IRON RANGES
OF MINNESOTA

Lake Winnibigoshish

ITASCA

Prairie River

Me...

CHISHOLM

HIBBING

KEEWATIN

NASHWAUK

TACONITE

D.M. & N.

G.N.

CALUMET

BOVEY

MARBLE

COLERAINE

Trout Lake

DEER RIVER

Great Northern

GRAND RAPIDS

G.N.

D.M. & N.

SWAN RIVER

G.N.

CASS

Soo Line

Great Northern

St. L...

AITKIN

Mississippi River

CROW WING

North...

CA...

Cuyuna Range

CUYUNA

CROSBY

RIVERTON

IRONTON

DEERWOOD

AITKIN

Soo Line

BRAINERD

Northern Pacific

Mille Lacs Lake

Burntside Lake

Shagawa Lake
ELY   WINTON

Vermilion Lake

Vermilion Range

SOUDAN

TOWER
& I.R.
Cu

Birch Lake

Pike River

BABBITT

River

Range

L A K E

BIWABIK

Partridge River

ON   VIRGINIA

D I& I.R.

EVELETH
D. M & N.

Embarrass R.

IRON JUNCTION

VERMILION TRAIL

D & I.R.

BEAVER BAY

Great Northern

D. M. & N.

S T.   L O U I S

Lake Superior

TWO HARBORS

OOK JCT. TON)

D & I.R.

DULUTH

SUPERIOR

N.P.

FOND du LAC
N.P.

TON

CARLTON
G.N.
Soo Line
G.N.

W I S C O N S I N

Soo Line

G.N.

Soo Line

LAKE SUPERIOR IRON RANGES

Vermilion

Mesabi

Two Harbors

M I N N E S O T A

Duluth

Ashland

Lake Superior

Marquette

M I C H I G A N

Marquette

Superior

Cuyuna

Gogebic

Escanaba

Menominee

W I S C O N S I N

IRON FRONTIER

The Discovery
and Early Development
of Minnesota's Three Ranges

David A. Walker

MINNESOTA HISTORICAL SOCIETY PRESS · 1979

2004 reprint of 1979 edition

Copyright © 1979 by the MINNESOTA HISTORICAL SOCIETY

Library of Congress Cataloging in Publication Data:
Walker, David Allan, 1941–
  Iron frontier.
  (Publications of the Minnesota Historial Society)
  Includes index.
  1. Iron industry and trade—Minnesota—History.
2. Iron mines and mining—Minnesota—History.
I. Title. II. Series: Minnesota Historical Society.
Publications.
HD9517.M6W34   338.2′7′309776   79-20626

International Standard Book Number: 0-87351-145-X

for Sharon, Sarah,
and Mark

PREFACE

THE FRONTIER played a major role in shaping the growth and development of the entire North American continent. Historians, beginning with Frederick Jackson Turner in 1893, placed various emphases on the nature and duration of this experience. Fundamentally the concept refers to a meeting point where several forces come together, a contact situation. For me the frontier is both place and process. It represents a geographical location where a specific activity — mining, lumbering, cattle raising, farming — occurred during a designated time period. Equally important, the frontier also describes the formative stages in resource exploitation and the changes affecting government and society. New institutions and attitudes were created by the unique opportunities that appeared.

The mining frontier attracted thousands of people to numerous isolated, undeveloped areas. It unfolded rapidly with an influx of capital, transportation, marketing facilities, and settlement patterns. Like many frontiers the search for mineral resources in northeastern Minnesota lured an extremely diverse group of people: miners, speculators, bankers, lawyers, merchants, saloonkeepers, and politicians. It was shaped by many forces, among them changes in technology, fluctuating economic conditions, and expanding social values. More than anything the iron range frontier stimulated expansion and development; it opened an area to new or restructured institutions, unique business opportunities, and challenging social interrelationships.

During the latter half of the 19th century the iron and steel industry was a key ingredient in building the economic strength of the United States. Iron had been a primary commodity in colonial manufacturing, and its use had expanded with the need for rails and rolling stock for the nation's growing transportation network. Technological improvements in steel production greatly accelerated demand for iron ore and stimulated its discovery and exploitation in the frontier region surrounding Lake Superior.

This book describes the early development of mining on the Vermilion, Mesabi, and Cuyuna ranges of northeastern Minnesota, the most extensive source of high-grade iron ore in the United States. It is not a social history of that region. Rather it focuses on people who were the economic decision-makers during the formative years when the potential wealth of the ranges was just beginning to be recognized. This mineral frontier attracted such diverse individuals as Charlemagne Tower, George C. Stone, members of the remarkable Merritt family, John D. Rockefeller, Henry W. Oliver, Andrew Carnegie, Cuyler Adams, and others. Their relationships were sometimes cordial, but more often hotly competitive, and some antagonistic feelings remain to this day. In his own way, each man was an entrepreneur, someone who recognized the commercial potential of the region and sought to design a means of profitably extracting its wealth.

The trend in business and economic history seems to be moving away from a focus on the individual toward analysis of inanimate institutions and corporations. In many ways I regret this new direction because it tends to ignore the central fact that men and women create history. Working separately or collectively, in harmony or in discord, they seem to me to be the key elements in understanding past events. It is my hope that this book illuminates both the people and events that shaped the early years of the iron ranges of Minnesota.

No historian can spend nearly eight years delving into various aspects of a topic without accumulating substantial debts, both personal and academic. The central core of this book first unfolded as a doctoral dissertation at the University of Wisconsin, where Allan G. Bogue, a warm, personable man, worked patiently with a struggling graduate student. I am grateful to John L. Loos of Louisiana State University, who introduced me to the serious study of the American frontier. His faith in my ability, his interest in my career, and his continued friendship are appreciated.

Special thanks must go to a group of Duluthians: to Glen J. Merritt and his son, Grant, who shared their family's views on the Rockefeller struggle in frank, hospitable interviews that added greatly to my understanding of these complex events; to Roy O. Hoover and his wife Marjorie who generously shared their expert knowledge and their valuable preliminary inventory of the Merritt Papers with researchers from the Minnesota Historical Society; and to Donald B. Shank and Franklin A. King of the Duluth, Missabe and Iron Range Railway

Company, who answered questions and kindly made available records that verified or expanded my understanding of the early years of the Duluth, Missabe and Northern Railroad.

I am also indebted to my former colleague, John Hevener, now at Ohio State University-Lima, who critically and perceptively read an early draft at a crucial time in its creation, and to my fellow historians at the University of Northern Iowa, especially Donald R. Whitnah, for providing a congenial atmosphere that encourages research and writing without sacrificing the faculty's primary obligation to students. Sandra Heller skillfully typed several drafts; I appreciate her kindness and her interest in the unfolding narrative.

Numerous librarians and archivists opened their collections in a genuine sense of scholarly co-operation. This is especially true of Eleanor C. Bishop, Kenneth Carpenter, and Robert W. Lovett of Harvard's Baker Library; Linda Edgerly and Joseph W. Ernst at the Rockefeller Family and Associates Archives; Judith A. Trolander and David Gaynon of the Northeast Regional Research Center at the University of Minnesota-Duluth; John D. Taylor and Trudy Byrum of the Northwest Area Foundation, custodians of the Louis W. Hill Papers, and especially Louis W. Hill, Jr., for permission to use them; Dale Carrison, Marilyn Lass, and Phyllis Roberts of Mankato State University; and Don Gray and Eva Bonney at the University of Northern Iowa.

Kindnesses were also extended by staff members of the State Historical Society of Wisconsin, University of Wisconsin's Geology and Engineering libraries, and Columbia University's Butler Library. Most important of all is the Minnesota Historical Society, especially its archives and manuscripts and publications and research divisions, where I was the recipient of many helpful courtesies. A special thank you to Mary D. Cannon and Ruby Shields. The book's design and its maps are the work of Alan Ominsky, the Society's production supervisor. All illustrations not specifically credited may be found in the Minnesota Historical Society's Audio-Visual Library.

I gratefully acknowledge the financial assistance provided by a Minnesota Historical Society Research Fellowship, a Kress Fellowship at Harvard, and by the National Endowment for the Humanities, Mankato State University, and the University of Northern Iowa.

Several aspects of the relationship between the Merritt family and Rockefeller were discussed in a paper I presented at the Missouri Valley History Conference in 1977 and at the annual meetings of the Economic and Business Historical Society in 1977 and the Minnesota

Historical Society in 1978. I wish to thank the former group for presenting me with the inaugural Charles J. Kennedy Award. The Lake Vermilion gold rush episode is more fully told in *Minnesota History* (Summer, 1974). A portion of that story is repeated here with permission.

June D. Holmquist, the Minnesota Historical Society's assistant director for research and publications, expertly transformed this manuscript from a thesis into a book. Her indomitable patience was put to the test on too many occasions. I deeply appreciate her candor, cordiality, and skill.

My wife Sharon lived with the project from its inception and carefully read more revised chapters than either of us cares to remember. Her patience, gracious attitude, and loving co-operation are irreplaceable.

Many have contributed to this endeavor, but as always the final product is mine. I assume total responsibility for its contents, both factual and analytical.

David A. Walker

WATERLOO, IOWA
JULY 6, 1979

CONTENTS

|1|

OPENING
THE LAKE SUPERIOR
HINTERLAND

NATURE buried the world's largest and richest deposits of iron ore deep in the heart of the North American continent. Then it tucked in lumps of copper, added a fringe of rich pine and hardwood forests, and sprinkled an abundance of fish and fur-bearing animals in the woods and waters. Ringing Lake Superior, the largest body of fresh water on earth, were six great iron ranges — the Marquette, Menominee, and Gogebic in Michigan and Wisconsin, and the Vermilion, Mesabi, and Cuyuna in Minnesota. Some of the copper was scattered near the surface of Isle Royale, Lake Superior's largest island, and around the shores of Michigan's Keweenaw Peninsula, which stretches like a finger into this coldest and deepest of the Great Lakes. Perched near the continental upland that separates the watersheds of Hudson Bay, the St. Lawrence, and the Gulf of Mexico, Lake Superior is connected to the Atlantic by an almost unbroken chain of landlocked waterways. In steplike progression the Great Lakes increase in length, breadth, and height above sea level from Lake Ontario on the east to Lake Superior on the west.

The story of North America's iron ranges is closely linked to the presence of this vast water stairway. Along it over the centuries moved a procession of people — ingenious copper miners of the prehistoric period, the later Ojibway and Dakota Indians, French and British traders seeking furs, military men and other representatives of the young United States, geologists and landlookers, miners and settlers, speculators and politicians, investors and entrepreneurs. In general, their movements progressed from east to west as the geography of the Lake Superior hinterland became known. Not until the last half of the 19th century — the period with which this book is largely concerned — did the drama of history focus on the opposite movement of minerals

1

from west to east, as the ranges one by one came into production and began to yield their long-buried ores.

Copper was the first mineral to be exploited, and its story is tied inseparably to that of iron in the Lake Superior region. Long before Europeans arrived on the shores of the lake, prehistoric peoples and later Ojibway Indians extracted pure copper with wood and stone tools, fire, and water. Indeed, a prehistoric culture known to archaeologists as "Old Copper" is unique to the western Great Lakes. Believed to have existed from about 5000 B.C. until 1500 B.C., it "represents the first known use of fabricated metal" by the peoples of either North or South America. On Isle Royale alone prehistoric mines covered as much as a square mile or more, and near Rockland on Michigan's Upper Peninsula the pits of these ancient miners "formed an almost continuous line for 30 miles." The soft metal was hammered into spear points, awls, fish gorges, ornaments, and knives, which have been found in archaeological sites located only in Minnesota, Wisconsin, Ontario, and Upper Michigan. Later Indian people used the less malleable "heavy stones" of native iron for striking fires. They also heated and mixed iron oxides with fish oil to make the durable red paint that can be seen to this day in the rock art paintings left by Indian artists in the Lake Superior basin.[1]

Like many other sections of the continent, the Superior hinterland of Minnesota, Wisconsin, Michigan, and Ontario owed its initial European development to a world-wide demand for furs and a rapidly expanding missionary zeal. The years of French control of the western Great Lakes also brought the area's mineral resources to the attention of European powers. Although Jacques Cartier mentioned western copper deposits as early as the 1530s and various accounts of native copper mines were published in France in the 1630s and 1640s, the first authoritative observations were probably those of Pierre Radisson and his brother-in-law Médard Chouart, sieur des Groseilliers, who learned of the mines on Isle Royale in 1659. Shortly thereafter Father Claude Allouez, a Jesuit missionary to the Ojibway Indians, further advertised the region's resources. An educated man of infinite curiosity about these new lands and their native peoples, Allouez typified the 17th-century Jesuit. Like many early clerics, he was a careful observer, recorder, explorer, and map maker. He noted the use of copper by the Indians and recorded the existence of a "great rock, all of copper" on the south shore of Lake Superior.[2]

The economic interrelationships between Frenchmen and Indians, originally based on furs, soon expanded to include copper. In the last decades of the 17th century French colonial administrators, aware of the potential commercial importance of Lake Superior copper, emphasized the economic stimulus mining could provide for New France. Interest in Montreal and in Paris, however, could not overcome the difficulties of transportation. Substantial boats were needed to carry the heavy ore across Lake Superior. The shipping season was short and the waters treacherous, given to sudden storms and high waves. The rapids in the St. Marys River between present Ontario and Michigan, where the channel drops nearly 20 feet in less than a mile, presented a formidable obstacle to navigation. Ships could not venture through them, and cargoes had to be unloaded at Sault Ste. Marie and hauled across by land. Because of this obstruction in the otherwise continuous water route to Lake Erie, the transport of ore required two fleets: one for the angry waters of Lake Superior and another for the lower lakes.[3]

The French did not begin commercial mining until the period between 1734 and 1740. Then Louis Denis, sieur de la Ronde, commandant of the French post on Chequamegon Bay in the present Apostle Islands at the western end of Lake Superior, constructed a ship to transport copper ore from several mines. The enterprise, although profitable, was cut short by La Ronde's death and war between the Ojibway and Dakota. "La Ronde," wrote historian Louise P. Kellogg, ignoring the much earlier native people, "may well be known . . . as the first practical miner on Lake Superior."[4]

Control of the Lake Superior basin passed from France to Great Britain in 1763 under the terms of the Treaty of Paris that closed the Seven Years War. During the next two decades British explorer Jonathan Carver, trader Alexander Henry the elder, and others called further attention to the copper deposits. Between 1767 and 1773 a British group called the Proprietors of Mines on Lake Superior, which included some of London's most influential politicians, made a concentrated effort to tap these mineral resources. Under the leadership of Alexander Baxter, Jr., and Henry Bostwick, the Proprietors erected Fort Gloucester at Sault Ste. Marie, constructed an assay furnace, and built a sailing vessel to transport the ore. But the Proprietors suffered one financial setback after another, and operations ceased within five years.[5]

In a treaty signed in 1783 at the close of the Revolutionary War, Great Britain officially gave up to the recently proclaimed United States its interest in the lands south of a vaguely defined border. Negotiating for the British was Richard Oswald, a wealthy London merchant, and for the Americans, Benjamin Franklin. A popular rumor, never substantiated, suggested that Franklin knew of the existence of minerals in the region west of Lake Superior. This knowledge, said to have been gained from Jesuit records and French acquaintances, supposedly guided him in the boundary negotiations. If so, it was an unspoken objective. The main question was not minerals, but inland navigation.[6]

The first proposal, extending the boundary of the new nation westward from Lake Nipissing, Ontario, to the then unknown source of the Mississippi, proved unsatisfactory to the British because it would have denied them full control of the lower Great Lakes and lost them southern Ontario. The following month the Americans suggested that a line be drawn at the 45th parallel of latitude, giving the Toronto area to the United States but leaving to Britain all lands north of the Falls of St. Anthony in present Minneapolis. The English delegation again expressed dissatisfaction, so the American representatives proposed a boundary that would follow the main channel of the Great Lakes and the principal fur trade route from Lake Superior west to Lake of the Woods along what is now the Minnesota-Ontario border. This solution, which was accepted by both countries, allowed each nation to retain at least a portion of the inland waterways, but it gave to the United States the unknown mineral wealth of northern Minnesota, Michigan, and Wisconsin.

The ambiguities of the 1783 treaty were not entirely resolved until a final settlement of the boundary west of Lake Superior was worked out in the Webster-Ashburton Treaty of 1842. It has been claimed that rumors of mineral wealth in the border area again influenced the negotiations. Secretary of State Daniel Webster considered the land "valuable as a mineral region," but his British counterpart probably did not share Webster's optimism concerning it. Neither iron nor copper was among the expressed goals of the negotiators, and the border was settled without disagreement. Its terms assured to the United States the possession of the iron ores of Minnesota's three ranges.[7]

During the more than 50 intervening years of uncertainty over the exact location of the boundary in the Lake Superior hinterland, the United States government sponsored several expeditions to acquire information about its wilderness empire. Although these explorations reached only the fringes of northeastern Minnesota's mineral districts, they sketched a harsh land beyond the frontier about which Americans knew virtually nothing. After Thomas Jefferson negotiated the Louisiana Purchase with France, American interest in its northern reaches quickened. The area was visited in 1805–06 by Lieutenant

Zebulon M. Pike of the United States Army, who ascended the Mississippi in an attempt to discover its source. In January, 1806, Pike arrived at Pokegama Falls near the present city of Grand Rapids, Minnesota. Although he made no reference to the presence of minerals there, the lieutenant had reached what was later identified as the western end of the Mesabi Range.[8]

In 1818 Congress assigned to Michigan Territory all the land north of the new state of Illinois between Lake Michigan and the Mississippi River. Two years later Lewis Cass, governor of Michigan Territory, led an expedition into a portion of his domain. He hoped to investigate rumors of mineral deposits near Lake Superior, map the western Great Lakes country, and establish good relations with the native people in order to divert their loyalty from the British. The Cass party traveled along the south shore of Lake Superior to the western tip near what is now Duluth, Minnesota, struggled up the St. Louis River, made the long portages over to the Mississippi, unsuccessfully sought its source, and descended the great stream to Prairie du Chien in present Wisconsin. From there the men made their way back to Detroit via the Wisconsin and Fox rivers to Green Bay and Lake Michigan.[9]

Although his final report made no direct mention of minerals in northern Minnesota, Cass's expedition piqued the curiosity of Henry R. Schoolcraft, a young man of literary ambitions who traveled with the party as its mineralogist. In describing the area surrounding Pokegama Falls, Schoolcraft specifically called attention to the "granitical and metamorphic rocks" as well as to the existence of quartzite. Twelve years later Schoolcraft returned to northern Minnesota hoping to locate the true source of the Mississippi, a project that had remained in his mind since 1820. In June, 1832, he visited and named Lake Itasca, correctly designating it as the river's source. Lieutenant James Allen, the United States Army topographical engineer who commanded Schoolcraft's military escort, was one of the earliest Americans to take note of a "low ridge" that tended "in a northeast and southwest direction," a ridge later known as the Mesabi Range.[10]

Four years later Joseph N. Nicollet, a gifted French scientist and map maker, traveled to the Mississippi headwaters southwest of the iron ranges. Though perhaps partially financed by the St. Louis fur company of the Chouteau family, Nicollet carried letters of introduction and astronomical equipment supplied by United States government officials. In July, 1836, he arrived at Fort Snelling, a military post now on the edge of the Twin Cities of Minneapolis and St. Paul. Obtaining a canoe and a crew of voyageurs, Nicollet continued up the

Mississippi to Lake Itasca, where he meticulously confirmed and ex-
panded Schoolcraft's sketchy observations of the river's headwaters.
After two more journeys to Minnesota under the auspices of the War
Department, Nicollet's final report was published by Congress in 1843.
It contained a brief description of the rock formations at Pokegama
Falls, noting the presence of quartzite, schist, and iron sulphide. It also
included the first accurate map of the land between the Mississippi and
Missouri rivers, on which was depicted the "Missabay Heights" — the
earliest printed indication of the land form that held enormous iron ore
reserves. The word, meaning "giant's hills" — which Nicollet translit-
erated "Missabay" and which was later spelled Mesaba, Missabe, and
officially Mesabi — was part of the Ojibway name for the upland region.[11]

A different kind of problem facing the federal government in the
years from 1826 to 1855 was ownership of the land these men explored.
Settlers, attempting to wrest control of the Lake Superior country from
its Indian inhabitants, as well as reports of the existence of minerals
near the lake — especially copper along the south shore — stimulated
the government to action. In the summer of 1826 Governor Cass and
Thomas L. McKenney, head of the Bureau of Indian Affairs, met in
council with nearly 600 Ojibway at Fond du Lac on the St. Louis River
near modern Duluth. The Indian delegates represented an area that
stretched from Sault Ste. Marie on the east to the headwaters of the
Mississippi on the west and from the Canadian border to southern
Wisconsin. For five days each side displayed all the pomp and cere-
mony its members could muster as McKenney attempted to persuade
the Ojibway of the benefits of allegiance to the United States.[12]

The Fond du Lac Treaty, signed on August 5, 1826, contained a
provision important to the future development of iron ore. Under the
terms of article 3, the Ojibway granted to the United States "the right
to search for, and carry away, any metals or minerals from any part of
their country." Both delegations firmly asserted, however, that this
privilege was to have no effect on "the title of the land, nor the existing
jurisdiction over it."

Twenty-eight years later in September, 1854, after copper and iron
ores had been found both north and south of Lake Superior, the Ojibway ceded some of their lands in northeastern Minnesota in a treaty
negotiated at La Pointe, an Indian agency and trading post on
Chequamegon Bay in Wisconsin. Ten bands of Lake Superior Ojibway
represented by 85 leaders signed the document allowing the United
States to acquire formal title. Five months later the Mississippi Ojib-

way surrendered another portion of northern Minnesota extending west to the Red River of the North. Small reservations set aside within the ceded territory provided restricted homes for the numerous bands residing there. The treaty specifically provided for the mixed-blood members of the Lake Superior Ojibway regardless of their residence. Each head of a family or single person over 21 years of age received a certificate entitling him or her to select 80 acres of public land, surveyed or unsurveyed, from any section. These certificates, known as Ojibway scrip, later created controversy when whites took advantage of loopholes in the law to use them fraudulently to acquire mineral lands.[13]

Meanwhile along Lake Superior's south shore, Michigan officials urged the federal government to confirm that state's borders and eliminate potential conflicts with its native people. In 1837 the state of Michigan was created with the mineral lands of the Upper Peninsula within its limits. A boundary was drawn between Michigan and Wisconsin Territory that, for the most part, followed the Menominee and Montreal rivers. United States acquisition of all unceded Ojibway land in the Upper Peninsula was provided for by a treaty signed at La Pointe on October 4, 1842. As copper prospectors rushed into the newly opened lands, the War Department erected Fort Wilkins on the northern tip of Keweenaw Peninsula near present Copper Harbor. These actions assured to Michigan possession of known copper deposits on the peninsula and the Ontonagon River as well as yet undiscovered iron reserves.[14]

Two factors that were to affect mining in the Lake Superior area for years to come hampered the early exploitation of copper on the south shore. The first was the necessity for large investments of capital. The land was rugged and inaccessible, and the metal had to be brought out in large quantities to realize a profit. The development of Michigan copper was due in large part to heavy investments by Boston financiers. The second factor was related: the difficulty of transportation. The Upper Peninsula was then accessible only by water, and the water route was effectively blocked at Sault Ste. Marie. During the 1840s a few steam-powered vessels were towed laboriously around these rapids on greased rollers for launching on Lake Superior, but the troublesome bottleneck remained. Shippers often spent four or five days unloading and reloading on either side of the rapids. As early as 1837 Michigan Governor Stevens T. Mason advocated a canal to bypass them, and construction was actually commenced in 1839. But disputes between the state and federal governments intervened, and the plan was temporarily put aside. In the face of all these difficulties, the earliest copper mines on the Upper Peninsula did not thrive. Although well-financed efforts later provided substantial returns and copper remained a major resource well into the 20th century, iron ore proved easier to extract. Many of the miners and entrepreneurs who had been lured to the area by copper soon diverted their efforts to iron ore instead.[15]

The discovery of iron ore in the Lake Superior basin was a direct result of the search for copper. Douglass Houghton, a native of Troy, New York, who had accompanied Schoolcraft to Minnesota in 1832, spent four years systematically exploring the Upper Peninsula as Michigan's first official geologist. By 1844 he had already confirmed the existence of substantial copper deposits, and he was combining a federal land survey with geological investigations in the hope of avoiding unnecessary confusion over mining claims. Unexpectedly, he discovered iron ore on what became the Marquette Range.[16]

On the morning of September 19, 1844, Houghton's crews were working along the upper Escanaba River just south of Teal Lake where present Negaunee stands. Surveyors under the direction of William A. Burt, United States deputy surveyor, noticed an unusual declination in their magnetic compasses. After a brief search they found pieces of almost pure iron ore along an outcropping about 80 miles southeast of the Keweenaw Peninsula. When winter closed the surveying season, the men returned to Sault Ste. Marie. Then news of their discovery spread rapidly.[17]

One of the first to act was Philo M. Everett, who had traveled from Connecticut to prospect for copper. On July 23, 1845, Everett joined other Jackson, Michigan, residents in organizing the Jackson Mining Company. They located iron ore near Burt's discovery and entered a claim for the Jackson Mine (in present Negaunee). The company then constructed a forge on the Carp River, where on February 10, 1848, Ariel N. Barney and William B. McNair hammered out the first wrought-iron blooms in the Lake Superior country. Financial difficulties forced the company to close both the mine and the forge in 1854, but Everett and his partners had inserted the opening wedge in a new mineral frontier.[18]

By this time the Marquette Range, located in two great troughs extending east and west between Negaunee and Lake Michigamme, had already attracted the attention of other eastern entrepreneurs. At first miners merely broke up the ore with pickaxes, sledge hammers, and crowbars and shoveled it into wagons. Using this unsophisticated method in July, 1852, the Marquette Iron Company, formed by Pennsylvania industrialists who had leased Everett's property, extracted and shipped six barrels of ore to New Castle, Pennsylvania — the first cargo of iron ore from the Lake Superior region. The barrels were used to ease the transportation problem at Sault Ste. Marie, but it was still a laborious and expensive task to bypass the rapids there. Two years earlier the Cleveland Mining Company, directed by Samuel L. Mather, had purchased and explored iron lands near the small settlement of Marquette, ten miles east of Negaunee along the lake shore. In 1852 this firm shipped about 70 tons of ore to Sharon, Pennsylvania. The following year Mather gained control of the Pennsylvania company, and in 1855 he constructed the first substantial dock at Marquette.[19]

The increased mining activity in copper and iron finally forced a solution to the bottleneck at Sault Ste. Marie. Michigan legislators appealed to Congress for financial help, and the government responded with a land grant of 750,000 acres to finance the building of a canal. The land would be forfeited, however, unless construction began within three years and was completed within ten. In February, 1853, state lawmakers passed a bill that specified the canal's size, authorized a five-man commission to prepare plans, and hired Charles T. Harvey as general agent to superintend operations. The first excavation was made on June 4, 1853, and in spite of delays caused by supply shortages and a brief cholera outbreak, the first steamer passed through the locks on June 18, 1855. Use of the locks increased steadily from the day the Sault Ste. Marie Canal opened, and its building was to provide a giant gateway unlocking the wealth of Lake Superior's hinterland.[20]

Further transportation improvements were necessary, however, before the Marquette Range could be fully developed. In 1851 Heman B. Ely, a Rochester, New York, native who later represented the Marquette district in the Michigan legislature, laid plans to construct a short-line track, to be known as the Green Bay and Superior Railroad, from the mines to Marquette. After surveying a proposed route, he failed to obtain the necessary financial support. The Cleveland Mining Company then constructed a plank road, replacing it in 1854 with a "wooden strap railroad," wooden rails attached to the planks and covered with a thin iron strap. The newly created Iron Mountain Railroad Company proved more efficient, but unfortunately it was more accident prone than the plank road. In September, 1857, Heman's brother, Samuel P. Ely, shipped a 25-ton steam locomotive from New

Jersey to Marquette that quickly began regular trips to the mines. This rail system eventually became part of the Duluth, South Shore and Atlantic Railway. Thereafter ore shipments to lower lake ports increased as new mining companies gradually developed the range.[21]

While Philo Everett and the Ely brothers were opening the Marquette Range in the early 1850s, settlers and speculators began to fan out along the lake shore. At the western end of Lake Superior the first of three settlements that were to endure was Superior on the Wisconsin side of St. Louis Bay. There Henry M. Rice, Minnesota territorial delegate to Congress, and his associates attempted to promote townsite development by touting the beneficial effects that construction of the Sault Ste. Marie locks would have on the economy of the western Great Lakes. Rice received congressional and financial support from such prominent individuals as John C. Breckinridge of Kentucky, Stephen A. Douglas of Illinois, and William W. Corcoran, a Washington, D.C., banker. In January, 1854, the Superior settlers began cutting a winter road from the new town to Taylors Falls on the St. Croix River, and in 1856 mail was carried over a federally financed trail linking Superior with St. Paul. During the first four years after settlement began in 1853, Superior experienced phenomenal growth. In 1854 its population was listed as 449; by 1857 it had grown to about 2,000. The city charter, obtained in 1858, confidently predicted that within five years Superior "would be the second city of Wisconsin."[22]

Across St. Louis Bay, the beginnings of Duluth were barely visible. It grew slowly from a log trading post on Minnesota Point in 1853 to a platted and incorporated city in 1857, with a few homes, warehouses, stores, a post office, a shipping dock, and a land office. Although Duluth did not begin to match Superior in this period, it did achieve a firm foothold along the rocky hillside.[23]

The third enduring settlement was Beaver Bay, located some 50 miles east of Duluth. It was one of more than a dozen embryo townsites which sprang up amid a flurry of copper claims on the north shore. During the boom that followed the signing of the treaty of La Pointe in September, 1854, "Men from the Ontonagon copper district were arriving by steamer, sailing vessel, row boat, canoe and even on foot, and hurrying out . . . to secure a copper claim." Founded by a group of men that probably included Justus Ramsey of St. Paul and Thomas Clark of Superior, Beaver Bay survived, thanks largely to the arrival of the Wieland family and others in 1856.

Other hopeful settlements along the north shore were not so fortunate. When the panic of 1857 hit suddenly, "the bottom . . . fell out of all the booms at Superior, and at all other points at the Head of the

Lake," wrote one early historian who witnessed it. "Three-fourths of the people left the country, by every means of exit that was then available. Some, with pack and gun 'shot their way out'; some that had families and were without means to pay their passage on boats, were taken out free by those generous and charitable captains of the few steamboats that in those days visited the Head of the Lake. Sound money — yes, or any money — was then very valuable; a corner lot in Duluth was then not worth a pair of boots." In 1860 the census enumerated only 34 families, totaling 71 whites and 9 Indians living in Duluth. Not until the mid-1860s did the economy of the western hinterland begin to revive, and as late as 1865 one pioneer reported that he could choose which house he wished to occupy from among the many abandoned ones in Duluth.[24]

Fundamental technological improvements in the iron and steel industry stimulated the rapid exploitation of the ranges on Lake Superior's south shore. In 1856 Henry Bessemer, an English ironmaster, effected a method of producing steel by blowing a stream of air through molten pig iron.[25] This created heat so intense that it burned off several impurities, mainly carbon and silicon, present in all iron ore. A huge, pear-shaped receptacle, called a Bessemer converter, performed these functions to turn iron ore into steel. Superior to wrought iron (pig iron with the oxygen removed), possessing greater tensile strength and hardness, and less brittle and more malleable, steel replaced iron for many industrial uses, especially for the production of rails and structural beams.

Unfortunately the Bessemer process could not be immediately utilized in the United States. At about the same time Bessemer was working in England, William Kelly, a Kentucky ironmaster, made a similar discovery. In 1845 Kelly had acquired the Eddyville Iron Works on the Cumberland River and had begun to manufacture kettles. By the mid-1850s he was attempting to decarbonize iron by introducing a current of air. Although the Englishman applied for an American patent in 1856, a short time before Kelly did, the commissioner of patents ruled that the Kentuckian had been the prior developer and sanctioned his method. Bessemer, however, succeeded in patenting necessary auxiliary equipment that compelled users of either process to infringe upon the other's rights. The demand for steel rails, rising rapidly after the Civil War, led to a compromise. In 1866 the legal patent rights were combined, and for several decades the Bessemer process dominated the industry.[26]

American attempts to produce steel in the Bessemer converter encountered an immediate problem. The Englishman had done his original experiments with low phosphorous ore from Wales. American steelmakers, using ore from Pennsylvania and Michigan — the major sources in 1870 — found that a high phosphorous content caused the steel to fragment and disintegrate under the slightest pressure. In 1879 two English cousins, Sidney G. Thomas and Percy C. Gilchrist, eliminated this obstacle by lining the converter with lime rather than acid. This permitted the phosphorous in the iron to combine with the lime and be carried off in the refuse, or slag.[27]

Before the Thomas-Gilchrist process was perfected and adopted, the demand for Bessemer steel forced American furnacemen to search for a supply of low phosphorous iron ore. The Marquette Range met that basic requirement. Consequently, during the Civil War northern manufacturers' dependence on south shore iron ore increased, and Marquette shipments more than doubled from 114,401 tons in 1860 to 296,713 six years later.[28]

During the late 1860s and early 1870s the development of new machinery for underground shaft operations allowed entrepreneurs to exploit the range's full potential. By the end of the 1868 shipping season the output of 13 mines surpassed 500,000 tons. Nine years later the Marquette Range established itself as a consistent annual producer of over 1,000,000 tons. Until 1895 the Marquette remained the largest shipper among the Lake Superior ranges.[29]

Led by Eber Brock Ward, steel manufacturers in Detroit were the first to utilize successfully Marquette ore. Working his way up from cabin boy on a schooner to owner of a large merchant fleet and several railroads, Ward also made successful investments in shipbuilding, lumber, and real estate. In 1857 he established the Chicago Rolling Mill along the banks of the Chicago River, a plant that on May 24, 1865, rolled the first Bessemer steel rails in the United States. Before his death in 1875 Ward had expanded his operations to include two large furnaces and a rolling mill at Bay View near Milwaukee, Wisconsin. His pioneer manufacturing efforts eventually were absorbed into the complex United States Steel Corporation. Harvard business historian Fritz Redlich wrote perceptively that Ward "was one of the greatest business men of his generation and was the first man in American iron business to become a captain of industry. . . . With E. B. Ward . . . begins the modern era of large-scale production in American iron and steel."[30]

Ward's plants on the western Great Lakes, combined with expanding eastern steel production, stimulated a search for additional sources of ore. Before the end of the 19th century these efforts were to bring about the development of two new south shore ranges — the Menominee and the Gogebic — as well as the ore resources in northeastern Minnesota. During the late 1840s geologists had discovered a second major concentration of iron ore in Michigan's Upper Peninsula. In November, 1849, Dr. Charles T. Jackson, director of the geological and mineral survey of federal lands in Michigan, reported an ore formation along a section of the Menominee River that formed the state's border with Wisconsin. He noted that the ore was "as rich as is desirable, and [it] will make the best kinds of cast and wrought iron." Two years later geologists John W. Foster and Josiah D. Whitney traversed the region near present Iron Mountain, situated about 45 miles directly south of the Marquette Range. They described "a deposit of iron of considerable extent. It is at least a hundred feet in breadth, and extends probably three-fourths of a mile. . . . The specimens of the ore show a very high degree of purity." In 1867 after more than a decade of inactivity, Hermann Credner, a German scientist, visited the Upper Peninsula and conducted a geological survey of the Menominee Range. His report, published in German two years later, described numerous iron ore deposits.[31]

In 1872 geologist Nelson P. Hulst directed the first thorough exploration of the Menominee district. Based on his findings, the Milwaukee Iron Company, which four years earlier under Eber Ward's direction had built Wisconsin's first rolling mill at Bay View, uncovered enough iron ore to warrant investing the capital needed for extensive mining operations. Transportation remained a problem until the summer of 1877 when the Chicago and Northwestern Railway Company completed a line from the port city of Escanaba on Lake Michigan to the eastern section of the Menominee Range. Three years later the company extended its rail system westward to the communities of Iron Mountain and Iron River.[32]

Menominee Range shipments through Escanaba began in 1877 when the product of two mines, a total of 10,405 tons, reached Lake Erie ports. Within four shipping seasons miners extracted more than 500,000 tons of ore from 11 mines, and by the 1882 season the range's annual output reached 1,170,819 tons. Thereafter the Menominee remained a major source of high-grade Bessemer ore. As recently as 1977 it mined and shipped 5,370,081 tons.[33]

As the nation's industrial capacity grew, demand for raw material accelerated. Lake Erie prices for iron ore stayed at or above $9.00 per ton from 1880 through 1882. With such incentives many market observers speculated that the two producing Michigan ranges would shortly be depleted by the rising demand. Mining companies began a fresh search of the Lake Superior hinterland, a search that would lead to the opening of a third range in Michigan and Wisconsin and a fourth on the iron frontier in northeastern Minnesota.[34]

The Marquette and Menominee soon faced competition from the third major south shore district, the Gogebic Range. Geologists studied the two sections of this deposit, lying along the Montreal River in both Michigan and Wisconsin, for many years before firmly establishing it as one continuous geologic structure. As early as 1848 Arthur Randall observed exposures of lean iron ore in Wisconsin between the present communities of Mellen and Hurley. Several years later Charles Whittlesey traced iron-bearing slates farther to the west in Wisconsin. In 1858 a group of Milwaukee speculators, hoping to discover enough ore to begin a mining venture, financed an exploratory survey by Increase A. Lapham, who later served as Wisconsin state geologist from 1873 to 1875. Lapham reported finding an immense bed of magnetic iron ore along Bad River southwest of English Lake.[35]

Although initially explored from the Wisconsin side, the Gogebic Range first yielded iron ore in marketable quantities near the town of Bessemer, Michigan. Seeking to exploit this new area in the early 1880s, the Penokee and Gogebic Development Company established the Colby Mine. Under the leadership of Joseph Sellwood, later a pioneer on the Vermilion and Mesabi ranges, miners in 1884 used a steam-powered shovel to load ore cars for the first all-rail shipment of 1,022 tons south to Milwaukee. In 1885 the Milwaukee, Lake Shore and Western Railway Company established direct rail connections from the mines to the port of Ashland, Wisconsin, on Lake Superior.[36]

Mining companies in this district benefited from the experience of pioneers on the other ranges and, as a result, the deposits were extracted at a rapid pace. During 1885 seven mines produced a total of 119,590 tons. Two years later, in its fourth season of operation, the range's 24 producing mines shipped a total of 1,322,875 tons of iron ore. In 1887 the Gogebic surpassed the annual production of the Menominee Range and ranked second to the Marquette.[37]

The growth of open-hearth steel production after 1885 was a major stimulus to continued development on the Upper Peninsula as well as to the opening of Minnesota's Vermilion and Mesabi ranges. William and Frederick Siemens, two Germans living in England, and Pierre Martin, a French ironmaster, first introduced the open-hearth method in 1866. The new process created extremely high temperatures in a gas-regenerating furnace that eventually allowed manufacturers to control the ore's chemical quality more effectively than they could in the Bessemer system. By 1900 open-hearth steelmaking, utilizing cheaper raw materials, including scrap, than the Bessemer process, produced higher quality steel of a more uniform grade. The shift from manufacturing rails to making more varied products hastened the obsolescence of the Bessemer converter. Throughout the first decade of the 20th century the Bessemer process slowly yielded to the open-hearth method, so that by 1909 the latter accounted for 60 per cent of the market. With the gradual adoption of the open-hearth method and the decline of the south shore ranges, the spotlight shifted to the Minnesota ranges atop the great divide at the western end of Lake Superior.[38]

|2|

THE
VERMILION RANGE
DISCOVERED

THE PRESENCE OF IRON ORE in northern Minnesota was known to both Indians and whites centuries before large-scale development took place. In the 1730s and 1740s French explorer Pierre Gaultier de Varennes, sieur de la Vérendrye, traversed the boundary waterway and established Fort St. Charles on Lake of the Woods in what is now Minnesota's Northwest Angle. In 1734 during a council at the fort with six leaders of the Cree and Assiniboine Indians, La Vérendrye inquired about iron. The chiefs told him that they "knew of several iron mines with ore of different colours," which they described in detail.[1]

Almost a century later in 1821 a British fur trader again recorded the presence of iron ore in northeastern Minnesota. He was Dr. John McLoughlin, who wintered at various posts in the Lake Superior hinterland in the years before he moved to the Oregon country. McLoughlin possessed a good working knowledge of the region as far south as Mille Lacs Lake. In his description, written about 1821, he commented that "The Country in general is not perhaps so fit for cultivation, as Canada, however there are very Extensive tracts of very fine land. . . . The only Mineral I have seen in the Country is Iron, which though very common I never saw in any large quantity."[2]

Awareness of the mineral resources of northeastern Minnesota increased gradually as the result of a series of systematic explorations. The first official recognition of the presence of iron ore there was buried in a large volume published by the United States Geological Survey in 1852. Carried out under the direction of David Dale Owen, the nation's leading western geologist of the period, this survey of Minnesota, Wisconsin, and Iowa was conducted in part by such able lieutenants as Joseph G. Norwood and Charles Whittlesey. Norwood, who was in charge of the northern Minnesota section, spent the three

16

seasons from 1848 through 1850 examining that region as far west as Red Lake. A well-trained geologist, he wrote the first detailed account of the physical and geological features of northeastern Minnesota, including discussions of its mineral, agricultural, and timber resources, which pioneers settling there soon utilized.[3]

While the earlier expeditions of Pike, Schoolcraft, and others had recorded many isolated facts concerning northern Minnesota, the work of Owen's survey was considerably more comprehensive and detailed. These men attempted a scientific examination of the country rather than a topographical reconnaissance. During its first year in the field in 1848, the Norwood party followed the St. Louis, Embarrass, and Pike rivers to Vermilion Lake, where the geologist called attention to rocks along the shore that contained "beautiful crystals of iron pyrites" and noted the possibility of other minerals associated with them. He cataloged one specimen as "Quartz — colour, reddish brown, crystalline; with yellow iron pyrites" — the first published references to the ores of the Vermilion Range. During its second year, the survey team moved along the north shore, making short ascents up the numerous streams that flowed into Lake Superior, before penetrating the interior via the Pigeon River route along the present Canadian boundary. Paddling west to Flint (now Gunflint) Lake at what was later believed for a time to be the extreme eastern end of the Mesabi Range, Norwood observed "an exposure . . . the seams and joints of which are filled with imperfectly agatized quartz, chalcedony, and iron ore." Although he crossed what he called the "Missabé Wachu" and found "thin layers of iron ore" at Gunflint Lake, Norwood devoted much more time and attention to the possibilities for copper mining on the north shore.[4]

The brief passages in Owen's report, which went largely unnoticed by the general public, constituted the first scientifically accurate statement of the existence of iron ore in Minnesota. Writing in 1872, a later geologist proclaimed that this report threw "the first real light, derived from the systematized science of modern times, on the geology . . . of Minnesota." Less scientific and less accurate but more widely read was the statement of J. Wesley Bond in a guide to Minnesota published in New York in 1853. Bond wrote that "A mountain extends all the way between the St. Louis River and Pigeon River. It evidently abounds in copper, iron, and silver. The terrestrial compass can not be used there, so strong is the attraction of the earth. The needle rears and plunges 'like mad.' "[5]

As for the Mesabi Range, it is possible that an attempt to mine there was carried out "soon after, if not earlier than, the Norwood report." In 1905 Dwight E. Woodbridge, a reputable Duluth mining engineer, maintained that James Whitehead, a fur trader representing St. Paul interests, sank a shaft in the vicinity of Pokegama Falls and the mouth of Prairie River where that stream cuts across the Mesabi near the present city of Grand Rapids. Woodbridge wrote that in 1876 Whitehead "showed friends a rotted and tree-covered log cabin near the Mississippi river, where he and associates had sunk a shaft for iron many years before, and endeavored to mine."[6]

It is certain, however, that Alexander Ramsey, the first governor of Minnesota Territory, early recognized the economic potential of the Lake Superior basin. In his message to the territorial legislature in September, 1849, three months after he arrived from Pennsylvania, Ramsey called for the construction of a road from St. Paul to Lake Superior in order to "open the mineral regions on the shores of that lake to the farm produce of our Territory, and lead to a trade mutually advantageous."[7]

Ramsey's statement immediately raises the question: Was he aware in 1849 of the Minnesota ores later officially revealed by the Norwood survey? While it is certainly possible that the governor possessed such advance information, it seems more probable that the "mineral regions" Ramsey had in mind were in Michigan rather than in Minnesota. Arriving in the territory on May 27, 1849, the newly appointed executive stepped into a controversy over the Minnesota-Wisconsin boundary which was complicated by the desire of politicians and investors to combine the development of the white pine in the St. Croix River Valley with the recently opened copper mines on Michigan's Keweenaw Peninsula. The governor was surely cognizant of these issues when he addressed the legislature.[8]

One of Ramsey's informants may have been Caleb Cushing, who was deeply interested in both timber and mineral lands. The two men had served together briefly in Congress in 1845 when the superintendent of mineral lands on Lake Superior made a report on "ores, sufficient in quantity and richness to justify the investment of a large amount of capital." Cushing, a Bostonian who later became attorney general of the United States, had visited the midwestern region as early as 1846 when he canoed from Lake Superior to the Falls of the St. Croix River near modern Taylors Falls. He invested heavily in water-power rights and lands at the Falls of the St. Croix, at the Falls of St. Anthony in what is now Minneapolis, and in copper and iron lands on the south shore of Lake Superior. He was eager to see the boundary of Minnesota Territory drawn as far to the east as possible on the theory that its leaders would be more amenable to his plans than would those of Wisconsin, which was already a state. It has been suggested that Cushing may have hoped to be governor of the new territory.

Ramsey, who was to prove an adroit political leader as well as an able advocate of the interests of his adopted home, remained a resident of Minnesota for the rest of his life. In 1860, again occupying the governor's chair as chief executive of the two-year-old state of Minnesota, Ramsey once more called the legislature's attention to the Lake Superior basin. He spoke of its agricultural possibilities and "safe and commodious harbors," called for "direct rail communication with the Lake," and stressed the presence of minerals, declaring that they were "not only of magnificent extent, but the ores, both of iron and copper are known to be of singular purity."[9]

The 1860 legislature responded by appointing Charles L. Anderson and Thomas Clark as commissioners to devise a plan for a state geological survey. In 1861 the two men recommended that the proposed survey be "full and scientific" and that its results be made public annually. They also suggested that Minnesota should profit from the mistakes of other states, which had allowed "*party prejudices* to interfere" by "rewarding a political leader with the office of State Geologist, and a liberal yearly salary, when he is totally incompetent for the task." Clark, who had purchased land at Beaver Bay on the north shore in 1857 and represented that area in the first state senate, was especially emphatic concerning Minnesota's copper and iron resources, which he said had been "privately" explored but kept secret by those who wished to exploit them. "To Minnesota belongs the furnishing of the entire Mississippi valley demand for copper, and the upper portion with iron," he wrote. "Geographically and physically Minnesota preeminently commands a position second to no interior State. Within our borders we have commercial command of [an] outlet to the competing Atlantic markets — an ample surplus of food to sustain a manufacturing population — an abundance of natural power . . . and inexhaustible material to convert to useful purposes."[10]

Ramsey was succeeded as governor by Henry A. Swift, who promptly in 1864 reminded the lawmakers that the portion of Minnesota bordering Lake Superior abounded "in precious ores, and has interests peculiar to itself which deserve the fostering care of the Legislature. Mining companies are operating here with good prospects of success," he remarked optimistically of the copper ventures then in progress on the north shore. At last convinced of the probable value of the region, the legislators in 1864 responded by authorizing Swift to select a "suitable person or persons" to conduct a geological survey, appropriating $2,000 to search "the mineral lands on the north shore of Lake Superior" as well as "all other mineral or coal districts of the State."[11]

Augustus H. Hanchett received the appointment as Minnesota's first state geologist. Described by a contemporary newspaper as "not much of a geologist himself," Hanchett engaged Thomas Clark to carry out the field work. A competent civil engineer, the former commissioner and his survey team devoted a summer to two areas. One was the north shore and the other was near Vermilion Lake. Hanchett and Clark reported that "Specimens of hematitic specular iron ore were obtained from a heavy deposit" just south of Vermilion Lake. Although they tried to reach this deposit, they said, low water and the lateness of the season prevented them from ascertaining the "precise percentage of commercially pure iron contained in this ore." They also reported "evidences of a deposite of magnetic iron north of Duluth," but deemed the funds appropriated for the survey too limited to explore its extent. They recommended that the work in the northeastern counties be continued. "From the abundant richness of the ore in that region," they advised, "it will be to the general interest of the State to do so at as early a day as practicable."[12]

Acting immediately on this suggestion, Governor Stephen Miller authorized the continuance of the survey along the north shore and in other "mineral bearing districts." Seeking a more exhaustive examination, the governor and the legislators directed the new appointee to analyze "metal-bearing rocks to ascertain their commercial value" and to draw section maps "showing the location of minerals examined." To carry out the work, the lawmakers specified that the funds "to pay the actual and necessary expenses incurred" could not exceed $1,000. Governor Miller appointed Henry H. Eames as director of the new effort.[13]

Assisted by his brother Richard, Eames established his headquarters in Duluth and spent the summers of 1865 and 1866 in northeastern Minnesota. Like his predecessors, the geologist found that time and limited funds confined his attention to a small portion of the state. When the survey team set out, the *Superior Gazette* announced on June 3, 1865, that Eames intended to "make a thorough examination of the iron range which crops out in the neighborhood of Vermillion [sic] Lake, and trace it eastward toward Lake Superior."

Although he was more interested in gold and silver, Eames con-
scientiously recorded the presence of iron ore. Near Vermilion Lake in
1865 he found it "exposed at two or three points between fifty and sixty
feet in thickness," and he included in his report an analysis of three
specimens varying from 65 to 80 per cent iron. "The ore is of the
variety known as hematite," he reported. Along the trail from Beaver
Bay to the lake "for some distance there are exposures of . . . magnetic
iron." Speaking of St. Louis and Lake counties in his 1866 report,
Eames wrote: "The varieties of iron ore occurring in these counties is
very extensive; not only that near Vermilion lake, but I am informed
that large masses were seen north-east of that place, also to the south-
east, both in place and of great elevation. That to the north-east is of
the hematatic variety, and that to the south-east magnetic." [14]

Eames stirred up considerably more excitement over gold, which he
thought he found near Vermilion Lake, than he did over its iron ores.
After he reported veins of gold- and silver-bearing quartz there, Governor Miller sent numerous samples to the United States mint and to
private assay firms. When they returned favorable reports, what became known as the "Vermilion Lake gold rush" got under way. During
1865 and 1866 no fewer than 15 mining companies were incorporated
by some of the leading politicians and financiers of St. Paul, Superior,
Chicago, and New York. Vermilion Lake quickly acquired signs of
settlement. The town of Winston was established, numerous cabins
were constructed, and water-power sites, a sawmill, and a boat yard
were developed. A post office, hotel, general store, and saloons made
their appearance during the spring and early summer of 1866. Much of
this feverish activity took place on land still claimed by the Bois Fort
band of Ojibway Indians on the grounds that it had not been ceded in
the treaty of 1854. The *Superior Gazette*, noting that "if the land has
not been ceded to the government" the gold seekers were in for a
surprise, said that the "Indians are determined to assert their rights,"
adding, "we think these Boisforts have just cause for complaint." A
treaty was finally signed on April 7, 1866.[15]

From the beginning, the availability of provisions and machinery
was recognized as essential to the survival of the mining companies. As
early as September, 1865, miners had begun to open a sled road from
Duluth to the new gold fields. Those involved in the project hoped to
provide an immediate winter route for transporting needed supplies as
a first step toward construction of a year-round stage and wagon road.
Late in November a party of ten men under the direction of Joshua B.
Culver set out from Duluth to blaze an 80-mile trail to the lake. Citi-

zens of St. Paul, Superior, and Duluth, aware of the economic benefits to be derived from such an enterprise, contributed funds to defray building costs. By the end of February, 1866, after the last half of the route had been completed by the largest of the mining companies, construction parties had widened the Vermilion Trail into an acceptable winter road.

Since the route, out of necessity, crossed rivers, creeks, and swamps, it could be used only after the ice froze firmly. During the summer, travel over most of the trail became either treacherous or downright impossible. Several mining companies strenuously advocated the construction of a year-round roadway, and after successful lobbying efforts, Congress in 1869 appropriated $10,000 to improve the winter trail. Secretary of War William W. Belknap appointed George Riley Stuntz, a pioneer resident of Duluth, not only to upgrade the Vermilion Trail but also to extend it northwest to the recently established Bois Fort Indian Reservation at Nett Lake. Early in July, 1869, Stuntz organized a party of ten axmen, purchased necessary supplies, and began work. By the end of the season, pestered by unusually heavy rainfall and swarms of mosquitoes and black flies, Stuntz and his men had slashed a 12-foot-wide roadbed through the forest from Duluth to Vermilion Lake. Beyond the lake, however, they merely located and surveyed the portion leading to the Bois Fort Reservation.

By the time this transportation link had been completed, the unproductive gold boom had subsided and interest steadily declined. During the late summer and early fall of 1866 many miners had abandoned the gold fields without locating deposits of value. Large companies with heavier capital investments remained throughout 1867, but they too returned empty-handed. The only individuals who realized a profit were teamsters freighting equipment, store owners handling mining supplies, and operators of boardinghouses and saloons. No sizable earnings resulted from investments in the 15 companies that organized, established claims, and sought gold-bearing quartz. But the Vermilion Lake gold rush had other results. It evoked a speculative excitement that renewed and redirected attention to the western end of Lake Superior. It briefly attracted national recognition to the region, and it opened the only existing road in northeastern Minnesota, a road later important to the profitable extraction of iron ore.

One of the ironies of history is the fact that hundreds of men flushed with gold fever tramped over the largely unnoticed iron ore riches of both the Mesabi and Vermilion ranges. One of the men who did notice them, however, was Ossian E. Dodge, a special correspondent of the *St. Paul Pioneer* who was sent north to report on the gold rush. Writing under the pen name of "Oro Fino," he speculated on October 25, 1865, that the quality of iron ore on a peninsula in Vermilion Lake could be favorably compared with high-grade Swedish ore. He reported that it existed in "immense beds hundreds of feet in height," containing "enough to supply for a hundred years all of the furnaces of the entire world." The quality of the ore, wrote the journalist, was "fully equal" to anything he had ever examined. As late as December 23, 1867, the *St. Paul Daily Press* reported: "Vast fields of the richest iron ore are said to exist upon the shores of this [Vermilion] lake, and some specimens brought down . . . shows that the report is not unfounded." Despite these endorsements, the Vermilion Range was to remain undeveloped for another 20 years.

The man who set in motion the complicated series of events that eventually led to its development was George Stuntz, who had been in charge of improving the Vermilion Trail in 1869. Stuntz, who is regarded as the first permanent settler in Duluth and St. Louis County, built a log cabin, trading post, warehouse, and dock on Minnesota Point in 1853. Born in rural Erie County, Pennsylvania, in 1820, he had studied mathematics, chemistry, engineering, and surveying. From 1840 to 1843 he taught school in several Illinois communities before working as an axman on a survey of Keokuk County, Iowa. Later he moved to Wisconsin, where in 1847 his Grant County neighbors elected him surveyor and shortly thereafter county sheriff. In 1852 young Stuntz arrived at the western end of Lake Superior as a surveyor in the employment of the federal government.[16]

It was Stuntz, by then a well-established merchant and government surveyor, who received the first samples of iron ore brought to Duluth. In 1863 North Albert Posey, a blacksmith hired by the Bureau of Indian Affairs to teach his skills to the Bois Fort Ojibway, carried several specimens to Stuntz for his opinion. Posey told the surveyor that Ojibway people had brought the ore to his cabin. Stuntz surmised that the samples had reached the Indians through trade with the tribes near the Marquette Range and jumped to the conclusion that they were thus of no interest.[17]

During the Vermilion gold rush in October, 1865, Stuntz and several partners joined in the search. In nearly two months of prospecting, they found only small quantities of gold-bearing quartz. They did, however, find iron ore, and Stuntz began to reconsider the samples Posey had

shown him two years before. Concerning his prospecting efforts near
Vermilion Lake at the spot where the Soudan Mine was later developed, the surveyor recalled that "we proceeded . . . to look for the
'mountain of iron' where Posey had taken his few specimens," and
"found the place, and on the same day, while wandering about alone, I
discovered the great iron outcropping, later known as the Breitung pit
at Tower." Later he wrote, "I took a few specimens from the first
discovery — some beautiful and novel specimens of banded quartz —
one, in the shape of an S, that was done surely by compression and not
by water. After coasting around and sketching in the country, I started
home," arriving in Duluth on December 7, 1865.[18]

Stuntz packed nearly 60 pounds of samples back to Duluth but was
unable to attract the interest of financial backers. Eventually he donated several specimens to the Smithsonian Institution and discarded
the rest. For nearly a decade thereafter, however, Stuntz continued to
boost northeastern Minnesota, firmly committed to the belief that
"When this country is developed, that big mountain of iron will do it.
When they get to hauling that iron out, they will haul in its supplies
cheap."

One man became interested in the rumors and in the reports of
Stuntz and others concerning the existence of iron in the range of hills
near Vermilion Lake. He was George C. Stone, a native of Massachusetts who had moved west with his family and entered the mercantile business, first in St. Louis and then in Muscatine, Iowa. Stone
then expanded into the banking field, where he soon overtaxed his
resources. When the Civil War began and news of the firing on Fort
Sumter arrived in 1861, his Iowa depositors demanded their money
and Stone fell into bankruptcy. Throughout the remainder of the decade the struggling businessman moved from Muscatine to Chicago to
New York City and to Philadelphia. In 1869 at the age of 47, he arrived
in Duluth as an assistant to George B. Sargent, the agent of Philadelphia financier Jay Cooke, who was at that time the country's foremost
banker.[19]

Beginning in 1866 Cooke and his partner, William G. Moorhead,
had acquired for speculation more than 44,000 acres of pineland near
Duluth as well as several possible water-power sites along the St. Louis
River. In 1868 Jay Cooke and Company purchased a small block of the
bonds of the Lake Superior and Mississippi Railroad Company, which
with the help of 1,686,000 acres of state and federal land grants intended to construct a line connecting the Mississippi River at St. Paul
with the head of Lake Superior. In June, 1868, Cooke personally vis-

ited Duluth to inspect the lands his agents had purchased and to form his own judgment concerning them. He stayed in a hotel at Superior and — decked out in a high silk hat — crossed St. Louis Bay in a small boat to have a look at Duluth and go fishing in the Lester River. Impressed "with the possibility of developing a harbor," Cooke selected this "village of but six or seven frame houses, a land office, and a schoolhouse" as the terminus of the Lake Superior and Mississippi. He was convinced that the land values, timber stands, and mineral possibilities of the area demanded his immediate attention. As a result, he had dispatched Sargent and Stone to the head of the lakes to look after his interests.[20]

Backed by Cooke, the two men quickly established the first bank in Duluth in 1869. Stone took advantage of the speculative nature of the economy, invested in real estate, and was elected city treasurer in 1870. That was the year the first cars of the Lake Superior and Mississippi Railroad, financed by Cooke, rumbled over the 156 miles of track from St. Paul. It was also the year the Northern Pacific Railway Company, in which Cooke had invested heavily, began the construction of its line from Duluth to the west. In 1871 the completion of a ship canal provided the convenient harbor entrance that enabled Duluth to become a great inland port. As a result of these events, the city's position as an economic center improved rapidly; its future appeared settled beyond a doubt. Duluth, with a population of 3,131, seemed destined to become the "Zenith City of the Unsalted Sea" and the major emporium for the surrounding hinterland.[21]

But the city's prosperity was short-lived. When the Lake Superior and Mississippi completed its track in August, 1870, the railroad was in a weak financial position. Cooke, who was now in control of the railroad company, had invested too much to let it fail. Connecting the Lake Superior and Mississippi's tracks to the Northern Pacific system, he tried unsuccessfully to capture the wheat-carrying trade from Milwaukee and Chicago.

He also took an interest in reports of iron ore deposits in northeastern Minnesota. Cooke had previously owned iron lands in New York and had invested in several iron-processing companies. With the great need for railroad iron following the Civil War, he set out to construct charcoal furnaces and rolling mills at Duluth to manufacture rails and other equipment. On April 21, 1869, the *Duluth Minnesotian* speculated that within two years thriving future towns near Vermilion Lake would be connected to Duluth by railroad. The city would then become a major iron-shipping port.

But in September, 1873, Jay Cooke and Company collapsed under the weight of its railroad investments, initiating the panic of 1873 and six years of depression. So thoroughly intertwined were Duluth's economic interests with the personal fortunes of the Philadelphia banker that the community almost became a ghost town. Superior also suffered; in less than two years its population fell from 2,500 to barely 500. George Stone, too, lost most of his holdings.[22]

Refusing to give up and eager to rebuild his fortunes, Stone recalled Stuntz's stories of iron ore in northern St. Louis County. Without ever traveling north of Duluth to see for himself, the enterprising Stone headed east in 1874, hoping to attract the support of men with capital to invest. He met with industrialists Orrin W. Potter of Chicago, Amasa Stone of Cleveland, and Eber Ward, the "steel king of Detroit." Unfortunately each financier rejected his proposals, complaining of the necessarily heavy expenditures that would be required to construct a railroad through the unsettled region to get out the ore.[23]

Stone's tenacity and his previous business connection with Jay Cooke's now-defunct Duluth bank finally brought him to the Philadelphia offices of Charlemagne Tower. Born near Utica in central New York, Tower graduated from Harvard University in 1830 and, after teaching in a private school and running his father's business, completed his legal training in 1836. In 1844 Alfred Munson, a wealthy iron manufacturer and coal-land speculator who lived in Utica, employed Tower's legal talents to investigate titles to coal lands in eastern Pennsylvania. A warm friendship developed, and the young lawyer eventually became Munson's partner. In 1851 Tower moved his family to Pottsville near the Pennsylvania anthracite fields, and after the elder Munson died in 1854 he formed a partnership with Alfred's son, Samuel A. In 1871 Tower sold his Pennsylvania holdings and sought to reinvest the fortune he had made.[24]

Among his investments were Reading Railroad bonds, which he purchased through Philadelphia banker Jay Cooke. This association also led Tower to acquire Northern Pacific Railway Company stocks, bonds, and lands in which Cooke was vitally interested. The lawyer fully supported the banker's attempt to develop rail connections west across the Great Plains and to establish Duluth as a major grain terminal. During 1871–72 Tower invested a total of $250,000 in Northern Pacific bonds, especially attracted by the firm's immense land grant. The following year he became one of 13 company directors, a post he held until 1879. Although the value of his holdings in the Northern Pacific plummeted during the panic of 1873, Tower confided to a close friend that "it has not injured me." Confident that the company would recoup its losses, he worked with other directors to continue the railroad's essential building projects. In 1875, as cotrustee under the Northern Pacific mortgage, Tower helped guide the company through the inevitable bankruptcy proceedings.[25]

Following the panic of 1873, Tower absorbed Cooke's interest in exploiting the Lake Superior iron ore region. Samuel Wilkeson, secretary of the Northern Pacific who handled much of the railroad's publicity and possessed considerable information on recent developments in the western lake country, had predicted that "There are mountains of iron ore" within the railroad's land grant "from which the Road will undoubtedly, and at an early day, supply itself with rails and other iron." Formerly employed as a journalist on Horace Greeley's *New York Tribune*, Wilkeson in 1874 relayed to Tower rumors and reports of an iron mountain northwest of Duluth in an area later considered a part of the Mesabi Range.[26]

Thus in April, 1875, when George Stone called on him, Tower listened intently to the Duluth banker's reports. As a lawyer Tower had frequently handled patent applications for new manufacturing processes, and as an investor he had kept abreast of changing industrial trends. He realized the exciting future possibilities of iron ore reserves. But the panic and depression were temporarily restraining demand. Bessemer ore prices at Lake Erie ports dropped steadily from $12 a ton in 1873 to $5.50 in 1878. Nevertheless Tower and his partner Samuel Munson were sufficiently interested in Stone's proposals to sponsor an investigation of the reported deposit in order to determine its potential value. They appointed Stone their general manager to gather the necessary men, supplies, and equipment in Duluth. He, in turn, hired Stuntz to guide Tower's expert into the hinterland.[27]

Early in the summer of 1875 Albert H. Chester, mineralogist, chemist, and geology professor at Hamilton College in Clinton, New York, arrived in Duluth to direct the investigation. In 1867 he had prepared the mineral exhibit for the Paris Exposition that may have included the Minnesota iron ore samples once donated by Stuntz to the Smithsonian Institution. Upon close examination Chester had seemingly assured himself of their quality, and he readily accepted Tower's offer to gather additional samples and analyze them in his college laboratory for $250 a month and expenses. Richard Henry Lee, Tower's son-in-law and a competent surveyor with a rudimentary knowledge of chemistry, accompanied Chester.[28]

On July 13, 1875, with Stuntz and five other men, Chester and Lee left Duluth and traveled slowly by canoe for ten days before reaching the Ojibway Indian agency on Vermilion Lake. After spending two days inspecting hematite exposures and planning another trip to the lake, Chester retraced his steps southward toward the Mesabi hills between the upper Embarrass and Partridge rivers. His instructions from Tower and Munson had been specific. He had been requested to "thoroughly examine" the "Mesaba Iron Range . . . from end to end," to ascertain the length, width, thickness, and depth of the ore as well as its purity and value, and to determine "the best and nearest route for carrying the Mesaba ore by Rail Road, or otherwise, to Lake Superior." Tower and Munson also expected Chester to provide them with "full information as to whether the ore is worth having of itself, and as to the cost of mining it, and of getting it to market, or at least to navigable waters."[29]

For nearly a month Chester and his party traveled over the eastern end of the Mesabi Range between Colby (formerly Partridge) and Birch lakes, an area which had been examined by federal government surveyors Christian Wieland and Winfield S. Humason in 1872 and 1873. The surveyors had reported poor agricultural land and frequent stands of burned-over timber. In addition Humason, in particular, had noted iron deposits that made it "next to impos[s]ible to run a correct line by the needle in that portion." Chester and Lee probed for such magnetic deposits at favorable locations and collected samples for future analysis.[30]

Characterized by a series of ancient eroded hills capped by granite outcroppings that rose abruptly to 400 or 500 feet in height, the region they explored exhibited one of the earth's longest geologic histories. Cedar and tamarack swamps dominated the lower slopes, while black scars left by forest fires marked the stands of hardwood. The iron ore there, as in the entire western Lake Superior basin, occurred in bedded formations varying in thickness from less than 100 to over 2,000 feet. Geologists define these deposits as chemical sediment consisting of layers of chert, a finely crystallized gray quartz. Approximately 7,000,000 years ago a vast inland sea covered the northern portion of Minnesota, Michigan, and Wisconsin. Successive flows of lava spread over the floor of the sea and built a formation several thousand feet thick. Iron and silica, derived from the decomposition of iron-rich volcanic rocks, accumulated in small basins or saucerlike depressions rather than as continuous blanket deposits. When solidified and compacted this material developed into what is referred to as an "iron formation." A thick body of greenish rock of volcanic origin, named Ely Greenstone by geologists, encased the deposits. These surrounding rocks, estimated to have existed for about two billion years, are among the oldest in the world.[31]

A great mountain-building movement accompanied by volcanic activity followed the period of deposition and raised the beds of iron formation (or taconite) and associated sedimentary rocks into solid land. This process folded or pushed together the rock layers, forming a chain of mountains. When the inland sea disappeared and exposed the newly formed mountains, rain, wind, streams, and other weathering agents eroded much of the land and uncovered the iron-bearing rocks. Water percolated through cracks in the rock layer, softening it and dissolving many valueless materials. Sand and silt from the old ocean bottom were washed away, leaving pockets of high-grade iron ore.[32]

Glaciers made the final impact on the geologic structure of Minnesota's iron ranges. Giant ice sheets pushed down from the north on four successive occasions, scraping and gouging out major landforms. Each time the ice melted and the glacier receded it left on the surface a layer of sand, rocks, and gravel, called glacial drift. After millions of years this covering was either penetrated by shafts on the Vermilion Range or removed by steam shovels on the Mesabi in order to exploit the ore.[33]

At the time of his initial search, Albert Chester failed to realize that he was investigating merely the eastern end of the Mesabi Range, the portion that contained only lean magnetic ores. The high-grade deposits of soft hematites lay to the south and west. As a result, the Mesabi samples Chester gathered showed an average iron content well below the desired industrial minimum. He expressed discouragement with the lack of uniformity in Stuntz's alleged "mountain of iron."[34]

Although he was not sufficiently perceptive to search out the soft marketable ores of the western Mesabi, the geologist did succeed in identifying the deposits of hard hematite on the Vermilion Range. On July 31 Chester dispatched Stuntz, John Mallmann, an experienced miner, and two Ojibway back to Vermilion Lake. For several weeks they searched along the eastern shore of what later became known as Stuntz Bay. The men drilled three holes each 42 inches deep — the first test pits on the Vermilion Range — and filled them with "about eighteen inches of black powder." The resulting explosion opened a crack 4 to 5 feet deep and nearly 40 feet long. From this beginning the crew worked by hand with sledge hammers and "ash wedges generously covered with pieces of soap" to expose the ore body. A second blast, which consumed all their remaining powder, exposed more than 60 tons of high-grade ore. Mallmann recalled these events in 1890, when he wrote: "I was the only miner in the party, and Mr. Stuntz asked me to put in a blast in the ore, which was the first blast ever put in the iron ore of the Vermilion range."

On August 29 Lee gathered the Mesabi samples to be analyzed, broke camp, and led the main party back to Duluth, while Chester went north to check on the work of Stuntz and Mallmann. He observed "many good exposures of hard rich hematite," collected additional specimens for subsequent analysis, and described one particularly interesting formation as "a solid cliff of pure hematite . . . twenty-five feet wide . . . standing at least thirty feet out of the ground, and with large blocks of the same rich ore scattered in profusion over the ground." [35]

On September 8 the professor returned to Duluth and shipped his samples to Hamilton College. His analysis of those gathered on the Vermilion revealed desirable ore for Bessemer furnaces, containing only a trace of phosphorous and assaying as high as 67.77 per cent iron. Chester condemned the Mesabi locations he had seen — a judgment for which he said he was later "set down" as a "'d--d' fool and one who did not know a good thing when he saw it" — but he recommended that the Vermilion area be subjected to "most careful and exhaustive examination." Tower and Munson accepted these conclusions, and, after dividing the total costs of $7,323.39, they wrote off the Mesabi and Stone's mountain of iron. Although they did not blame him for Chester's lack of success on the Mesabi, Tower told Stone: "The expense of mining what ore there is . . . and of getting it to market forbids us entirely. The Vermilion district is more promising." Three months later, on March 14, 1876, Tower wrote Stone, "The Mesaba territory we would not touch at all at any price . . . and as for the iron ore, that is there, we would not give anything for it. In regard to Vermilion, we are not adverse to learning more about . . . that region." [36]

Throughout the last half of the 1870s Stone repeatedly offered to aid Tower and Munson in purchasing land near Vermilion Lake. Because of the lingering effects of the depression, both initially refused, agreeing that the distance of the Minnesota iron frontier from markets and the difficulties of transporting the ore hindered the development of a profitable operation. On June 10, 1876, Stone wrote from Duluth rather desperately to Tower, pointing out the value of the land, telling the financier that "Stuntz is *sure* of it," and urging him to "take up all these Iron Lands as cheap as possible." Stone added, "My necessities compel me to make you this offer . . . if it is not taken I cannot remain here." In a letter of June 16 Tower offered Stone little immediate encouragement, but he did not completely rule out the possibility of future action. "I have no doubt that the Vermillion Iron District contains good iron ore, and a great deal of it," Tower wrote. "It lies so far out of the way however, that even if it were more easily to be acquired than it now seems to be, and at quite a small price, it is not inviting to either Mr. Munson or myself. We don't feel disposed to make a purchase there, or any further explorations in that field, for the present."[37]

By 1880, however, market conditions had improved, and the available supplies from Michigan ranges seemed nearly depleted. Bessemer ore prices rose from a previous low of $5.50 per ton in 1878 to $9.25 per ton two years later. Tower and Munson heard rumors of the exhaustion of the Marquette Range, the greatest south shore iron ore producer. Moreover, the revived Minnesota Geological Survey began to publish information on the ores near Vermilion Lake. Its 1878 annual report called attention to the "costly exploration" Tower and Munson had made there "a few years since," which revealed ore varying "in metallic iron from 61 to 66 per cent." It was clear that if the easterners did not move rapidly, the extent and richness of the Vermilion Range would soon become common knowledge.[38]

The state effort, discontinued after the 1866 season, had resumed six years later. On March 1, 1872, the legislature revived it as the Minnesota Geological and Natural History Survey under the direction of Newton Horace Winchell as state geologist, a post he was to occupy until 1900. For 28 years Winchell would provide inspired leadership for one of the most complete and detailed state surveys in the nation. He directed the publication of 24 annual reports as well as a monumental multivolume *Final Report*. He helped organize the Minnesota Academy of Science and served as its president in 1879 and again in 1897. He was one of a small group of geologists in the United States who perceived the need for a strictly professional journal to provide a forum for new information. In 1888, with his brother Alexander, he established the *American Geologist*, a periodical he edited for the next 18 years.[39]

Under Winchell's energetic direction the state survey delineated the geologic boundaries of Minnesota's ore reserves. During the early years, he concentrated on the older, more settled parts of the state, but beginning in 1878 Winchell turned his attention to northeastern Minnesota. In that year he reported, "there is no actual mining being done at any place in the state," but he noted outcroppings of iron ore that resembled the high-grade ores of Scandinavia and Russia. "For making steel these ores excell," he wrote, adding "It is highly probable that these iron deposits will not lie long undeveloped."[40]

Newton Winchell's prophecy was soon fulfilled. In 1880 Tower and Munson decided that the time had come to develop the Vermilion Range. With Stone's help, they began to acquire land in the areas Chester had recommended. Accumulating the large blocks of land they wanted was the first and the easiest step in the development process. In 1872 Congress had established mining laws for the United States that set the price of mineral land at $2.50 per acre for placer deposits and $5.00 per acre for lode mining. During the next session, however, Michigan Senator Zachariah Chandler pushed through legislation exempting Michigan, Minnesota, and Wisconsin from the previously established mining codes. Thereafter mineral lands in the northeastern counties could be purchased under the same terms as agricultural or timber acreage; that is, through pre-emption, homestead, or cash payment.[41]

In December, 1879, Stone wrote to Tower outlining how the Vermilion ore lands could be cheaply obtained by taking advantage of the provisions of various laws passed by Congress for the benefit of settlers and individual states. At first Tower wanted to acquire the land after it had been surveyed by the government at public sales conducted by the federal land offices, but he abandoned the idea when Stone convinced him that a direct purchase by a single buyer would tip their hands. Instead the agent made use of pre-emption, which allowed supposedly bona fide settlers to claim 160 acres by living on it for a specified period of time and improving it. The various pre-emption laws passed by Congress over the years generally required that the land must be surveyed before it could be offered for sale, that settlers had to "prove up" and pay a minimum of $1.25 per acre at a public sale, and that they must swear they had not settled on the land for the purpose of selling it for speculation. The Homestead Act of 1862 was similar but even more generous; homesteaders could obtain land free by fulfilling the requirements and paying only small fees rather than a set price per acre.[42]

Stone also told Tower that he wanted the surveys to be made by Stuntz, "who knows this country better than any man in U.S. & whom I do not wish to have make swamp lands of any 'Iron Lands.' Winter is time to make the surveys for the men with him will see nothing but snow." In true robber-baron style, Stone continued, "Now this matter has all got to be fixed with Surveyor Genl — Local Land office, & Stuntz — and I can manage them all — am on the *best terms* with all of them. . . . It is in some respects a *delicate matter* to handle." The agent suggested that Stuntz could select 2,500 to 3,000 acres for Tower while doing the survey for the government. If Professor Chester were to return, Stone pointed out, he could reach the Vermilion region from the north via the Canadian Pacific Railroad, Thunder Bay, and Rainy Lake, and "no one in Duluth would know it."

Not surprisingly the laws passed to encourage the rapid settlement of the United States were often subject to abuses. Tower acquired much of the land fraudulently from "settlers" placed on it by Stone, a practice frequently used to circumvent the intent of the law. In so doing, he and other men of his day simply took advantage of the least expensive means to obtain land, even though the original law had been intended only to aid settlers in buying small homesteads. Tower was convinced that the land around Vermilion Lake could not adequately sustain agriculture, a view soon reinforced by the field notes of government surveyors. The region could best be exploited, he argued, through the acquisition of large tracts by entrepreneurs willing to risk substantial amounts of capital in order to build railroads and market the ore. Thus the price per acre in iron regions, originally placed at a low figure to aid small purchasers, had the effect of allowing speculators to secure vast holdings at reduced rates, defeating the original spirit and purpose of the legislation.[43]

On January 26, 1880, Tower and Munson signed a contract with Stone to purchase at least 8,000 acres near Vermilion Lake. Stone also agreed to direct the supplying of the men and equipment needed to establish mining operations. He left Philadelphia, where he had been residing, and returned to Minnesota to establish headquarters in St. Paul and Duluth. During the early months of 1880 the land in the immediate vicinity of Vermilion Lake remained unsurveyed and thus legally unavailable either for purchase or pre-emption. Although federal surveyors, including George Stuntz, had been at work on the north shore since 1856, there was as yet, nearly 30 years later, little pressure to open a frontier so universally regarded as undesirable for settlement. As late as 1883 William R. Marshall, Minnesota's former governor and railroad commissioner, told Tower that "The whole country north of Duluth, to the international boundary, is an uninhabited wilderness, there not being a single settler on the road from Duluth to Vermilion Lake. . . . To my mind there is not a quarter section of this land that is worth the Government minimum." Despite such negative views from not entirely disinterested observers, Tower continued his preparations, however slowly, for land acquisition.[44]

Still closely associated with Stone, government surveyor Stuntz measured several townships south and east of Vermilion Lake between June, 1880, and October, 1882. He began with the most important location, Township 62 North, Range 15 West, which had earlier gained Chester's praise, the only township where the amount of swampland was so small that the surveyor did not mention it in his summary description. Stuntz described every surrounding tract as containing "extensive swamp" or "largely covered with swamps." But his field notes characterized the prime area that became Tower-Soudan as showing "veins of iron ore [that] appeared to be of excellent quality."[45]

His failure to list any swamp and overflow lands was significant, for it brought into play yet another series of laws useful to speculators. Beginning with the swampland acts of 1849 and 1860, Congress willingly donated to the states, starting with Louisiana and Arkansas, millions of acres of swamp, overflowed, relatively inaccessible, and seemingly unwanted land when it appeared that these tracts would not produce revenue. Supporters argued that the transfer would allow states to finance drainage projects, construct levees, and reclaim the land by turning it into tillable acreage. In 1860 these statutes became valid in Minnesota, and two years later the legislature established a state land office. A provision dealing with the selection of swamplands in Minnesota specified that "surveys on file in the surveyor general's office are hereby adopted as the basis upon which will be accepted the swamp lands granted to the state." By not designating swampland in the key area, the obliging Stuntz assured that it could be purchased from the federal government rather than reserved to the state.[46]

Unfortunately, as happened so often throughout the history of the United States, the theories behind a particular policy quickly disappeared in its practical implementation. It soon became apparent that states were selecting tracts that could hardly be classified as swamp or overflow lands. In many areas, including Minnesota, most of the acreage was transferred to individual speculators for projects already under way or in the planning stages. Historian Paul W. Gates concluded that "Few problems have absorbed as much of the time and attention of Land Office officials, have created so much ill feeling and friction with the states, and few acts have accomplished so little of the purpose for which they were adopted as the original swampland acts of 1849 and 1860."[47]

Stone wanted to take advantage of this statutory opportunity and of Stuntz's favorable survey in order to obtain all the potential orelands and forestall any immediate competition. Beginning in August, 1880, he methodically acquired acreage by using Ojibway scrip and carefully selected homesteaders and pre-emptors, all of whom subsequently transferred their titles to Tower and Munson. Although legal challenges frequently threatened, Stone very quickly gathered in more than 20,000 acres in northeastern Minnesota.[48]

Late in 1879 before Stone began his purchases, Tower and Munson had again employed Albert Chester to make a second expedition into the iron country north of Duluth accompanied by his laboratory assistant, Herbert M. Hill. This time Chester received $100 per week plus expenses as full compensation. During the winter of 1879–80, in anticipation of Tower's actual possession of any land, George Stuntz and Nathaniel Youngblood contracted with Stone to transport supplies from Duluth to Vermilion Lake. Using sleighs, they had by the end of March hauled several tons of provisions, groceries, and tools to sustain 20 men for two or three months. In addition, Stuntz erected a small log house, which later served as a storehouse and office on the lake shore.[49]

Professor Chester arrived in Duluth on July 3, 1880. After his men celebrated the "Glorious Fourth," which left them in a condition "not worth much for work," they started north two days later. The geologist and his assistant were accompanied by three experienced miners, a cook, and two Ojibway guides. Stuntz traveled with them, although he was at the time in the employ of the federal government charged with the task of surveying the most promising townships. Nearly a dozen miners under the leadership of William H. Bassett met Chester on the upper Embarrass River. The combined exploration and surveying crews reached Vermilion Lake on the afternoon of July 16, 1880.[50]

The men quickly established a daily routine of "trenching, drilling, blasting, measuring and sampling" the iron formation. Chester spread them out along the southern and eastern lake shore and spent his time visiting each of the small work camps. He also traveled for three days with Stuntz northeast toward Burntside Lake but failed to substantiate the existence of "solid cliffs of ore." After following this schedule for nearly six weeks, the camps closed down on August 21, and most of the miners returned south via the Vermilion Trail. At the same time Chester, Hill, and several Ojibway packed ore samples into two canoes and also headed for Duluth. Traveling from the first light of dawn until darkness, they arrived on Sunday, August 25. During the next few days

Chester and Hill sorted and repacked the physical evidence of a summer's effort and totaled up their expenditures. Both men were back in New York on September 4. Stuntz was disgusted with Chester's work. On September 11, writing to Stone in a letter marked "Private," he accused the professor of having "schemes of his own. . . . With a splendid outfit[,] a good crew, and the most favorable season approaching, he suddenly abandons the whole work, and hurrys out of the country."[51]

Chester quickly completed the laboratory analysis of the samples and made to Tower a second report dealing with the Vermilion Range. The professor found two principal deposits where iron ore existed in well-defined strata covered by glacial drift. Both formations followed a general east-west direction along the southern edge of the lake. That closest to the shore line stretched for "nearly a mile" in length and varied from 11 to 58 feet in width. Samples obtained from it registered between 50.27 and 61.84 per cent iron; one contained 69.17 per cent. The second deposit measured up to 120 feet wide but extended for only a half mile. Although streaked with quartz and assaying only an average of 57.04 per cent iron, this deposit included such small amounts of phosphorous that Chester was prompted to describe it as "a true Bessemer ore." The geologist believed that the ore bodies he had seen represented an extension of the Michigan and Wisconsin formations, especially those of the Marquette Range. He concluded that "it is safe to predict the development . . . of an iron district of immense value and importance." Years later Newton Winchell wryly described Chester's contribution as "the first step which bore fruit, toward the economic development of the Minnesota ores, if not the first toward the elucidation of their geology."[52]

In spite of Chester's glowing report, Tower and Munson, aware of the unforeseen difficulties that can ensnare any new mining operation, proceeded with the utmost caution and restraint. In order to utilize effectively the results of the professor's labors, Tower re-emphasized the need to acquire additional ore lands. Effective on December 3, 1880, the financier signed a new three-year contract with Stone, agreeing to pay him an annual salary of $5,000 plus traveling expenses provided he did not engage in any other business enterprise. In return Stone was to assist Tower and Munson by creating opportunities to purchase "with reasonable promptness, and as cheaply as practicable, any and all lands, Iron Ore, and other minerals, in Minnesota and elsewhere."[53]

Before Tower could take advantage of these new arrangements, personal tragedy struck. On May 26, 1881, Samuel Munson, his business partner and close friend, died of Bright's disease. Not wanting to abandon the mining venture, Tower eventually purchased for $20,000 Munson's share of the Vermilion operation from his joint heirs, Cornelia C. Munson, his wife, and Mrs. Helen E. M. Williams, his sister.[54]

With the help of Stone, the entrepreneur then enlisted the aid, knowledge, and financial support of Edward N. Breitung, a graduate of the College of Mining in Meiningen, Germany, who had immigrated to the United States in 1849. After spending several years learning English by clerking in general stores, the young German moved in 1855 to Marquette, Michigan, where he became a merchant. Four years later Breitung settled in Negaunee, operated several iron furnaces, invested in ore deposits, and emerged as a leading figure in the developmental stages of the Upper Peninsula's ore industry. He served as a spokesman for the iron mining companies in the Michigan legislature from 1873 to 1878 and in the United States House of Representatives from 1883 to 1885. The new partner purchased for $110,000 a 40 per cent interest in all of Tower's Vermilion property. The two men also agreed to share proportionately the costs of additional land acquisition, railroad construction, and harbor development.

In addition to bringing Tower and Breitung together, Stone performed an important service of a different kind in November, 1881, when he successfully lobbied through the Minnesota legislature a special tax bill "to encourage mining in this state by providing a uniform rule for the taxing of mining property and products." The law affected all corporations "now organized or hereafter organized" that attempted to mine, smelt, or refine copper, iron, or coal. In lieu of all taxes and assessments on its "capital stock, property, income or real estate," a mining company was required to pay an annual royalty of one cent for each ton of iron or coal and fifty cents for each ton of copper shipped. Half of the revenue received would go onto the state's general fund; the other half would be credited to the appropriate county government. With the aid of Representative James Smith, Jr., a St. Paul attorney, the new tax law passed the Minnesota House by a vote of 65 to 4 and breezed through the Senate 31 to 0. Stone obviously hoped to free Tower from levies on his iron lands until the mines, once in operation, could provide profits to meet obligations.[55]

By the end of 1881 Charlemagne Tower was fully committed to developing the iron ore resources of the Vermilion Range. The appropriate townships had been scientifically explored, sympathetically surveyed, and cheaply acquired. The immediate future would see additional expenditures, legislative debates, financial battles, and the construction of a railroad. But a solid foundation had been laid, and Charlemagne Tower was ready to proceed with the opening of the first of Minnesota's three iron ranges.

GEORGE R. STUNTZ, Duluth pioneer, found iron ore on the northern Minnesota frontier in the 1860s and tried to promote its development. Photograph about 1880.

GEORGE C. STONE, Duluth banker, traveled East in 1875 in search of men with money to invest in ore. C. C. Andrews, *History of St. Paul* (1890).

CHARLEMAGNE TOWER, SR., a Pennsylvania capitalist, acted on Stone's idea. He financed the opening of the Vermilion Range in 1884 and the building of the Duluth and Iron Range Railroad. Lake County Historical Society.

ALBERT H. CHESTER was hired
by Charlemagne Tower to investigate the Minnesota iron
ranges. He condemned the eastern Mesabi and recommended
the Vermilion. Hamilton College.

CHARLEMAGNE TOWER, JR., (right) as pre
and Richard H. Lee (left) as chief engineer
the construction of the Duluth and Iron Ra
road from Two Harbors on Lake Superior
the first Minnesota iron mine at Tower-So

CAPTAIN ELISHA J. MORCOM
led the first miners and their
families from upper Michigan in
the spring of 1884 and supervised the opening of what became the Soudan Mine. Tower-
Soudan Historical Society.

THE FRONTIER CITY of Duluth as it looked about 1870, the year Philadelphia banker Jay Cooke began to build the Northern Pacific Railroad west from the future iron port on Lake Superior. Photograph by W. H. Illingworth.

BEFORE the fire of 1887 frame buildings and wooden sidewalks characterized the muddy new mining town named for Charlemagne Tower. St. Louis County Historical Society Collection, Northeast Regional Center, University of Minnesota — Duluth.

ONLY a few log and frame buildings were to be found on Ely's Second Street before the opening of the Chandler Mine and the arrival of the railroad spurred its growth in 1888. Photograph by G. A. Newton. Grace Lee Nute Collection, University of Minnesota Library — Duluth.

THE VERMILION RANGE and the townsite of Spalding on Shagawa Lake were boomed in this advertisement in the *Duluth Daily News* of April 4, 1887. Spalding died quickly when Ely sprang up nearby on the Duluth and Iron Range Railroad.

42

AN EXPLORATION CREW digging a test pit at the Pioneer Mine near Ely in the late 1880s. Note the evergreen forest in the background. Nute Collection.

HAND LABOR and a skip hoist were used to remove ore at the Chandler Mine near Ely in 1895. St. Louis County Historical Society Collection.

THE SOUDAN, the first mine in Minnesota, began as a series of open pits, one of which is pictured above. The ore was worked down into these cuts from which it was taken by wheelbarrow or mule cart to the skip hoist in the background.

LARGE CREWS of relatively unskilled workers were required to mine ore in the 19th century. This group, employed at the Soudan Mine about 1890, used air hammers, picks, shovels, and quantities of dynamite to loosen the hard ore.

UNDERGROUND MINING quickly replaced open pits at the Soudan as shown in this diagram of its operations in 1927. By 1962 when the mine closed, its shaft extended down 2,400 feet with drifts running about a mile to the east and west. The mine is now in Tower-Soudan State Park. Duluth and Iron Range Railroad, *Transportation of Iron Ore* (1927).

AN ORE TRAIN leaving the Pioneer Mine near Ely in 1889. The headframes of shafts no. 1 and no. 2 can be seen at left, and a typical "location" of miners' houses is clustered at right. Nute Collection.

AT THE SOUDAN MINE in 1884 a small, diamond-stacked Baldwin engine (left) switched wooden ore cars into the loading pocket beside the stock pile. During the winter months ore was stock-piled at the mines for shipment when Lake Superior was free of ice. Tower-Soudan Historical Society.

RESTORED to its original appearance, the "Three Spot," the first locomotive used by the Duluth and Iron Range Railroad, is preserved at Two Harbors.

VETERANS of the Duluth and Iron Range Railroad posed before the "Three Spot" in 1925. At left are William Pettibone and Thomas Owens, the engine's first fireman and its first engineer, respectively.

THE STEAMER "Hecla" and the schooner "Ironton" loading the first cargo of Minnesota ore at Two Harbors in August, 1884. The ore was shipped from the mine at Tower-Soudan over the newly completed Duluth and Iron Range Railroad.

THE "ELLA G. STONE," the tug that towed the first locomotives and cars from Duluth to Two Harbors in 1883, continued to serve the Duluth and Iron Range Railroad for many years. Duluth, Missabe and Iron Range Railroad Collection.

|3|

MINES AND A RAILROAD ON THE VERMILION

BEFORE any mining operation could profitably develop the hard hematite near Vermilion Lake, Charlemagne Tower and his associates had to devise an efficient way to transport the ore to Lake Superior for shipment to eastern furnaces. They chose to build a railroad. In the 1860s Henry M. Rice, the founder of Superior who was then representing Minnesota in the United States Senate, had correctly predicted, "When we have communication by railroad to the head of lake navigation, the most skeptical can not overrate the mineral wealth that will be developed, nor the commercial advantages that will inure to the state."[1]

Well aware of the heavy investment of capital that would be necessary to lay tracks across difficult terrain, Tower was hopeful that the Minnesota legislature would provide a grant of state swamplands to help offset his expenditures. Late in January, 1881, he incorporated the Duluth and Vermilion Railroad Company, proposing the "survey, location and construction" of a standard-gauge track from Duluth to Vermilion Lake. Concerning the carefully selected name of the proposed road, Tower wrote Stone on January 20, 1881: "I concede you must use 'Duluth' as part of the corporate name, it looks as if you could not well avoid doing so. . . . Spell Vermilion with one 'l' according to the dictionaries." But before Minnesota's legislators could react to the financier's request for a land grant, the state's voters on November 8, 1881, approved an amendment to the constitution stipulating that in the future state swamplands should be sold like school lands, or, in other words, they should no longer be handed over free to railroad companies. This brought to a halt Tower's plans for the survey and construction of the Duluth and Vermilion.[2]

Since the acquisition of a new land grant seemed impossible, Tower

49

turned to the most promising alternative — obtaining one already approved and previously assigned to another railroad firm. The one he fixed on had been given to the Duluth and Iron Range Railroad Company, which had been formed by a syndicate of Duluth and Ontonagon, Michigan, businessmen, including George Stone. Ontonagon, located in the copper district of the Keweenaw Peninsula, was for many years the largest commercial center on Lake Superior. Merchants there had frequent business across the lake in Beaver Bay, where they were well acquainted with the Wieland brothers who kept them informed concerning the rumors and reports of mineral deposits in northern Minnesota. On December 21, 1874, the ubiquitous George Stone and eleven others, including three Duluth men, had incorporated the Duluth and Iron Range Railroad Company, proposing to construct a line from Duluth to ore lands in Township 60 North, Range 12 West, on the Mesabi Range. Through the active support of Charles H. Graves, a state senator from Duluth who later became United States minister to Sweden, the Minnesota legislature on March 9, 1875, had awarded the railroad company a generous swampland grant of ten sections for each mile of track built. The land could be selected within ten miles of either side of the proposed right of way, and the company could make up any deficiencies by choosing other state swampland elsewhere in St. Louis or Lake counties. To aid in offsetting the costs of construction, which was to be completed within five years, the law further provided that none of the land granted would "be subject to taxation until the expiration of five years from the issuance of the patent . . . unless previously sold or disposed of by said railroad company." [3]

The Ontonagon and Duluth investors, however, for reasons that will become clear in the next chapter, never built a single mile of track. Between 1875 and 1882 numerous events transpired to complicate the firm's title to its land grant. The original legislation, as amended on February 17, 1876, gave the railroad until February, 1879, to finish its track and receive all of its swamplands. In 1877 William W. Spalding, a wealthy Duluth businessman who was the railroad's president, organized a new firm called the Duluth and Winnipeg Railroad Company and proposed a line running from Duluth to some unspecified location on Minnesota's northern border between the Red River and Lake of the Woods. On March 9, 1878, the Minnesota legislature "transferred and vested" the Duluth and Iron Range land grant in the Duluth and Winnipeg "in case of forfeiture by the said Duluth & Iron Range" — a seemingly sensible provision that was to have unforeseen results. The lawmakers also allowed the new recipient of the grant until 1888 to complete its track.[4]

Noting the ambiguity of the "in case of forfeiture" phrase, George Stone, acting on Tower's behalf, asked Representative Smith, who had earlier been so helpful, to determine whether the 1878 transfer to the Duluth and Winnipeg was valid or whether the original 1875 legislation giving the grant to the Duluth and Iron Range took precedence. He raised the question because the time specified in the 1875 law had not expired when the 1878 statute was passed. Counsel Smith concluded that the 1875 legislation held precedence until it expired, and that additional legislative action would be needed because the transfer to the Duluth and Winnipeg had been passed before the expiration of the old law. Buoyed by this news, Tower sought to gain control of the older railroad company and its land grant.[5]

With the help of Stone, Tower set out to purchase a majority stock interest in the Duluth and Iron Range from its original incorporators. Because the stock had never been publicly subscribed, it was held by the Duluth and Ontonagon syndicate, of which George Stone had been a member. Tower's plan was a simple one. As a legitimate member of the original group, Stone persuaded seven of the eleven other incorporators to join him in petitioning for a meeting of the stockholders. When they gathered in Duluth on the afternoon of March 1, 1882, Stone proposed that the company's books be opened in order to sell at least 5,000 shares for a cash payment of only 1 per cent, the remainder to be paid in undesignated future installments. Despite vigorous and heated protests from Spalding, Stone's resolution passed by a vote of 8 to 2. The majority then divided the Duluth and Iron Range stock into shares worth $100 apiece and, as previously arranged, each man subscribed for one share. Then Stone subscribed for 1,000 shares each in the names of Tower, his son, and Richard Lee, his son-in-law — a total of $500,000 worth of stock. He took the remaining 1,993 shares himself, giving the Tower associates firm voting control over all but 7 of the 5,000 shares for a cash outlay of only $4,993. The investors immediately elected a new board of directors who, in turn, chose Stone as the railroad company's president and named Lee as chief engineer. In one afternoon Tower had gained control of the Duluth and Iron Range and, he hoped, its swampland grant.[6]

As the next step in his plan to develop the iron ore near Vermilion Lake, Tower incorporated the Minnesota Iron Company on December 1, 1882, with headquarters in St. Paul. He named himself president, Edward Breitung vice-president, George Stone general manager,

Charlemagne Tower, Jr., treasurer, and Thomas L. Blood, Stone's son-in-law, secretary. Tower divided the $10,000,000 worth of capital stock into 100,000 shares at $100 each. Stockholders possessed one vote for each share owned, and they annually elected a board of directors who selected the company's officers from within their own group. The eastern entrepreneur assured himself of complete control, however, by providing the president with the power to appoint an executive committee and all other committee chairmen. In addition he insisted that only the president and the treasurer — Tower and his son — could oversee the transfer of stock in the firm's Philadelphia branch office.[7]

By the end of 1882 Tower had completed the framework of the two key companies he needed to mine Vermilion ore and transport it to Lake Superior by rail. Then he turned to the practical problem of determining the best route for his railroad. After completing several surveys, Stuntz, Lee, and John B. Fish, a civil engineer, reported their findings to the new Duluth and Iron Range stockholders. They recommended that "about 70 miles" of track be constructed from Agate Bay on the north shore of Lake Superior inland to the Vermilion Range, rather than from Duluth to the Mesabi Range as originally proposed in the 1875 legislation. They also suggested Duluth as a desirable secondary lake terminus to be reached by a 26-mile extension along the north shore from Agate and Burlington bays, the site of the future town of Two Harbors. Tower was willing to accept these revised plans if the legislature would authorize the new route to the Vermilion instead of the Mesabi, extend the period allotted for actual construction, and, most importantly, renew the reorganized Duluth and Iron Range's claim to the land grant.[8]

The prospect of obtaining new legislation providing for these changes brought the Tower forces once more up against the opposition of William Spalding, who held firmly to the position that the land grant belonged to the Duluth and Winnipeg. To thwart this contention, Representative Smith, Stone's friend and counsel, introduced a bill in the Minnesota legislature in January, 1883, that would permit the new controlling interests of the Duluth and Iron Range to "re-locate and change" the proposed terminals and that would reaffirm its right to the land grant provided for in the 1875 law. Smith also asked his fellow legislators to extend the time period for completing the track for an additional five years from the date of the new law's passage. The bill sailed through the state House of Representatives on February 13 by a substantial margin of 61 to 14, but the Senate was not going to be so easy. Tower's associates realized that a struggle with Spalding's forces awaited them in the upper house.[9]

When the bill reached the Senate, Stone secured the active support of Charles D. Gilfillan, a prestigious senator and a respected businessman who owned the building in which the Duluth and Iron Range's St. Paul offices were located. Gilfillan argued that the Duluth and Iron Range deserved legislative assistance because it would benefit the entire state by profitably developing the Vermilion iron resources. William W. Billson, a young Duluth lawyer and former United States attorney for Minnesota, vigorously defended the position of Spalding and the Duluth and Winnipeg aided by Cushman K. Davis, an influential St. Paul lawyer who was a former governor of the state. Billson and his colleagues persuaded the Senate Judiciary Committee to recommend that the bill be postponed indefinitely. Gilfillan fell ill, and Stone through sheer force of argument lobbied the committee members so successfully that enough of them changed their votes, permitting the bill to come before the Senate during its last session before adjournment on the evening of March 1.[10]

In a dramatic entrance as debate began, Gilfillan was carried onto the Senate floor. Too ill to stand, he contended that, unlike the Duluth and Winnipeg, the Duluth and Iron Range now possessed vigorous leadership with capital to spend on immediate construction. His personal prestige and his rhetoric prevailed only after the legislative clock had been stopped at 11:30 P.M. The Senate approved the bill by a vote of 25 to 15 and adjourned at 3:45 in the morning. Tower had successfully obtained the right to the land grant as well as the right to relocate the line.

As spring turned to summer in 1883, only a few minor organizational and financial difficulties remained to be overcome before actual construction of the railroad between the Vermilion Range and Lake Superior could get under way. Tower transferred to the Minnesota Iron Company not only full ownership of the road, but also 17,666 acres in St. Louis County and 2,840 acres in Lake County, which Stone had acquired for him by means of questionable homestead and preemption claims. Stone resigned as president of the Duluth and Iron Range because of ever-increasing responsibilities as general manager of the parent iron company. As a result, Tower's 35-year-old son, Charlemagne, Jr., moved to Duluth as the railroad's chief executive. The senior Tower insisted that the railroad's board of directors vote to issue $2,500,000 worth of 20-year bonds in order to finance construction, and they did so on May 30, 1883. Both the railroad and the iron companies then mortgaged their properties as well as the future swampland grant of some 600,000 acres to the Fidelity Insurance Trust and Safe Deposit Company of Philadelphia.[11]

One final project involved the awarding of the construction contract for the rail line. "In response to advertisements in the newspapers twelve contractors submitted sealed bids which Lee, as chief engineer, examined painstakingly," wrote Tower's biographer. "The lowest bid, $12,519 per mile, came from John S. Wolf and Company of Ottumwa, Iowa. Wolf was a well-known railroad builder in the West, having performed contracts on the Chicago, Burlington, and Quincy and on various other roads in Iowa, Illinois, and Wisconsin. . . . A contract between the Duluth and Iron Range and Wolf and Company was signed as of June 20, 1883. Wolf and Company agreed to complete the railroad from township 62-15 to Agate Bay on or before August 1, 1884, or forfeit a penal[ty] bond of fifty thousand dollars." The younger Tower wrote his father that the contractors "look like good men . . . we are satisfied to have them." Seemingly, no serious problems now stood in the way of building Minnesota's first ore-carrying railroad.[12]

During the summer of 1883 Stuntz, Fish, and Lee completed rough surveys of the proposed right of way. Chief engineer Lee had earned degrees in civil and mining engineering and had gained practical field experience in northeastern Minnesota as a member of Albert Chester's first exploring party. Lee established an office at Two Harbors but kept in close telegraphic contact with the younger Tower in Duluth. The work on the railroad began at the Lake Superior end of the line. The small lake port that was soon to be formally named Two Harbors hummed with activity as men unloaded quantities of provisions and equipment during the warm summer months. Since there were no roads from Duluth to Agate Bay, all the supplies had to be carried by boat along the north shore.[13]

The most hazardous journey involved the Duluth and Iron Range's first locomotive, the "Three Spot," a wood burner built by the Baldwin Locomotive Works at a cost of $12,000. William A. McGonagle, a civil engineer from Pennsylvania who was to spend the rest of his life in the service of Minnesota ore-carrying railroads, directed the moving of the locomotive from Duluth to Two Harbors. His workmen carefully strapped the 60-ton engine on a specially built scow towed by the "Ella G. Stone," the company's tugboat which had been named in honor of George Stone's daughter. On July 8, under the command of Captain Cornelius O. Flynn, the tug steamed out of Duluth on a clear and calm

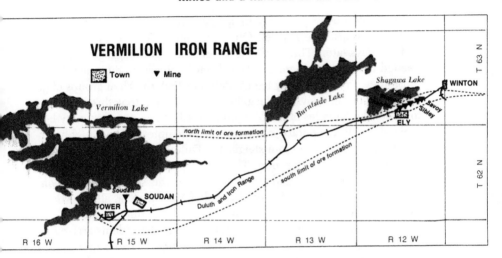

lake. But about eight miles south of Agate Bay near the mouth of Knife River, a storm blew up suddenly. The men pumped furiously as Flynn pulled alongside with axes ready to cut the scow free if it began to sink. But it did not sink, and that night the "Three Spot" puffed safely ashore ready to aid in building the rail line to Vermilion Lake.[14]

Two more locomotives were delivered that summer, and a wooden ore dock on the lake shore was partially completed. Some 600 laborers were "actually at work upon the road" in August, 1883, following Lee's survey, blasting and clearing the forest, grading the roadbed, filling in the muskegs and swamps, and laying track. Working into December, even though ice had to be chopped from the roadbed, Wolf's crews completed nearly 20 miles of track before stopping for the season.[15]

During the winter months, the company's attention turned to Vermilion Lake and the development of Minnesota's first iron mine. In order to commence stock-piling quantities of ore in the spring of 1884, work crews, using sleds and horses, hauled tons of supplies over the snow-covered Vermilion Trail. To obtain the experienced miners he needed, Tower planned to transport them from northern Michigan. Through the efforts of Stone once again, he hired Elisha J. Morcom, a Cornish mining superintendent, to recruit whole families in order to ensure a stable core of workers. Morcom, who had served in the Michigan legislature in 1882, registered some 350 men, women, and children, largely at Quinnesec, Michigan, where he had recently served as a mining captain, and at Iron Mountain and Vulcan on the Menominee Range.[16]

As winter slowly gave way to spring, Morcom divided the people
into two groups. The first left Quinnesec on March 10, 1884; the second ten days later. Three railroad coaches and two baggage cars were
needed to transport them overland from Michigan to Superior. From
there they crossed the bay ice by sleigh to Duluth. Then groups of 30
people set out in relays for Vermilion Lake. Since it took three full days
to travel the 80 miles of the Vermilion Trail, the mining company had
erected small temporary camps at 12-mile intervals. In highly organized fashion three or four sleighs left Duluth in the morning and the
same number each afternoon. The initial parties stopped at the first
camp for a noon meal and then traveled to the second way station,
where they spent the night. By the end of March, 1884, all the Michigan miners and their families were settled at Vermilion Lake.

Tower had left little to chance. Before the first residents arrived on
March 17, John Owens of Escanaba, Michigan, had supervised a
crew of carpenters who erected a sawmill and 28 very plain but warm
one-room frame houses. The Minnesota Iron Company opened a well-
stocked general store in the new settlement, which was soon fittingly
named for Charlemagne Tower. The company store went out of business, however, as soon as private merchants established themselves.
The eastern entrepreneur had no desire to create a tightly controlled
company town. George Stone expressed the opinion of the firm's management when he said, "my policy is to get out of stores, boarding
houses, sawmills — everything, but mining at the earliest day."[17]

Some preliminary work at the mine site had been done by a small
crew of men in 1882 and 1883. Now Elisha Morcom was ready to begin
larger-scale operations in what became a total of nine open pits. By the
beginning of April, 1884, 140 men were already working at the mine,
and by early June Morcom had nearly 540 men operating hand drills
and sledge hammers. They stripped away the surface glacial drift,
blasted out giant boulders, and exposed several very promising veins.
As the slow, backbreaking labor deepened the cuts, the miners used
horses to hoist the ore to the surface for stock-piling.[18]

Morcom controlled the mining activity with a firm hand, while also
concerning himself with the general welfare of the miners and their
families. From March 1 to December 31, 1884, he reported that he had
employed an average of 384.6 men "in and around the mines." Skilled
men received less pay from the Minnesota Iron Company than did
their Michigan counterparts — $1.79 per day instead of $2.09. In
contrast, unskilled labor earned $1.60 per day compared to $1.40 to
$1.50 in Michigan, where the work was done in underground shafts

instead of outdoors in the harsh climate. March, 1884, brought unusually good weather, and by the end of July Morcom's men, having stock-piled sufficient ore, were awaiting the completion of the tracks to the new settlement. Eagerly anticipating the end of isolation, Franklin Prince, the company's mining engineer, wrote Tower, Jr., on July 10: "The R.R. is only about 12 miles from us and I hear very encouraging reports of the progress. . . . We hail with pleasure the coming of the road and to my ears, no music can sound as sweet as the rumbling of an ore train as it threads its way through this wilderness."[19]

Construction of the Duluth and Iron Range Railroad between the mine and Two Harbors proved to be a major undertaking. John Wolf's company faced a terrain that varied from hard rock ridges to bogs and tamarack swamps. Problems of faulty bridge erection, a snowy April, and deserting laborers plagued the effort. After inspecting the work in May, 1884, Samuel P. Ely wrote confidentially to Tower that finishing "the railroad by the 1st of August is still practicable with a sufficient force; but it is a question of force. It will take *1500* men to do it and there are not more than 400 men now." A bonus of $50,000 awaited Wolf if he met the designated deadline of Friday, August 1, a penalty of $50,000 if he failed. But when Tower realized that a widespread superstition among the miners regarded Friday as bad luck, he substituted Thursday, July 31. Wolf took advantage of several weeks of excellent weather early in July, employed 1,400 men working in shifts around the clock, and completed the road on schedule.[20]

Company officials planned a celebration and declared the day of the first shipment a holiday for the miners and the railroad gang. Thomas Owens was the engineer in the cab of the "Three Spot" as it pulled ten empty wooden ore cars built at Stillwater, Minnesota, and a caboose containing the young railroad president, Charlemagne Tower, Jr., and his party. Leaving Two Harbors early in the morning of July 31, the train arrived at Tower to the cheers of the miners, their families, and most of the Ojibway population from the nearby reservation whom the company had invited to the celebration. Wolf wheeled the first load of ore up a platform and dumped it into car number 406, one of the 20-ton, eight-wheeled cars to be loaded. In turn each person attending threw in a chunk of ore for good luck. Then someone placed a small fir tree upright in the first car, a tradition that continues to the present upon the opening of each new mine. About midafternoon Owens cautiously started the train down the steep grades, assisted by hand brakes on the overloaded cars. Because the track through the wilderness had no water tanks or pumps, the crew stopped frequently at streams and ponds to take on water for the tender tank. The first shipment of ore from the Vermilion Range reached Two Harbors at 11:00 P.M. that night, eight hours after it left Tower and one hour before the deadline set for completion of the railroad.[21]

At the port workmen stored the cargo in the 46 pockets of a partially finished wooden ore dock, which as yet had a storage capacity of only 3,000 tons. The first lake shipment left Two Harbors on August 19 aboard the steamer "Hecla" (1,427 tons of ore) and the barge "Ironton" (1,391 tons). It was consigned to George H. and Samuel P. Ely of Cleveland, but it was unloaded and stored on the Carnegie Brothers & Company's docks at Cleveland, Ashtabula, and Sandusky, Ohio. By the end of its first season, the Minnesota Iron Company had shipped 62,124 of the 94,643 tons of ore mined by December 31, 1884, and the Vermilion Range was launched on its long career as a major producer.[22]

Unfortunately the first Vermilion ore was shipped in the midst of a depressed market accompanied by shrinking stock values and numerous bank failures. Eastern furnacemen, the major consumers of Lake Superior ore, constantly complained that mineowners demanded high payments for their products. In 1884, however, the furnacemen took advantage of a sluggish demand for steel and their full inventories of iron ore and organized a pool to force ore prices down.[23]

This situation presented grave problems for Charlemagne Tower, who had invested a substantial share of his personal fortune in developing the Vermilion. By the time the "Three Spot" reached the Two Harbors ore dock his expenditures for railroad construction and equipment totaled more than $2,000,000. Rather than halt mining operations, as his son suggested, Tower decided to continue shipping, wait for an improved market, and borrow the necessary additional capital. Opinion in the financial world placed railroad speculation near the top of a list of causes responsible for the depressed economy, a state of affairs that offered little encouragement to investment in a new venture located in the hinterland of northeastern Minnesota. But Tower never seriously considered putting his stocks and bonds on the open market. Instead he sought a financial partner willing to invest capital and allow him to retain control of the entire operation. He had earlier limited Stone's negotiations with Andrew Carnegie, James J. Hill, and others by warning, "We don't want anybody concerned in that ore property but ourselves."[24]

After Edward Breitung offered a financial plan Tower would not

accept, the eastern financier turned to George and Samuel Ely, prosperous Marquette Range mining promoters. These natives of Rochester, New York, had abandoned a milling operation to invest in mining
and railroad construction. In 1883, then living in Cleveland, they purchased several thousand shares of Minnesota Iron Company stock from
Breitung. Samuel Ely was especially interested in rapidly developing
the mine and completing the railroad. As early as February, 1883, he
told George Stone that "the most important thing is to get the line
begun and finished at the *earliest practicable moment.* . . . We shall not
make much impression on the minds of consumers until we can say that
a railroad is building, and *when* our ore is coming into market." However, Stone told Tower on February 19, 1883: "You will readily see that
no time is to be lost . . . nothing would please Mr. Breitung better than
for us to stop all work at Vermilion and shut up the saw mills until it
was settled that we had a railroad."[25]

During the summer of 1884 the Ely brothers, after accepting an offer
to act as exclusive sales agents to market Vermilion ore, established a
branch office of the Minnesota Iron Company in Cleveland. Even with
their years of experience, knowledge, and prestige in the iron and steel
world, however, they were unable to counteract poor eastern market
conditions. Discovering that furnacemen were reluctant to purchase a
new and untried ore, the brothers gloomily advised Tower on August
15, "It is the dullest ore market there has ever been." The uneven
quality of early shipments from the Vermilion apparently complicated
the use of the small amounts of ore the Elys were able to sell in
Chicago. Of the 62,124 tons shipped in 1884, only 523 tons were sold
that year.[26]

Through persistent effort and individual agonizing, Tower and his
associates attempted to secure sufficient capital to meet their major
obligations. In desperation Tower turned to close personal friends and
former business colleagues. Working through Joseph M. Shoemaker, a
banking agent, he approached Samuel Munson's widow and sister for a
loan of $400,000. Mrs. Munson and Mrs. Williams both expressed a
deep regard for their old friend, but after reviewing a confidential
statement of Tower's financial situation they considered it "unwise" to
make the loan.[27]

On November 11, 1884, with their backs to the wall, Tower, Jr.,
Stone, Breitung, and Samuel Ely met to confer in Chicago. Eleven
days later they entered the office of banker Lazarus Silverman. A
Bavarian emigrant, Silverman had accumulated his fortune as a merchant, real estate speculator, and banker. After driving a "hard bar-

gain" the Chicagoan loaned his visitors a total of $150,000 in six notes dated November 26, 1884, at 8 per cent interest. As collateral Silverman held $150,000 in railroad bonds and another $150,000 in iron company promissory notes. In addition he received a bonus of 800 shares of iron company stock from Stone, Breitung, and the Ely brothers, and 3,000 shares of Breitung's stock in a Michigan mine.[28]

Several weeks earlier Breitung had agreed to raise an additional $150,000 if Tower could supply $200,000. As a result, on November 12, Tower left Philadelphia by train to make a personal appeal to Helen Williams at Utica, New York. He emerged from this visit with a loan of $300,000. In return Mrs. Williams received a promissory note, payable on demand, at 12 per cent annual interest, a cash bonus of $10,000, and $600,000 in Duluth and Iron Range Railroad bonds as collateral security. Tower and his associates had paid a heavy price, but they managed to keep their venture afloat through the winter of 1884–85.[29]

The following year witnessed a marked improvement in the nation's economy. *Iron Age*, a major trade journal, cautiously editorialized that steel rail manufacturers had abandoned their differences over pool allotments. As a result, the eastern market began to accept the new source of raw material. The high iron content of Vermilion ore, combined with its low percentage of phosphorous, made it increasingly attractive. By August the Elys had sold 113,000 tons, and the following month they made their first large sale of 45,000 tons to Carnegie Brothers & Company of Pittsburgh. By the end of the year they secured additional substantial orders from several Chicago steel manufacturers and improved on the Carnegie contract.[30]

During 1886 the position of Vermilion ore on the eastern market continued to advance. The Bessemer price at Lake Erie ports had stabilized at $9.00 per ton during the first shipping seasons of the 1880s. In 1884 it dropped to $5.25 and the following year it fell another $.50 per ton, but by 1887 the rate rose to $6.00. The non-Bessemer price did not fluctuate so drastically, although it opened the decade at $8.00 per ton, fell to $4.75 in 1883, and remained within $.50 per ton of that figure for the remainder of the 1880s. Tower used the financially successful seasons of 1885 and 1886 not only to liquidate completely all the debts accumulated during the first year, but also to raise the wages of his employees. In April, 1886, the daily rate for experienced, skilled miners went up to $1.95; unskilled laborers now earned $1.60 and $1.70.[31]

The year 1886 also saw the completion of a rail line from Two Harbors to the Duluth terminus specified in its land grant legislation.

There is some evidence that the threat of competition from the Duluth and Northern may have moved the Duluth and Iron Range to finish the extension, although Lee believed that the rival proposal was "gotten up for the sole purpose of inciting the Duluth & Iron Range Railroad Company to build the uncompleted portion." Wolf and Company employed nearly 600 men to construct the 26-mile Lake Division. In spite of the numerous bridges which had to be built, favorable weather conditions, relatively level terrain, and little deadline pressure allowed the firm to finish track laying by December 1. When the extension opened for traffic the following spring, the Duluth and Iron Range owned 95.7 miles of track connecting the mines at Vermilion Lake with the nation's railroad system at Duluth.[32]

That port city was not, however, destined to play a major role in the marketing of Vermilion Range ore, which was shipped almost exclusively from Two Harbors. Named for the adjoining Burlington and Agate bays, the site of Two Harbors had been occupied only by a single rough log cabin built by one Thomas Saxton when William McGonagle and his crew arrived early in 1883 to begin clearing the railroad right of way. By September Tower, Jr., presided over a town meeting that elected "two justices of the peace, two constables, a town clerk, a treasurer, and overseer of highways." The single shanty had been replaced by a town of nearly 400 residents. Population continued to increase, and on March 9, 1888, the village was incorporated, and the seat of Lake County was moved to it from Beaver Bay.[33]

Charlemagne Tower planned and controlled Two Harbors' early physical development, demanding wide streets, public squares, a reliable water supply, and the complete elimination of real estate speculation. Lots were reserved solely for "men who will build and live there." This atmosphere prevailed even after the Pennsylvania financier's direct involvement ceased. For example, on December 23, 1890, the *Duluth Daily News* reported that Two Harbors contained several large hotels, a substantial courthouse, a fine gravity-pressure water system, and "a church for each saloon in town."[34]

In order to operate more efficiently the Minnesota Iron Company constructed a brick machine shop, railroad car repair facility, foundry, and roundhouse. The initial wooden ore dock stood 50 feet above the water, extending 1,200 feet into Agate Bay. When completed in the summer of 1884, its 162 pockets possessed a total storage capacity of 16,000 tons. The firm built a second ore dock of equal length the following summer. It contained only 141 pockets but could handle

18,000 tons of ore. By the end of 1890, claiming a population over 1,200, Two Harbors also had several general merchandise docks and a very efficient coal dock.[35]

Approximately 69 miles northwest, along the southern shore of Vermilion Lake, community development was even more intense. By 1885, 1,431 people resided at Tower, the business center, and at Soudan, the residential center for miners located a mile to the east. (Soudan was humorously named for the tropical Sudan region of Africa, whose hot weather provided a contrast to the extremely cold winters of northern Minnesota. The mine and the town eventually assumed the same name.) A variety of ethnic backgrounds composed its settlers. Barely one-quarter had been born in the United States, principally in Minnesota, Michigan, or Wisconsin. Iron ore had been the major attraction, and the large proportion of the inhabitants from Cornwall, England, and Scandinavia, especially Sweden, reflected the presence of experienced miners. Father Joseph F. Buh, a Slovenian missionary who established the Catholic church at Tower, reported in 1888 that the community had "20 Irish families, 12 French families and 300 Krainer (fellow countrymen, Slovenes from the Krain district in Jugoslavia)." From the beginning a family orientation was evidenced by the fact that women composed nearly one-third of the total population.[36]

The interests, concerns, and goals of these people were similar to those of residents in many new, often isolated frontier towns. Drawn to the area because of its mineral and timber resources, they gradually devised the institutions and services necessary to community living. Among them was a volunteer fire department, for fear of forest fires was widespread. Tower suffered its first serious fire in 1887. Eight lives were lost when a three-story saloon and boardinghouse burned to the ground. After that, a volunteer fire department was organized and trained, hundreds of feet of hose were purchased, and pumping facilities to utilize lake water were quickly established. With the threat of fire in mind, merchants set out to construct substantial buildings. A yard, with a daily capacity of 40,000 bricks, carried on a brisk trade.[37]

By mid-1888 the principal ingredients for a thriving community had been established with a public school, several churches, and the *Vermilion Iron Journal*, the first weekly newspaper north of Duluth. On July 19, 1888, the paper boasted: "Tower today has a genuine boom — brick blocks, elegant residences, and costly mercantile houses are being erected." The five-year-old town claimed 5,000 residents, six general stores, two hardware and two furniture stores, a bank and a building and loan association, men's and women's clothing shops, two drugstores, jewelers, barbershops, and three photographers among other businesses. Many of the unmarried miners and employees of the two sawmills were willing customers of the six hotels, numerous boardinghouses, and restaurants.[38]

The local editor frequently urged Tower residents to improve their community and solidify its economic foundations. By 1889 more and more newspaper space was devoted to advertisements, many from Duluth firms as well. Two years later a street railway system about two miles long was built, making it more convenient for Soudan residents to shop in Tower. In October, 1891, electric street lights replaced the kerosene lanterns of previous years. Within another year the city had a newly constructed three-story, 40-room hotel, described as one of the "neatest buildings and most modernly equipped hostelries in Minnesota."[39]

When a 12,000-barrel-capacity brewery opened, the *Vermilion Iron Journal* of April 21, 1892, tried to find some economic benefit in this newest addition to Tower's businesses. "The Iron Range Brewing Company has no uncertain future before it. . . ," the paper said. "With the foreigners that constitute a large per cent of the population in this locality, there is sure to be a market for the company's product. . . . If the people will have beer — and it is evident that they will — why not manufacture it here and let the city derive the benefits of an employment giving institution."

Like people in many similar communities, Tower residents demonstrated a genuine interest in baseball, boxing, and drama. There was even a brief attempt to construct an opera house. But all social activities were not so enthusiastically supported. The iron ranges of northeastern Minnesota were no different from the gold and silver mining towns of the West when it came to an exploitive social atmosphere. The men and women who came to "mine the miners" in Tower incurred the wrath and outright hostility of many citizens. The *Vermilion Iron Journal* of November 17, 1887, provided a classic Victorian description of what the reporter regarded as an urgent concern: "a *Journal* scribe . . . dropped anchor within the amiable walls of a rotten den — the most depraved house in St. Louis county. Inside were blisters — bums, burglars, sluggers, toughs, poor whiskey, old chippies, dogs, and other animals to[o] numerous to mention . . . a blear eyed pimp, and two half decent citizens of this city sat in close conju[n]ction to the mass of whiskey and disease, [they] was covered by a wardrobe that would be considered exceptionally warm in the hereafter — if their destination was, as it ought to be hell. The old female carbuncle with her sieve-like voice kept on singing her chant; which was as harsh as the ravenous braying of a mule about day break. . . . In the bar-room the same state of affairs existed — miners spending their hard earnings. . . . The *Journal* pities the fallen female wretch . . . but would be only too glad to bless the hour that wiped the pimp and his following from this green earth."

On April 9, 1889, the community of Tower completed its incorporation as a city, but the initial period of prosperity and expansion was soon to end for its founder, Charlemagne Tower.[40] Three years earlier the *Minneapolis Tribune* of August 17, 1886, unknowingly forecast his future difficulties when it reported: "At present the only mine . . . is that owned by the same parties who built the road, but those who realize the tremendous future development of the mining industry of northern Minnesota, look for a number of new mining interests in a short time."

New York and Chicago industrialists had maintained a close watch over the success of Tower's operations. In the spring of 1887 John Birkinbine, chief engineer-reporter for *Iron Age*, traveled to the northern Minnesota frontier to describe for readers the Vermilion mines and the railroad. The result was an article in an April issue of the influential national weekly praising the startling purity of the ores mined in the five open pits then operating and the "efficient management" of the whole enterprise, which the journalist ranked as the nation's third largest mining company.[41]

Even as Birkinbine wrote, a move was well advanced to seize control of the desirable source of iron ore Tower had developed. Pig iron, Bessemer, and open-hearth steel production more than doubled in the 1880s. Improvements in machinery and technology, lower production costs, and more efficient rail and lake transportation facilities encouraged the formation of direct links between manufacturers and their supplies of raw materials. In three short years the Vermilion Range had established its position as a valuable source of high-grade ore, and it loomed as a desirable acquisition for any major steel producer.[42]

The move to loosen Tower's grasp on his mine and his railroad came from Henry H. Porter, a director of both the Illinois Steel Company and the Pullman Palace Car Company, who headed a syndicate that included among others John D. and William Rockefeller of Standard Oil, Marshall Field of the Chicago mercantile firm, Cyrus H. McCormick the farm implement manufacturer, and Jay C. Morse of Union Steel Company, which had been an early customer for the Vermilion ores sold by the Ely brothers. A Maine native, Porter was educated in public and private schools before beginning to clerk in a rural general store. At the age of eighteen in 1853, he moved to Chicago and slowly accumulated a fortune by investing in lumber, banking, stockyards, and several railroads. In 1882 Porter sold some of his railroad holdings and sank the profits into the Illinois Steel Company, a consolidation of several manufacturing firms. [43]

Looking for a source of high-grade iron ore, Porter's syndicate acquired 25,000 acres adjoining Tower's property to the northeast near the present community of Ely with the intention of squeezing him out. Through Samuel Ely, the financially powerful group opened negotiations in 1886 to purchase the Duluth and Iron Range Railroad by threatening to construct a competing line from their ore locations to Lake Superior. Tower and his son had no immediate intention of selling. By encouraging the development of new ore deposits, they believed the D&IR would haul ore so cheaply that the Porter group could afford to do nothing but build connecting branch lines. Tower, Jr., did not think that "these gentlemen will ever find it to their advantage to build a railroad to Duluth, or anywhere else. We have done that and have done it successfully." But he expressed a willingness to meet with Porter "to go over our own policy and agree upon the plan of action that we are willing to take." [44]

Samuel Ely had worked with Porter and Morse for several years and knew them well. With a direct connection to the Minnesota Iron Company, he was a natural intermediary between the two interests. Ely had initially proposed that Tower could avoid destructive competition by selling the D&IR to a nationally established railroad network. He believed it inevitable that these larger railroad firms would appear on the Minnesota iron frontier just as they previously had moved onto the Marquette and Gogebic ranges. He suggested that the eastern financier take full advantage of the booming iron ore market and sell the company and all its property, including the railroad and ore docks. In 1887 Ely emphasized that Tower could not ignore the pressure from the Illinois group. "If we sell the railroad we are safe; if we do not, we can be frozen out," he wrote. [45]

The idea of selling the complete holdings of the Minnesota Iron Company had previously been proposed by Lazarus Silverman, the Chicago banker who had loaned Tower $150,000 in November, 1884. During late August, 1886, he approached the company, perhaps as an unofficial agent of the Porter group. Tower, Jr., consulted with his father and then told Silverman they would not be "willing to sell it all, under any circumstances." [46]

By early 1887, however, Charlemagne Tower had re-evaluated his position. A Chicago meeting between the younger Tower and several Porter lawyers in February directly stimulated that reassessment. The syndicate attorneys informed young Tower they possessed evidence to prove that Breitung and Stone had acquired land from fraudulent pre-emptors and homesteaders. Porter did not want to take the matter to court, they said, but he was convinced that the land claims were invalid and open to challenge. The young mining executive initially wanted to stand and fight. He wrote to his father: "The land is ours. We shall never give it up until we are obliged to do so by the decisions of the courts. . . . The thing for us to do is to stand right at the door and defend ourselves." But at the age of 78 Tower, Sr., "had no desire to engage in a transportation war and a simultaneous defense of his land titles," as his biographer put it. His old partner, Edward Breitung, died on March 3, 1887, and George Stone warned that they were up against "a strong and unscrupulous crowd. We have got to make the best deal we can and get out," he wrote.[47]

In April, 1887, Donald H. Bacon, a Porter associate who was later to be president of the Minnesota Iron Company, visited the Vermilion Range and reported to the syndicate that the Tower company would shortly require additional heavy and expensive equipment to convert the Soudan Mine to underground operation. He optimistically estimated that its future annual production of from 400,000 to 500,000 tons would provide substantial profits.[48]

Tower and Porter were both under pressure to reach a final settlement. Beginning in April, 1887, Tower, Jr., Lee, and Stone met in Chicago on several occasions with Porter, Samuel Ely, and others. Ely was especially instrumental in carrying through the negotiations. He pointed out that the Minnesota Iron Company owned 26,800 acres of ore, timber, and lake front property. After three years nearly 9,000 acres of ore land was under development. The Duluth and Iron Range Railroad had a state swampland grant of 600,000 acres. In addition the railroad had completed 95.7 miles of track, owned 13 locomotives, 340 ore cars, and other equipment, had "no old debts," and a total operating budget surplus, as of December 31, 1886, of $82,111.22.[49]

The parties completed their negotiations on May 3, 1887, although the final agreements were not signed until June 14 in the offices of the Fidelity Insurance Trust and Safe Deposit Company of Philadelphia. Tower sold "not less than 65,000 and not more than 71,000 shares of Capital Stock of the Minnesota Iron Company, [and] $1,300,000 of bonds of the Duluth & Iron Range Railway Company . . . upon the entire issue of $2,500,000." Porter and his associates paid $51.50 per share for the stock — $13.50 less than Tower's initial demand — and $110.00 each for the D&IR bonds. In what the *New York Times* of May 17 described as "the largest transaction in the way of a cash sale that has ever taken place" in Philadelphia, Tower received $6,000,000 in certified checks and $2,500,000 in cash. Regretfully he wrote to his cousin Albert, "I sold; because it was necessary for . . . [Porter] to control and manage the whole Minnesota Iron Company. . . . I did not want to sell . . . but I had to."[50]

Porter immediately reorganized the Minnesota Iron Company, subscribing personally for $750,000 worth of stock. He gathered in a total of $10,000,000 from 81 subscribers that included Heber R. Bishop, who was named the new president of the Duluth and Iron Range, $500,000; Tower, $500,000; John D. Rockefeller, $500,000; Marshall Field, $300,000; Silverman, $250,000; William Rockefeller, $200,000; Stone, $110,000; and Cyrus McCormick, $100,000. When Tower resigned from the board of directors, he asked that his son be retained in an executive capacity, but young Tower's association with the new firm was short-lived. He resigned as president of the D&IR on September 20, 1887, although he remained a member of the executive committee until 1889 when he was replaced by Marshall Field. Returning to Philadelphia, he accepted a vice-presidency in the Finance Company of Pennsylvania, and later he went on to a diplomatic career. Lee was retained in spite of his belief that Porter's "intentions are, and always have been, to introduce new people into the management of both Companies, just as rapidly as this can be done, without giving too great offense." Lee was allowed to stay on as a railroad vice-president, and in June, 1889, he was elected to the board of directors and the executive committee.[51]

The Minnesota Iron Company grew and prospered under its new management. In 1888, applying for a security listing on the New York Stock Exchange, President Jay Morse reported among its assets more than $8,258,512 in real estate, including 14,270 acres of developed mining property; $4,400,000 in Duluth and Iron Range securities; and $2,285,803 in ore, supplies, and bills receivable. The company had issued a total of $14,000,000 in capital stock and registered a profit of $473,873.[52]

Under Tower's leadership the Minnesota Iron Company had operated only one mine — the Soudan two miles northeast of Tower. It began production in July, 1884, as a series of shallow open cuts at a spot initially probed by George Stuntz in 1865 and again ten years later. The pits, which were named for Stuntz, Stone, Ely, Tower, Breitung, and Lee, were worked from the surface to a depth of 50 to 100 feet, and the ore was removed with the help of dynamite and nitroglycerine. From 1884 to 1885 shipments increased from 62,124 to 225,484 tons. By 1888 it became necessary to go deeper by utilizing underground drifts and shafts. The hard character of the Soudan formation proved especially advantageous for underground operations because it formed solid pillars "and seldom requires timbering to support it." With the walls surrounding the ore body so well defined, miners extracted comparatively small amounts of waste with the ore. (Some of the refuse became fill for the roadbed of the D&IR.) As the shafts were extended, production rose steadily, reaching 457,341 tons in 1888 and 535,418 tons the following year.[53]

Because the ore formation possessed a vertical rather than a horizontal depth, the company, by test pitting and diamond drilling, delineated it, locating marketable ore as deep as 2,000 feet below the surface. (By the time the Soudan closed in 1963, it was the deepest mine in Minnesota, with shafts extending down 2,400 feet.) To develop the Soudan as an underground operation, crews opened a main hoisting shaft and as many auxiliary timbered shafts as dictated by the extent of the ore body. At right angles to these shafts the miners ran a main hauling or tramming level. The ore body to be mined was divided into a series of horizontal strata at 60-, 80-, or 100-foot levels, depending on the mining methods used, with a main tramming level on the bottom of each layer. The miners picked the ore loose by hand whenever possible, but in the Soudan they also used an "enormous amount" of small dynamite charges to break it up.[54]

Throughout most of the formative period of mining on the Vermilion, the Tower- and Porter-controlled companies divided the ore formation into standard "square-set rooms." Underground laborers used birch, spruce, and tamarack, cut from the surrounding countryside during the winter months, for the timbering needed. The ore mined from these rooms was then sent down through chutes to the main level, where hand-operated or mule-pulled trams hauled it to the hoisting shaft.

By 1900 the "slicing and caving" system was replacing the square-set in some mines. Previously, the larger square-set rooms had frequently caved in with little warning and required expensive repairing and bracing. When a large room collapsed, surface water often flooded the area and plugged the drifts with mud. In the new system, miners opened smaller rooms, or sublevels, laid boards down as a floor, and reduced the engineering costs through less expensive timbering. After they removed one or two slices across the ore body, the crews blasted out the supporting timbers or the "hogbacks" (pillars of ore) and let the surface sand and remaining ore cave in, filling the slices. The miners repeated this process as they worked their way down extracting the iron ore. In addition to the savings in the time and the cost of timbering, the slicing and caving system provided a safer and more compact working situation.

Using these methods, Vermilion Range companies faced little preliminary investment for stripping and disposing of thousands of tons of glacial overburden before exposing marketable ore. Virtually all parts of a deposit could be accessible to miners who were able to work year-round extracting ore while developing the mine. Thus ore could be marketed quickly to help defray initial expenses and pay the interest on the capital invested in hoisting machinery, pumps, electric lighting systems, and tram cars. In 1890 a total of 1,770 men were employed in the iron ore industry in Minnesota. Of these, 976 underground workers on the Vermilion averaged $1.96 to $2.55 per day for 257 days. These wages compared favorably with the $1.73 to $2.23 per day received by 8,450 Michigan miners and the $1.81 to $2.00 per day paid to 1,110 Wisconsin men. But the Porter-controlled syndicate benefited even more than the wage earners. With its integrated organization, its directors' administrative skills, no destructive competition, and the refinement of underground techniques, the daily productivity per man rose from 1.80 tons in 1890 to 2.32 tons a decade later.[55]

The Porter syndicate was also responsible for the development of a second famous mine on the Vermilion Range — the Chandler. In February, 1883, Martin and William H. Pattison, two brothers from Superior, Wisconsin, were searching for pineland between Vermilion Lake and the Canadian border. They had been encouraged to probe there by reading the field notes of government surveyor Stuntz. His observations, recorded from October 23 to November 20, 1880, depicted a "surface hilly & rocky" with good stands of Norway pine, birch, spruce, and aspen. While camped near present Shagawa Lake, the Pattisons gathered hematite ore specimens but could not trace the iron formation. On a second try in May, 1883, they found the outcropping at what became the Pioneer Mine. They returned the following year, still primarily interested in timber, and in the fall of 1884 they located the ore body, on land acquired by Porter, that would shortly be developed as the Chandler Mine near modern Ely.[56]

The Chandler, operated by the Minnesota Exploration Company which was closely affiliated with the Minnesota Iron Company, was the first of the two to ship ore. It opened in 1888 with Joseph Sellwood as superintendent and John Pengilly as captain. Unlike the Soudan operation, the Chandler quickly instituted underground techniques during its second year, utilizing the caving system. Between 1888 and 1901 it produced a total of 7,027,830 tons, or 43.9 per cent of the output of the entire Vermilion Range. Four other nearby Vermilion Range mines also began shipping ore before the turn of the century — the Pioneer in 1889, the Zenith in 1892, and the Savoy and Sibley in 1899. Smaller mines were also operated by independents for various lengths of time. None of these, however, individually or collectively came close to matching the steady production of the Chandler, which shipped nearly every year through 1930 and closed in 1942.[57]

Realizing the potential of the new district 21 miles northeast of Tower-Soudan, the Duluth and Iron Range Railroad began construction of a branch line in the fall of 1887. No serious obstacles were encountered by the nearly 800 men employed at $1.50 to $1.75 per day, except the difficulty of organizing a series of construction camps during the cold, snowy, winter months. A road was cut from Tower-Soudan, and stagecoaches and freight wagons hauled people, supplies, and lumber from the end of the railroad overland to Shagawa Lake until the tracks were completed in July, 1888.[58]

The first settlement at the eastern end of the lake was the townsite of Spalding, named for Tower's old antagonist, William Spalding, who with F. W. McKinney was one of its proprietors. In spite of a promotional pamphlet touting Spalding as "the coming town of the Vermilion Iron Range" and offering to supply "a good class of citizens" with free timber for building on lots that could be purchased for one-third down and the balance in six months, its few residents abandoned Spalding as soon as iron ore development got under way. They moved a mile or so to the west on the south shore of Shagawa Lake closer to the mines and to the new terminus of the Duluth and Iron Range extension. There an enduring new village of "rude-looking but substantial log buildings" was named in honor of Cleveland promoter Samuel Ely or his brother Arthur.[59]

When it was platted in 1887, Ely had a population of 177 persons. A year later after the railroad was completed "a noteworthy 'Boom' was in progress and the town soon began to take on quite a metropolitan appearance," wrote one early historian. "The first grocery store was opened in a small log building by a man named McCormick [J. D.

Cormack?] who had started business a year or two earlier at Spaulding [*sic*]. A. J. Fenske built the first frame building in the fall of '87 and opened a hardware and furniture store bringing the stock as well as material for the building on sleighs. . . . The Pioneer hotel was built the same fall by R. B. Whitesides [*Whiteside*] at the corner of Sheridan Street and Fourth Avenue. . . . Dr. [Charles G.] Shipman, the first physician, came in 1888. . . . The first school was opened in January, 1889, in a small frame building on Second Avenue" and "attendance reached 112 during the first season." A volunteer fire company was organized in 1889; a newspaper, the *Ely Iron Home*, made its appearance; a bank was opened; and a Slovenian Catholic congregation led by Father Buh erected the earliest church.[60]

At the first village election in May, 1888, Captain Pengilly of the Chandler Mine was named council president, and after Ely was incorporated in 1890, he also became the city's first mayor. The federal census that year counted 901 people in Ely, with another 702 residing in Morse Township. The isolation of the communities of Ely and Tower-Soudan — the first three towns to be settled on the Vermilion — was depicted in the *Duluth Daily News* of April 20, 1890. Hiking near Soudan through an "almost illimitable snowy waste of wooded wilderness and frozen lakes," the reporter saw only miles of virgin pine. "It must be borne in mind," he wrote, "that from Two Harbors to Tower we had ridden by train nearly seventy miles through what was practically an unbroken forest of pines, birches, spruces, and tamaracks. Now and then there were a few houses and an occasional deserted claim shanty and a burnt over clearing. That was all — naught else to break the unbroken forest."[61]

Under the direction of the Porter group the Minnesota Iron Company not only continued to expand the Soudan Mine, but also fully exploited its transportation monopoly over the other mines near Ely. Although the Vermilion regularly ranked at the bottom among the five Lake Superior ranges in total yearly production, it nonetheless provided a consistent source of high-grade ore. Beginning in 1892 the Vermilion annually, with only eight exceptions, shipped more than 1,000,000 tons of ore every year for nearly six decades. The first exceptions were the years of economic depression in 1893 and 1894, when even the Soudan Mine suspended for a time. After the depression hit northeastern Minnesota, hundreds of men lost their jobs, wages declined, and the total ore output of the range dropped sharply. From 1,770 employees in 1890 the Minnesota Iron Company employed only 1,131 miners and laborers in 1895. Miners' wages in the latter year had fallen to $1.75 to $1.60 a day, while those of underground laborers were only $1.40 to $1.25. By 1900 prosperity had returned, and the Vermilion miners extracted more than 1,600,000 tons of ore for $2.40 to $2.20 per day, while laborers received $1.90 to $1.85 daily. The Soudan Mine's all-time annual record of 592,196 tons was registered in 1897, while that of the Vermilion Range as a whole occurred in 1902 when 2,084,054 tons left the docks at Two Harbors.[62]

The dismal years of the 1890s also saw what the *Vermilion Iron Journal* of June 23, 1892, called the Soudan's "first labor strike," the earliest miners' strike on the Minnesota ranges. The paper reported that the "mine, employing over 1,400 men, has been shut down and the commonwealth thrown into a state of some excitement. It was not the strike of organized labor," but rather apparently a walkout caused by "a large number of miners and laborers" taking off to celebrate Corpus Christi Day on June 16. As a result "the company saw fit to lay off for fourteen days 315 of these men, mostly Austrians and Italians" and "trouble broke out that evening . . . when the night shift came on for duty and were met by a crowd of these men, who declared themselves strikers and refused to permit any one to go to work." A posse of sheriff's deputies from Duluth as well as several companies of militia under the command of the state adjutant general were dispatched to the scene, where they stood guard until June 22. By that time, things had calmed down.[63]

Charlemagne Tower did not share in the problems of the 1890s, nor did he live to see the development of the mighty Mesabi he had spurned. On July 19, 1889, he suffered a severe paralytic stroke and died five days later. He had contributed to America's growing industrial strength by providing eastern furnacemen with a valuable new supply of low phosphorus iron ore. Without ever setting foot in Minnesota, he developed the state's first iron range, and he left to others the problems and rewards of operating the mines and the railroad his vision had brought into being.[64]

❙4❙

THE MESABI — A NEW EL DORADO

ALTHOUGH the Vermilion Range provided Minnesota's first marketable iron ore in the 1880s, it was the opening of the Mesabi in the 1890s that moved the state into a position of commanding leadership among the world's iron mining districts. On New Year's Day, 1894, the *Duluth News Tribune* remarked: "From the day when the first hardy settlers pushed their way into the great Northwest and clustered about St. Louis bay there were stories of a new el dorado which lay only a little way beyond. . . . Day after day explorers tramped over the rocks of the Vermilion and Mesaba [*sic*] ranges and in their search for gold ignored or neglected the wealth of iron over which they walked, yet it was the greater which they neglected for the less. . . . The wealth of the great iron field is still immeasurable."

Stretching across portions of St. Louis and Itasca counties, the low granite ridge of the Mesabi rises only 1,300 to 1,600 feet above sea level. Yet it forms a three-way continental divide between those streams flowing east to the St. Lawrence River, north to Hudson Bay, and south to the Gulf of Mexico. Its timber and soil are characteristic of northeastern Minnesota. Pine, spruce, cedar, balsam, and tamarack dominated the forest, although birch, maple, and other deciduous trees were scattered throughout. In places a heavy cover of glacial drift contained numerous large boulders that hindered initial mining operations and discouraged attempts to clear the land for agricultural purposes as well. Although a Pittsburgh reporter commented that "it scarcely looks like a mining country at all," geologists and mining experts determined that the Mesabi iron district extended for approximately 100 miles from Pokegama Falls, where it was first noticed, northeast to Birch Lake. The iron formation, varying in width from two to ten miles, paralleled that of the Vermilion Range, although it lay south-southwest of Tower's mine.[1]

73

Because the mining frontier largely followed the axes of the lumberjacks on the Mesabi, "Old choppings, windfalls, fires, underbrush, and swamps . . . combined to make the scene a desolate one," wrote a geologist. Edmund J. Longyear, the man who sank the first diamond drill hole on the range in 1890, later said of his first visit: "Both white and red, or Norway, pines of majestic proportions as well as considerable hardwood, covered the area. I remember seeing partridge, deer, and one moose on our trip. Beyond the logging camp our way lay through a region that had been logged for the most part . . . I am glad that I saw those beautiful trees before they vanished. For anyone who knows only the modern appearance of the Mesabi Range, it would be impossible to form a true mental picture of the original. . . . Few areas in the United States have been so completely altered by man."[2]

The harbinger of a succession of men who played roles in altering the region over the years was Henry Eames, the Minnesota state geologist responsible for initiating the gold excitement at Vermilion Lake. In 1865 Eames canoed up the St. Louis River to Embarrass Lake on the eastern Mesabi. He reported sighting "immense bodies of the ores of iron, both magnetic and hematatic," in some of which "iron enters so largely into its composition as to effect [sic] the magnetic needle." Eames also ascended the Prairie River on the western end of the Mesabi, and after analyzing several specimens of iron ore, acknowledged their good quality. Thus in one year he had unknowingly hit upon the eastern and western extremes of America's richest iron ore deposit.[3]

One of the geologist's guides in 1865 was Christian Wieland of Beaver Bay. A surveyor and civil engineer, Wieland noticed that strong magnetic attraction made his compass useless, and he concluded that large deposits of iron existed on the eastern end of the Mesabi near what is now Babbitt. He returned with samples that he exhibited in Duluth and in the Michigan copper-mining town of Ontonagon, where the Wielands often traded lumber from their sawmill for supplies.[4]

Attracted by these samples, the Ontonagon Pool or Syndicate, a loosely organized group of Minnesota and Michigan speculators, proposed about 1870 to explore mineral lands on the Mesabi and, if promising, to acquire them. Although membership remained fluid, William Willard, an Ontonagon merchant, served as the pool's leader. Other early members included Ontonagon residents Peter Mitchell, an experienced woodsman; Samuel Mitchell, mine operator; Louis S. Langpre, saloonkeeper; and Lewis M. Dickens, James Mercer, and Linus Stannard, merchants or warehouse owners. William Spalding, the Duluth booster and hotel owner who battled Charlemagne Tower for control of the Duluth and Iron Range Railroad, remained an active participant for many years. Other Duluth residents were George Stone, Josiah D. Ensign, and Joshua B. Culver, all of whom were later involved in Minnesota iron mining. Alexander Ramsey, former Minnesota governor and then United States senator, was also in the syndicate.[5]

These prospective capitalists hired fellow pool member Peter Mitchell to examine Wieland's findings before committing themselves to land purchases. Born in Lancaster, Pennsylvania, in 1828, Mitchell had moved with his family to a farm in Illinois. In April, 1847, tired of farming, he joined a small group traveling to the copper region of Michigan's Upper Peninsula. There Mitchell made his home at Ontonagon until the village burned in 1896. In the spring of 1870 he arrived at Beaver Bay, and for the next three years he periodically studied Minnesota's iron formations, put down some of the first test pits on the Mesabi, gathered ore samples, and recorded the magnetic attraction. He carefully traversed the whole eastern end of the range and on one occasion traveled north to the Vermilion, frequently carrying specimens back to Duluth or Ontonagon for examination by his interested employers.

The Ontonagon Pool, ready to acquire many of the tracts Mitchell had explored, realized that none of the land in this area north of Duluth had as yet been surveyed by the federal government. Senator Ramsey secured the necessary appropriations, and in 1872 Christian Wieland was hired as a deputy surveyor to examine Township 60 North, Ranges 12 and 13 West near present Babbitt. As a result, the following year the syndicate acquired nearly 9,000 acres of eastern Mesabi land. In January, 1875, Mitchell, Spalding, Culver, Dickens, George Stone, and seven others incorporated the Duluth and Iron Range Railroad Company in order to transport their prospective ore to Lake Superior. No right of way was cleared, however, and, as we have seen, Charlemagne Tower subsequently acquired the Duluth and Iron Range charter and its land grant.[6]

The pool's mining operations on the eastern Mesabi Range also failed to progress beyond the organizational stage. The panic of 1873 and the death of Willard, the syndicate's principal supporter, in 1874 discouraged further exploration. But in June, 1882, members of the pool made one final effort to develop their mining properties by establishing the Mesaba Iron Company with Ramsey as president of its board and Spalding as secretary and treasurer. Headquartered in Duluth, the firm issued $3,000,000 in capital stock divided into 100,000 shares at $30 each. Although it was incorporated five months before the Minnesota Iron Company, the Mesaba directors never successfully developed their property; eventually they sold the land to various promoters. To Peter Mitchell and his associates, however, goes the credit for carrying out the first intensive search for iron ore on the Mesabi Range.[7]

That giant range did not quickly give up its secrets. The failure of the Mesaba Iron Company and the negative assessment of Albert Chester, who examined Mitchell's test pits for Tower, influenced opinion for nearly another ten years. Chester's report, published in 1882, noted the magnetic character of the iron in layers that proved to be neither thick nor continuous. He believed that 44.68 per cent was a fair average of the iron content in the surrounding rock formation of the eastern Mesabi, and he was correct. That portion of the range, which did indeed contain only low-grade ores, again attracted brief attention in 1887 when James B. Geggie, exploring on the Vermilion, noticed evidences of iron south and west of the "Red Pan Cut" near Mesaba, a station on the Duluth and Iron Range Railroad. The following year state geologist Newton Winchell persuaded John Mallmann, the experienced guide who had earlier accompanied Chester and Stuntz, to explore the area. Using supplies obtained from the Minnesota Iron Company at Tower, Mallmann with a crew of 25 men dug several shafts near Mesaba. He found iron-bearing ore but not in large quantities.[8]

Prospectors were still confining their efforts to the poorest section of the range, as Chester did in the previous decade, but Winchell believed that Mallmann's work was worth while. Writing in 1898, he said: "From this point [*Mesaba Station*] explorations were extended by other parties still farther west, and although there was not good success for a year or two, every test-pit that was sunk to the bedrock confirmed the idea that the Mesabi Range rocks were not only iron bearing, but that they were a different set of rocks from those containing the ore at Vermilion Lake. In the latter part of 1890 and the first part of 1891 the great discoveries were made which have brought the Mesabi range into the front rank of the iron districts of the world."

More than any other individuals, members of the Merritt family were responsible for the successful discovery and initial development of the Mesabi Range. As part of the westward sweep of the American nation, the Merritts migrated from Connecticut to New York — first to Onondaga County where Lewis H. Merritt was born in 1809 and then to Chatauqua County. There in 1831 he married Hephzibah or Hephzibeth Jewett, a young lady of 19 from Massachusetts, and there six of the family's eight sons were born — Jerome in 1832, Napoleon B. in 1834, Lucien F. in 1835, Leonidas J. in 1844, Alfred R. in 1847, and Lewis J. in 1848. Apparently about 1849 the family moved to Warren County, Pennsylvania, where two more sons were born — Cassius C. in 1851 and Andrus R. in 1853. Restless once more, Lewis then pursued his trade for two years as a carpenter and millwright at Austinburg, Ohio, where he chanced to meet Edmund F. Ely, a Methodist missionary stationed at the head of Lake Superior. Ely persuaded Lewis to go to Minnesota and supervise the construction of the first sawmill on the site of present Duluth. In July, 1855, Merritt arrived at Oneota, a small settlement on the Minnesota shore of Lake Superior now a part of West Duluth. The following year his wife and six of their eight sons, ranging in age from 22 to 2 years, moved into a newly constructed plank house on a quarter section Lewis pre-empted there.[9]

Once again the Merritt family was part of the opening wedge of settlement, this time in the Minnesota hinterland. Each of the three oldest sons carved out varying careers. Jerome, who had attended the Grand River Institute at Austinburg and, for a short time, Oberlin College, became the first teacher in a Duluth district school. He provided an education for some of his younger brothers and other children in the Oneota community in a schoolhouse built by his father. Napoleon, like his father a millwright by trade, apparently also inherited his urge to move frequently. He had followed Lewis on his initial trip to the head of the lakes in 1855, but he returned to Ohio and Pennsylvania in 1859. From there he moved to Missouri in 1866, where his parents as well as three brothers — Jerome, Lewis J., and Andrus — joined him in the 1870s. After Jerome and the elder Lewis died in 1878 and 1880, Hephzibah and the others gradually returned to Duluth, where the men later joined other family members in several mining and pineland ventures. Lucien, who was 21 in 1856, did not travel to Minnesota with the rest, remaining in Ohio to study for the Methodist ministry. After he "served various pastorates" in Pennsylvania, he, too, joined the family in Oneota in 1887.[10]

It was Leonidas, Alfred, and Cassius, however, who remained in Duluth and became the most visible family leaders there. Lon, as he was known to family and friends, had also briefly attended the Grand River Institute before moving to Oneota as a boy of 12. In 1864 after a two-year sojourn in the Pennsylvania oil fields where Lucien was then

THREE GENERATIONS OF MERRITTS

Lewis Howell Merritt (1809–80)
m. Hephzibah Jewett (1812–1906)

Jerome (1832–78)
m. Emily Walker (d. 1906)

Hattie May (1861–1950)
Wilbur J. (1863–1939)
Mary J. (1865–67)
Alfred R. (1865–1918)
—— Thomas H. (1869–1938)
Everett W. (1870–1953)
John J. (1872–   )

Napoleon B. (1834–1924)
m. (1) Jennie Holman (d. 1881)
(2) Matilda C. Tanner
(d. 1914)
(3) Margaret Shaw

Eugene T. (1858–1935)
—— Franklin W. (1860–1933)
Thomas A. (1871–1928)
Fred H. (1873–   )

Lucien F. (1835–1900)
m. Mary J. Richardson

Annis (1839–45)

Eugene (1842–51)

John E. (1862–1932)
Ada H. (1864–1941 or 1942)
—— Alta A. (1867–88)
Louis H. (1872–1948)
Alva L. (1875–1947)
Leonidas C. (1882–1945)
Mary E. (adopted)

Leonidas J. (1844–1926)
m. (1) Elizabeth Wheeler
(1851–1902?)
(2) Cordelia R. Sprague

Ruth (1875–1963)
—— Lucien (1877–1945)
Harry (1882–1970)

Alfred R. (1847–1926)
m. (1) Elizabeth Sandilands
(1854–82)

Lewis H. (1877–1940)
—— Thomas J. (1880–81)
Elizabeth S. (1882–1965)

(2) Jane A. Gillis (d. 1949)

Jessie T. (1886–1922)
Alta H. (1888–1973)
—— Ernest A. (1890–1968)
Glen J. (1894–   )
Clark C. (1897–1970)
Merna Jean (1906–   )

Lewis J. (1848–1929)
  m. Eunice Annette M. Wood
    (d. 1924)

Annice L. (1870–72)
Hulett C. (1872–1956)
—— Bertha H. (1876–   )
Lewis N. (1880–1940)
Ada Evelyn (1889–   )

Cassius C. (1851–94)
  m. Eliza M. Felt

Carolyn E. (1877–1951)
Mary H. (1878–   )
Charles C. (1880–   )
—— Wilbur C. (1885–   )
Olive L. (1888–1977)
Harriet L. (1890–[?])

Andrus R. (1853–1939)
  m. (1) Susan Bullock (d. 1880)

    (2) Elizabeth Clark

    (3) Jessie L.

James C. (1877–   )
—— Thomas A. (1879–   )

—— John Wesley (1884–   )
Louis W. (1885–   )

living, he volunteered for cavalry duty and served during the Civil War with Colonel Alfred B. Brackett's battalion in the Dakotas. After the war Lon and Alf, who were only three years apart in age, formed a team to work together on numerous projects. In the winter of 1867–68 both men were employed as chainmen on a survey crew for the first railroad to be constructed by Jay Cooke between the Mississippi River and Lake Superior. Shortly thereafter they bought a small scow and engaged in the lake-coasting trade. On May 8, 1869, they advertised in the *Duluth Minnesotian* that they were ready to "transport Lumber, Stone, Wood" and deliver "all kinds of Freight to any point of the Bays."[11]

The following winter, having absorbed a measure of their father's building skills, Alfred and Lon constructed the first sailing vessel built at the head of Lake Superior, the two-masted, 49-ton schooner "Chaska," which they used to haul stone for the foundations of a government pier at Superior. Their resulting debts and the stiff competition from steamboats convinced Lon to dissolve the partnership and join Frank W. Eaton in buying and selling pinelands. The firm of Eaton & Merritt prospered, and by the 1880s Lon was in comfortable circumstances.[12]

Alfred turned his considerable talents to a variety of enterprises. He sailed the Great Lakes and engaged in the lumbering business, logging the shores of Lake Superior and the St. Louis River in winter and towing great log rafts to sawmills in the spring and summer. In 1873 he hired a crew that included Andrus, the youngest brother, and spent most of the summer and fall cutting a road from Siskiwit Bay on the western end of Isle Royale inland to a copper mine. He finished the work the following summer and then cut trails and dug test pits on the island for the North American Exploration Company. About 1876 he and Lon built another sailing vessel, the "Handy," but they sold it when lake steamers made sailing unprofitable. Later Alfred and Thomas Sandilands, his brother-in-law, purchased and operated a tug-boat named the "John Martin."[13]

By the late 1880s the Merritt brothers warranted inclusion among Duluth's leaders. Both Alfred and Cassius held public offices. Alfred was elected a St. Louis County commissioner in 1886, a post he held until 1892. Cassius served as deputy county treasurer, deputy auditor, and deputy register of deeds in the 1870s. As lumbermen and dealers in pine and mineral lands, Lon and Alfred had also acquired a first-hand familiarity with much of northeastern Minnesota. Nevertheless it was Cassius, then a man in his 30s, who launched the family on its pursuit of iron ore. After operating a grocery in Superior for some years, Cassius, too, took to the woods, becoming an expert timber cruiser and estimator. While cruising for Eaton & Merritt, he had also acquired valuable pineland holdings. In 1887 he was in charge of the crews of William K. Rogers and Matthew B. Harrison, who hoped to build a rail line from Duluth to Winnipeg. Near present Mountain Iron in Township 58 North, Range 18 West, section 5, Cassius picked up on the surface what Alfred later described as "a boulder of iron ore . . . the first chunk of pure ore taken off" the Mesabi Range.[14]

Andrus recalled the event this way: "The estimator makes his way from point to point, carefully observing the quantity and quality of the pine, making notes as he goes. In order to do this he must keep looking upward, measuring the timber by his eye. But on this day, the date of which was never recorded . . . as Cassius was making his observations, he fell headlong. But, because he was an observing man, . . . this fall was instrumental in bringing to light a piece of iron ore about as big as his two fists. . . . Picking up the ore, he observed that the corners were sharp. From this he gathered that it had not been carried far, as that would have led to the corners being rounded off. . . . He stored it carefully in his pack and when he had returned to Duluth took it to be assayed. It proved to be high-grade Bessemer ore. From this the boys, as they talked it over, decided that there must be valuable deposits of iron near at hand."[15]

That fall the Merritts discussed the possibility of conducting a systematic mineral exploration of both the Vermilion and Mesabi districts. Their intimate knowledge of the terrain, Cassius' discovery, and the Minnesota Iron Company's profitable Vermilion Range operations supplied the incentives. Many years afterward the brothers also alluded to the influence of their late father, who had traveled to Vermilion Lake during the gold rush of 1865–66. Alfred wrote that "Father was not boomed on the gold fields, but while he was out there, North Albert Posey, who was the Indian Blacksmith, showed Father a chunk of iron ore, and Father told us boys that some day there would be great mines there. . . . These words perhaps influenced us in later years."[16]

The Merritts decided to undertake a "systematic reconnaissance" from Gunflint Lake on the Canadian border southwest to the present town of Grand Rapids on the Mississippi River. By using small surveying parties to map and prospect in designated areas, they hoped to obtain an accurate chart of the surface iron formations and to record the places of magnetic attraction. Beginning in the spring of 1888 they sent out two-man exploring parties that remained in the field throughout the season. When a crew arrived in the area selected for examination, the men immediately went to work with a standard compass and a dip needle, a compasslike device developed in Sweden in which the needle swings on a vertical instead of a horizontal plane. At regular intervals the survey team, by comparing the amount of downward attraction with the deflections of a standard compass, could plot the rough outlines of iron-bearing formations on their charts.[17]

Andrus wrote of this work on the Mesabi: "Most of the territory was covered with thick pine timber interspersed with tamarack and cedar swamp. There were no roads through this wilderness and . . . it became the rule for two men to go out together with packs upon their backs, these packs containing a thirty-day supply of food and the necessary tools. These packs would weigh a hundred pounds or more and the rule was to carry them throughout the day while the prospecting was being done, as the camp moved forward from night to night.

"The method employed in deciding whether further investigation would be justified or not was to carry an instrument known as the dip-needle. With this, the prospectors having selected their 'forty' would go in about ten rods on the land and there go south across the 'forty', keeping track of the distance by the counting of paces — monotonous business but essential to the success of the enterprise.

This needle was observed every few paces and a record made where an attraction for the needle was shown. By going back and forth in this way over the 'forty' a fair idea is obtained of any ore-deposit on the land. . . . This is such exacting labor that only those whose muscles have been inured to toil and their spirits strengthened by abundant hardships are able to hold their course."[18]

At the end of the first summer's work, the Merritts realized that their crews had located iron ore on state-owned school lands — sections 16 and 36 of each township which had been granted by the federal government to the states for educational purposes. They were aware that the Minnesota constitution allowed them to purchase school lands, but they did not know if such acquisitions conveyed title to the mineral resources as well. Lon and Alfred counseled with Moses E. Clapp, then state attorney general and later a United States senator, and with his influential support and that of State Auditor William W. Braden, they lobbied in the legislature for a bill permitting the lease as well as the sale of school lands.[19]

In April, 1889, the Minnesota legislature approved a bill regulating the sale and lease of mineral lands belonging to the state. Later known as the Braden Act, the law authorized the commissioner of the land office to "execute leases and contracts for the mining and shipping of iron ore from any lands now belonging to the state, or from any lands to which the state may hereafter acquire title," and permitted, but did not require, him to reserve the mineral rights to the state. An individual lease could not exceed 160 acres, for which the state assessed the applicant $25. If two individuals sought to lease the same tract, the person willing to pay the larger amount received official approval. The owner of a mineral lease could then prospect for iron on his land for a period of one year. At any time during that year the leaseholder, in order to carry out mining operations and extract ore, had to sign a binding contract with the state. Valid for 50 years, that document called for a payment of $100 annually to the treasurer's office until shipment of the first 1,000 tons of ore was made. Thereafter, working on a minimum of 5,000 tons annually, the leasee paid the state a royalty of $.25 per ton. Under state supervision and inspection, the railroad companies were required to weigh the cargo; the mining companies submitted monthly statements and made quarterly payments to the state auditor.[20]

Like many other men, the Merritts within four years acquired, by purchase and lease, extensive holdings of Mesabi mineral lands. In

1890 alone, the first year under the new law, Leonidas obtained 142 leases, while four other brothers gathered a total of 13. Lon, Alfred, Cassius, and John E., Lucien's oldest son, sold their timber holdings and reinvested in mineral lands, while other family members contributed additional assets and exploring skills to the mining venture. Convinced of the value of their systematic exploration efforts, the Merritts' enthusiasm for the enterprise increased. After four long summers of persistent work, the family possessed an accurate map covering some 500 square miles of Itasca, Cook, Lake, and St. Louis counties, as well as detailed field notes delineating iron outcroppings and the lines of magnetic attraction.[21]

As early residents of Duluth, the Merritts held large tracts of timberland, many of which lay close to the city limits. While exploring and acquiring mining properties, family members profitably engaged in various real estate transactions that supplied them with sufficient capital to meet their initial exploring expenses. Indeed, it was agreed that Lon and Alfred would finance the venture, and the others would carry out the work. Throughout the 1880s and 1890s northeastern Minnesota timberlands increased in value as logging in the region approached its peak. The advent of mining boosted the demand; in 1891 logging began at Tower and shortly afterward at Ely. It was not unheard of for buyers of timberland to acquire inadvertently ore deposits of great value; men such as John S. Pillsbury, Minneapolis miller, Charles H. Davis of the Wright and Davis Michigan lumber firm, and David M. Clough, lumberman and Minnesota governor, were among such fortunate landowners.

The logging activity combined with the development of mining stimulated a speculative real estate boom and rapid population growth in Duluth. The city's economy was then dominated by logging and wheat shipping. With the opening of the Red River Valley and the Dakota prairies, Duluth's port facilities and grain-storage capacities were expanding. Situated between the Vermilion and Gogebic ranges, the Zenith City had not as yet developed into an ore-shipping port, but it benefited from land speculation, mercantile demands, and construction and mining equipment sales. To meet the new financial needs, the number of banks rose from 3 in 1885 to 12 in 1889; loan agents increased from 6 to 53. The legal profession also profited from the problems of land and title disputes. In 1885 when the city's population totaled 17,685, only 24 lawyers were practicing there. Five years later 33,115 people lived in Duluth and the services of 72 attorneys were available. By 1890 Duluth's population was almost three times that of Superior, its economic competitor across St. Louis Bay, and the city was optimistically contemplating a rosy future.[22]

The Merritts constantly received discouraging reports and advice from mining men, who were of the opinion that the Mesabi Range possessed no profitable ore deposits and that, in fact, the region's geologic structure made the presence of iron impossible. Only Minnesota geologists Newton and Horace Winchell consistently believed that the "nearly horizontal strata" of "the Mesabi was likely to yield large quantities of good ore." The Winchells favored "the Mesabi, against the opinion of nearly all the actual mine operators and scientists who had examined it." The generally unfavorable opinion limited the amount of credit available to the Merritts. As they reinvested their funds in numerous mineral operations, the family was forced to seek outside financial support, but their increasing involvement in Mesabi exploration seemed too risky for most local bankers, who would loan only on the brothers' personal accounts.[23]

Despite adverse opinion and the difficulty of attracting new investors, the Merritts continued prospecting. In March, 1889, Alfred left Tower with six men and hand-drawn sleds. They followed Pike River and passed along the shore of Big Rice Lake looking for the area where Cassius had picked up the first ore sample. Unaware that they were too far north, the crew dug numerous test pits in Township 59 North, Range 18 West, section 34. At first the "only method available to us was what is known as test-pitting," wrote Andrus many years later. "In this from ten to twenty men were employed with pick and shovel in the effort to locate the ore. They would dig a pit about four feet square and timber it up as they went down. If rock was struck, it was necessary to use powder for blasting. These pits were mostly dug to the depth of about 100 feet, . . . the dirt being removed with the help of a windlass. Work was done upon two or three of these pits at the same time, two shifts keeping the work going day and night. The test-pitting was begun, as was proved later, too far north and kept working farther south until the money was exhausted." Alfred continued this work off and on until the summer of 1890. By that time his search had taken him several miles south into sections 3 and 4 of Township 58 North, Range 18 West.[24]

Based on favorable reports from Alfred and Cassius, Lon acquired most of the potentially valuable parcels of land in Township 58, Range 18, either in fee or under lease. On July 10, 1890, he headed a group of three investors who incorporated the Mountain Iron Company. Headquartered in Duluth, the firm was capitalized at $2,000,000 and issued 20,000 shares of stock. Late in 1890 Lon transferred ownership of these lands to the organization in return for stock. As a result, the new company focused its activities in Township 58.[25]

Hoping to broaden the Mountain Iron firm's financial base, the three investors attracted the support of Kelsey D. Chase, a southern Minnesota banker, who was to figure prominently in the fortunes of the Merritts. A New Yorker by birth, Chase had moved to Rochester, Minnesota, in 1860. He entered the banking and mercantile businesses and speculated in real estate, railroad construction, and mining; in 1905 he would open the Chase State Bank in Faribault. With the help of Chase and other southern Minnesota businessmen, the Mountain Iron Company raised sufficient capital to purchase additional land, clear a tote road some 27 miles from their property to Mesaba Station on the Duluth and Iron Range Railroad, and undertake initial mining development.[26]

According to Lon, the Merritts considered themselves "explorers and discoverers and developers," not miners. As a result, they hired 15 men, under the supervision of James A. Nichols, a competent German mining captain, to conduct the search for a specific ore body. Influenced by other successful mining operations on the Vermilion and in Michigan, Nichols expected to dig to considerable depths before reaching marketable quantities of ore. Lon exaggerated only slightly when he recollected that "if we had gotten mad and kicked the ground right where we stood we would have thrown out 64-per-cent ore, if we had kicked it hard enough to kick off the pine needles."[27]

The geologic structure and composition of the Mesabi ore bodies varied from those previously encountered on the Vermilion and most Michigan ranges. All of northeastern Minnesota had experienced similar geologic movements and had, on several occasions, been covered by ocean waters. Yet nature had produced very dissimilar ore bodies. In the Vermilion district the iron formation was narrow and deep. This hard ore usually stood at a high vertical angle and, as a result, lent itself to underground mining. In the Mesabi district, on the other hand, soft powdery ore lay in beds ranging from a few inches to several hundred feet in thickness and occupying broad, shallow troughs between layers of quartz and slate. A mantle of glacial debris, composed of material differing in size from fine sand to boulders weighing several tons, enveloped that range. This covering, varying in thickness from 10 to 200 feet, usually had to be removed before a surface of clean ore could be scooped up from open pit mines.[28]

The oldest rocks of the Mesabi district, the Archean series, were of igneous origin. Geologists have found evidence that slow earth movements mashed, bent, and broke these rocks. The encroachment of ocean waters accompanied and followed surface erosion. In time layers of sediment hardened and metamorphosed, or changed structurally, and became the sidewalls and underlying rock. Organic agents such as iron salts, or silicates, deposited the iron formation in comparatively shallow coastal waters. Subsequently rather gentle earth movements slightly inclined and fractured the rock layers. This fracturing allowed the percolation of ground waters that leached, or extracted, the silica and left behind a rather porous, spongy residue of enriched iron deposit. Because of the perforated nature of the ore body, sagging or slumping of the formation took place, creating the trough or basin structure that generally characterized the Mesabi. The extracted ores ranged in texture from large crystalline masses, requiring the use of a crusher, to a fine, soft dirt that could be picked up by steam shovel.[29]

Nichols and his crew began test pitting with picks, shovels, and perhaps a diamond drill. Edmund Longyear recalled that "To get a road built so that the Merritts could move their expected diamond drill west of Mesaba Station [in 1890], the county commissioners had authorized the construction of a road from the bridge over the Embarrass River, near the site of Biwabik, to Birch Lake. This became the first segment of the famous — or shall we say notorious — Mesabi Trail."[30]

Over it the Merritts sent men and supplies to Nichols and his crew throughout the summer. The first drill holes put down revealed "nothing of importance," but late in October a new pit indicated a valuable find. After an expenditure of about $20,000, Nichols on November 16, 1890, encountered the first body of soft ore discovered on the Mesabi Range only "12 or 14 feet" below the surface. Recalling the event 40 years later, John E. Merritt said: "I remember just how beautiful that ore was, glinting blue there under the deep green of the white pines. But I am unable to describe to you just what this Pit No. 1 meant to us. It was a dream come true, the fulfillment of a hope long deferred, an urge to greater effort, a satisfying fact that at last nature had yielded to us the great secret she had guarded through all the ages. Above all this was the thought that this pit on the Mountain Iron would forever set aside any doubt in the mind of a skeptical world that here was the birth of a new era in the affairs of men and commerce. And would be proof absolute of the supremacy of a great range of high grade ore. . . . From Pittsburgh, from Cleveland, from all the iron world came men to scoff and many to believe."[31]

Although news of Nichols' strike at what became the Mountain Iron Mine attracted an influx of explorers, speculators, and knowledgeable men with capital to invest, the Merritts had not been alone in their attempts to locate ore on the new El Dorado. Other pioneer prospectors like John Mallmann continued to search the range. Since no roads had as yet made their appearance, most of these early landlookers reached the eastern end of the Mesabi after 1884 on Charlemagne Tower's railroad, which crossed the range at Mesaba Station, "the earliest and easiest point of access to the new Mesabi range." Edmund Longyear, who appeared with his diamond drill in May, 1890, to open a new era in prospecting, made his first headquarters near the station. For the Longyear Mesaba Land and Iron Company of Jackson, Michigan, which had purchased property in the southern part of Township 59 North, Range 14 West, the young graduate of the Michigan Mining School at Houghton sank the first diamond drill hole on the range about one and a half miles from Mesaba on June 3, 1890.[32]

The drill invented by Longyear revolutionized exploration, for it enabled operators to ascertain the extent and nature of an entire ore property before beginning to mine. Drilling quickly replaced laborious pick-and-shovel test pitting as the dominant exploring technique. A drilling crew systematically surveyed an ore property, divided it into 100- or 200-foot squares, and sank several drill holes. If this effort located promising ore, additional holes were put down on the adjacent corners of each square. Some prospectors used a churn drill, a Mesabi invention sometimes credited to John Mallmann. This device drove water down a rod and out through perforations near the bit, clearing away the chopped debris and returning it to the surface inside the casing pipe. The churn drill, which operated as much by percussion as it did by rotation, could travel rapidly through the overburden and soft ore, but if boulders were encountered, they had to be blasted out with small sticks of dynamite. The drill, extra rods, boiler, and pump frequently cost as much as $2,000.[33]

Because the churn drill was expensive and somewhat ineffective, many mining companies adopted Longyear's diamond drill. Its tubular bit, which was nearly two inches long, had a cutting edge set with industrial diamonds. Such a drill could bore long distances through the hardest rock and return a core sample of the strata it had passed through. This work was often carried on by special, three-man crews who sampled the core every five feet to obtain a complete record of the deposit. In the formative period on the Mesabi, the contract price for drilling ranged from $3.00 a foot in drift to $6.00 a foot in rock. A drill helper was paid $2.25 per day, while experienced operators averaged $3.00. A competent diamond setter, essential to keep the drills going, could earn between $90 and $125 per month by servicing several drilling crews and resetting five or six bits in a ten-hour shift. Usually a crew required at least two fresh bits a day. Thus the total expense of operating one drill, diamond or churn, for the usual ten-hour shift averaged $12 a day.[34]

Of his first trip to the new range, Edmund Longyear wrote: "I got off at Mesaba Station. Today Mesaba is a ghost town, but in 1890 it was booming. Here the grading of the railroad had made the 'Red Pan Cut' in the rocks, so that anyone could detect iron in the local magnetic ore formation. Accordingly, the area of the cut had already become the point of departure for scores of iron prospectors. They not only dug holes everywhere in the immediate vicinity, but they also struck off from this railroad station into the wilderness of the Mesabi Range. . . . Mesaba, therefore, was rapidly developing as the main outfitting point on the Mesabi Range, and crude, frontier hotels, stores, and saloons were springing up in ever-increasing numbers. For the next two years the Mesabi Trail penetrated on and on, west from this point, first as a packers' trail, then as a bridle path, and, finally, as the most execrable tote road imaginable, miles of torture for horses and wagoners alike, beset with mosquitoes and black flies in the spring and by heat and flies in summer." Another early engineer said that the trail "could not be described," adding that a publisher at Tower "refused to try to describe it and when he could not find language no one cared to try."[35]

An early historian of the area wrote that the little settlement at Mesaba Station grew "almost overnight into a place of fifteen hotels" with "very active and comprehensive general stores" that catered to the needs of "countless bands of explorers delving into the range." Mesaba became the first incorporated community on the Mesabi in July, 1891. Too new to have been included in the 1890 census, the village by 1895 had a population of 159; more than half the residents were foreign born and the leading occupation was "lumbermen and woodsmen," of whom there were ten. But Mesaba's existence was short-lived. Its demise was chronicled by the *Duluth Evening Herald* as early as 1898: "Within a short time the village of Mesaba, the oldest settlement on the Mesabi range, will be nothing but a memory. There is little left of it now except . . . a few families . . . no money on hand . . . . 75 to 100 houses."[36]

Some 27 miles to the west on the Mesabi Trail another village at first named Marfield and then Grant was incorporated in 1892 as Mountain Iron — the "Gem of the Mesabi" — on the same day that the first train-load of ore was shipped from the nearby mine. What had been an exploration camp reached by the miserable trail from Mesaba began in 1891 to assume the appearance of a community as a public school, general store, bank, and a Methodist church were built. These augmented "a number of tents and several log 'shacks' " that housed "a considerable population representing many nationalities and gathered from various parts of the world." Edmund Longyear explained that most range "communities were, first of all, exploration camps, then they grew into mining camps; and finally they ended as villages or 'cities.' " Tent settlements "persisted until there were sufficient indications of ore to warrant the erection of a log house for crew members," he continued. The "usual drill camp of the day . . . was erected on the lines of logging camps. . . . The crew slept in tiers of bunks in a bunkhouse and ate in a cook shanty, usually connected with the sleeping quarters by a sort of breezeway, though that appropriate term was not yet used. As the mine expanded, more houses were required. Then the wives and families arrived, and the step from drill camp to mining camp or 'location' was made when separate houses for the men's families appeared." Mining camps, according to one newspaper, were "constructed of durable timbers . . . built with a view to their long occupancy by . . . the mining companies." Once such camps were established, Longyear pointed out, "enterprising capitalists seized the opportunity, bought land near the diggings, and platted a townsite. Lots were sold, streets were laid out, and a new Mesabi community began its career."[37]

To reach Mountain Iron by the Mesabi Trail, "One had to have great powers of endurance even to ride horseback over it," wrote Longyear. And in the early years "horses were the chief means of transportation, other than shank's mare. A carting business of such proportions as one can scarcely imagine made immediate use of every section of this Mesabi Trail as it was opened up. Jostling and jolting over corduroy and muskeg, picking their way through the stumps left everywhere in the 'road,' the great vans rattled, carrying everything, except logs and lumber, that built Mountain Iron."[38]

With the completion of a railroad connection in the fall of 1892 and the further development of the mines nearby, the municipality grew rapidly. Early in 1893 a sawmill began operation, and additional businesses soon opened their doors — five hotels, three boardinghouses, four restaurants, real estate offices, saloons, barber shops, a blacksmith shop, and a livery stable. A weekly newspaper, the *Mountain Iron Manitou*, began publication, and a Duluth editor predicted: "Mountain Iron's future seems full of promise, and she is without a doubt destined to set a pace that will distance many of her competitors" — a prophecy that almost immediately turned out to be untrue.[39]

Two weeks later fire destroyed most of the village, and the townspeople had to begin again. Spurred by expanding mine production, they erected substantial brick buildings to replace the burned rubble. By 1895 Mountain Iron contained 443 people. It was unique at the time as the only range town with a majority (51.9 per cent) of residents who had been born in the United States, largely in Minnesota, Michigan, or Wisconsin. Among the foreign born, Swedes, Finns, Canadians, and "Austrians" predominated. Most of the men were employed as laborers, mechanics, and engineers in the mines. In this burgeoning community "the great Mesabi range had its beginnings," stated a reputable mining journal, and the work of the Merritts first "bore fruit."[40]

Never content to place their efforts in a single venture, the Merritts continued to seek additional ore deposits. Wilbur J. Merritt, Jerome's 27-year-old son, and John E., his 29-year-old cousin, working with Captain John G. Cohoe at the head of test-pit crews, superintended the exploration work. On October 9, 1890, Lon and two West Duluth businessmen, John J. Wheeler and Edwin H. Hall, filed the incorporation papers of a second firm — the Biwabik Mountain Iron Company with headquarters in Duluth. They organized it to obtain the capital required for continued exploration and acquisition of mineral lands. It was originally capitalized at $2,000,000 (20,000 shares at $100 each) with Lon and Alfred Merritt and Kelsey Chase as the largest stockholders. They were, however, joined by 16 others, including Cassius, Lewis J., Napoleon, Andrus, and John E. Merritt. Initially the family controlled 69.7 per cent of the shares, with Lon and Alfred between them owning nearly 59 per cent. After Cohoe discovered marketable quantities of ore in August, 1891, the company's capitalization was increased to 30,000 shares. The Biwabik Mine, located in Township 58 North, Range 16 West, proved to be the family's second important discovery on the Mesabi, and John E. became its first general manager.[41]

The initial Merritt discoveries at Mountain Iron and Biwabik substantiated the existence of ore at only two sites. Test pits and 7,133 diamond drill holes sunk by Edmund Longyear alone at various locations below the granite ridge and on additional properties by the Merritts and others decided the future of the range. Gradually accumulating knowledge of its geologic structure, miners and entrepreneurs concentrated their exploratory operations in the basins, but neither they nor the developers of townsites who followed them encountered any uniform measure of success.[42]

For example, three men active on the Minnesota ranges invested in land near Biwabik and platted the townsite of Merritt in February, 1892. Joseph Sellwood of the Chandler Mine on the Vermilion, and O. D. Kinney and James T. Hale of Duluth developed the town named for the Merritt family along the shore of Embarrass Lake about 12 miles east of Mountain Iron. Although the Merritts were not pleased with the "honor" and are said to have avoided the place whenever possible, the community had 217 persons in March, 1892. By May it also had a weekly paper, the *Mesabi Range*, and it was the second community on the range to vote for incorporation. In spite of these assets, however, Merritt declined when a branch of the Duluth and Iron Range Railroad (see map, page 92) bypassed it, leaving the town isolated.[43]

A new settlement named Biwabik was then established on the rail line, and the residents of Merritt moved many of its buildings to the new location. Those which were not moved were destroyed when the disastrous forest fire of June 18, 1893, virtually wiped out not only Merritt but also Mountain Iron and Virginia. Biwabik was more fortunate. It had organized a volunteer fire company and as a result "was saved by the strenuous efforts of its inhabitants," according to a local historian. Although spared from the fire, the young village started slowly. According to one account, "Promoters hesitated at first to build on the present site of Biwabic, supposing iron existed under the whole valley." But by October, 1892, the *Biwabic Iron Ore*, a "seven-column four-page paper [with] two local pages," began publication. The peripatetic Father Buh helped organize the new Catholic church in 1893; at the first service he held, the Slovenian priest addressed the congregation "in English, Slovene, French, and finally in Ojibway." Three years later the community had 1,011 residents, 59 per cent of whom were foreign born, a familiar pattern on the Mesabi.

The next village to be established was McKinley, about four miles southwest of Biwabik. It had 189 residents in 1892, but settlement of the area had probably begun in 1890 when the Mesaba Lumber Company established a sawmill there. The mill was owned by Arthur G., Duncan, and William McKinley, Duluth real estate dealers and lumbermen who, with a fourth brother, John, became interested in iron

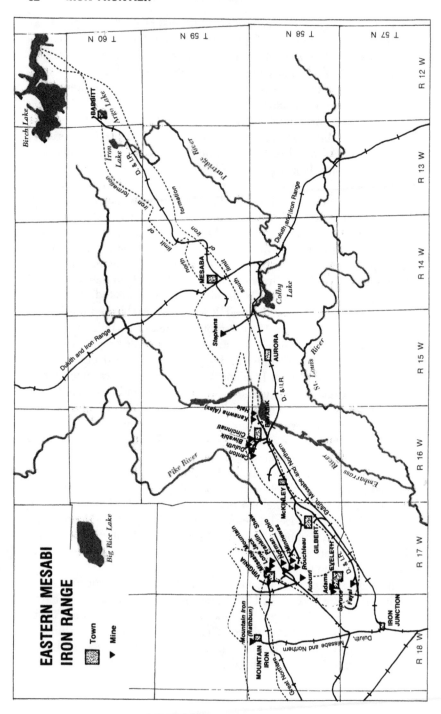

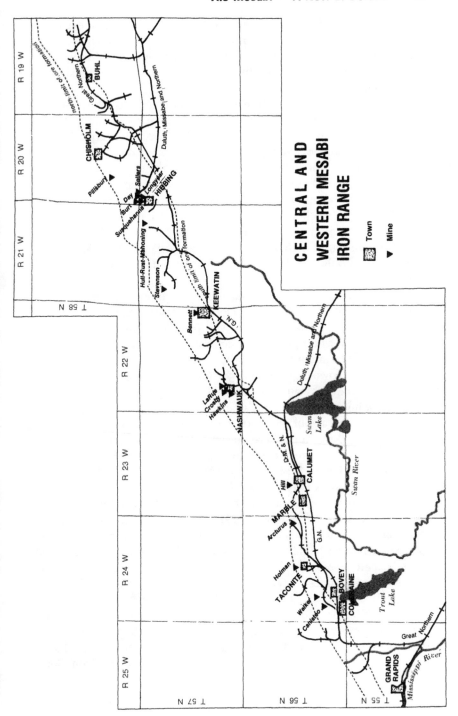

CENTRAL AND
WESTERN MESABI
IRON RANGE

▩ Town
▼ Mine

lands in August, 1891, when John McCaskill, an independent explorer, discovered traces of ore clinging to the roots of a tree overturned in a recent storm. McCaskill's find was located on property leased by John McKinley near what became the village of Biwabik in Township 58 North, Range 16 West.[44]

McCaskill informed the Merritts of his discovery, and they secured from McKinley a ten-day option on several quarter sections, agreeing to pay him $10,000 initially, an additional $20,000 in two equal installments within one year, and a royalty of $.30 per ton on any ore mined. Within ten days Merritt test-pit crews, under the direction of Wilbur, uncovered an enormous body of iron ore, and the family exercised its option to lease the mineral rights. McKinley agreed to take his remaining payments in stock rather than cash. Eventually this find led to the development of an important series of mines that included the Cincinnati, Canton, Duluth, Hale, Kanawha, and Williams. Near McKinley, the Corsica, Roberts (Atlas-Emmett), and Elba mines were developed beginning in 1897.

In the years from 1891 to 1894 numerous other explorers located valuable deposits on the new El Dorado, attracted investors, and incorporated. Only a few were successful. Among them was Frank Hibbing, the man who opened the central Mesabi. Working 15 miles southwest of Mountain Iron in 1892, he identified the first of the great ore bodies in Townships 57 and 58 North, Ranges 20 and 21 West. Born Franz Dietrich von Ahlen in the German province of Hanover in 1856, Hibbing took his mother's family name when he immigrated to Wisconsin at the age of 18. In subsequent years he worked on a farm and in a shingle factory there, briefly studied law, cruised timber, prospected for iron ore, helped plat the town of Bessemer on Michigan's Gogebic Range, and finally in 1887 arrived in Duluth. The next year he began to explore on the Mesabi.[45]

While others were concentrating their efforts between Mountain Iron and Biwabik, Hibbing was looking farther west near the community that now bears his name. After a thorough search and many test pits confirmed his belief that large bodies of iron ore existed there, the experienced woodsman leased numerous tracts of land from four Michigan pine-timber speculators: Ezra Rust, Wellington R. Burt, George L. Burrows, and Gilbert B. Goff. In 1890 Hibbing, along with John and William McKinley and Alexander J. Trimble, a well-to-do iron mining operator on the Gogebic Range, formed the Great Northern Iron & Steel Company. Two years later, again with the financial support of Trimble, Hibbing incorporated the Lake Superior Iron Company. The work of this firm led to the exploration and development of the Hull, Rust, Burt, Sellers, and Mahoning mines, which, when combined, became the largest and most productive open pit in the world.[46]

In 1893, with the aid of his partner Trimble, Frank Hibbing also platted the townsite that later justifiably claimed to be "the Iron Ore Capital of the World." The *Duluth News Tribune* of June 4, 1893, called the platting of Hibbing "the event of this week" and predicted it would become the "main place" on the central Mesabi and the "chief trading point" for both the mining and lumbering industries. The following month residents formed a village government and adopted articles of incorporation. One contemporary account stated that by September the community boasted 167 voters. Building materials were "brought in by wagons" until a railroad reached Hibbing in October, 1894. The settlement's first depot was a boxcar. The German-born entrepreneur constructed and operated the town's first hotel, sawmill, and bank; he also loaned the county $3,000 to build a road to the Mahoning Mine. In addition, he partially financed the water and electric light systems in the town that was named for him. During the winter of 1893–94, despite "dull, poor times," the settlement acquired three hotels, a restaurant, two stores, and eight saloons. Within two years it attracted a sizable population of 1,085 to become the second largest village on the Mesabi. As economic prospects brightened, the price of lots on Pine Street, which had been selling for little or nothing, rose to $300. Like most range communities in this formative period, Hibbing had a foreign-born majority of 67.4 per cent, with Finns and Swedes comprising two-thirds of that total. Acquiring the trappings of a somewhat raw but prosperous mining center, the town in 1895 had 35 merchants, 13 cooks and bartenders, and one photographer among its businessmen.[47]

Less than 10 miles north and east of Hibbing a heavily timbered area in Township 58 North, Range 20 West, was first explored for ore in 1891 by Frank Hibbing. The community which sprang up there as a result of ore discoveries took its name, however, from Archibald M. Chisholm, a Canadian-born mining man whose early experience had been gained on the Gogebic Range and in 1888 at the Chandler Mine on the Vermilion. Six years later he moved to Hibbing where he worked as a banker and entrepreneur in mining and real estate investigations. In 1896 Chisholm discovered his first mine, the Susquehanna, which did not enter the shipping lists until 1906. Other finds in the area led to the establishment of the community that bears his name. The Chisholm, Clark, and Pillsbury mines were all open by 1901 when the town was incorporated with a population variously listed as 250 and 496. By 1905 it had increased to 4,231, and Chisholm was well established.

In the 19th century, however, the leading Mesabi community was Virginia, the "Queen City of the Range," some 20 miles east of Hibbing near the Missabe Mountain Mine. None of the other burgeoning towns surpassed it in size or activity until the early 20th century. Named for the home state of one of its promoters, A. E. Humphreys, Virginia came into being in the fall of 1892 when David T. Adams, O. D. Kinney, and others incorporated the Virginia Improvement Company, which laid out the townsite, erected a few wooden buildings with lumber brought over the tote road from McKinley, and organized a local government. The construction of a branch rail line from Mountain Iron in December encouraged the establishment of many businesses, including a hotel, a bank, a sawmill, and in February, 1893, a newspaper, the *Virginia Enterprise*. Unfortunately drought conditions prevailed throughout the spring and early summer of 1893, and Virginia was destroyed by the fire that devastated two other communities in June.[48]

Like the others, Virginia rebuilt. By 1894 it had incorporated, becoming one of the first Minnesota municipalities to embrace the commission form of government. A year later its population reached 3,647. Although mining furnished employment for a great many of its workers, the lumber industry provided "occupation for as many or more," according to one historian. The city was unusual on the range because it was not a one-industry community. As a sawmilling center, it boasted for a time the largest white-pine mill in the world operated by the Virginia and Rainy Lake Lumber Company. One of the most prosperous mining ventures commenced in February, 1892, when Lon and John Merritt with Kelsey Chase incorporated the Missabe Mountain Iron Company capitalized at $3,000,000 for the purpose of "buying, selling, leasing and dealing in mineral and other lands." The Merritts formed the company to take advantage of a strike made by Captain Cohoe in March, 1892, in Township 58 North, Range 17 West, section 8, near Virginia.[49]

As was the case in other range towns, more than 60 per cent of Virginia's residents were foreign born. Finns and Swedes were most numerous, but substantial numbers of Norwegians, Canadians, "Austrians," Irish, Russians, and Poles were present. Six Chinese men also lived there in 1895; five operated laundries and one had a restaurant. Men outnumbered women by only three to one, but the hazards and uncertainties of life in a developing wilderness attracted many individuals who came to "mine the miners." In 1895 Virginia was the most wide open of the range towns. Fifty-nine young women, listed in the census as "demi-monde," were living in four separate residences. In each case the nearby buildings housed bartenders and saloonkeepers. Nineteen gamblers also plied their trade "and many forms of behavior were tolerated, if not directly sanctioned, that would have been condemned and outlawed elsewhere," wrote one serious student of pioneer mores on the Mesabi.[50]

The successful entrepreneur most directly involved in the platting of Virginia was David Adams, who is said to have struck his first good ore in August, 1891, on the same day Cohoe located the Biwabik Mine. A persistent but perhaps apocryphal story related that "Mr. Adams had worked down from the ridge, which seemed to mark the formation. Capt. Cohoe and Captain Nichols had dropped down to lower ground. Adams was working on [what became] the Cincinnati [Mine], Cohoe on the Biwabik and Nichols on the Mountain Iron. . . . It was on a day in the fall that Mr. Adams found the high-grade ore that he had been looking for on the Cincinnati and on the same memorable day Cohoe made his strike [on the Biwabik]. . . . Adams was on his way to the Merritt camp where Cohoe was working to apprise his neighbors of his good luck, when Cohoe hailed him with the announcement that he had found merchantable high-grade ore."[51]

A native of Illinois, Adams had spent two years on the Menominee Range near Crystal Falls and Iron River, Michigan, before moving to Duluth in 1882. As both a timber cruiser and ore explorer, he traveled throughout the Vermilion and Mesabi ranges during the next decade. In 1886, while employed by a timber firm in the Prairie River area, he concluded that this portion of the extreme western Mesabi possessed no ore of marketable quality. Undaunted, he spent the next few years working his way to the center of the range. By 1893 Adams had organized 17 explorations, and the following year he published the first Mesabi map compiled exclusively from field observations. In January, 1893, with Peter L. Kimberly of Sharon, Pennsylvania, and John T. Jones of Iron Mountain, Michigan, he incorporated the Adams Mining Company, modestly capitalized at $1,000,000. Within the next few years these entrepreneurs opened and developed numerous mining properties in the heart of the range; they were also responsible for the townsite of Eveleth.[52]

Platted by Adams four miles south of Virginia, the "Hilltop City" of Eveleth was incorporated during the summer of 1893 when Neil

McInnis, Adams' longtime associate, described it as having "a small start" with a boardinghouse, a hotel, and a saloon. At the time of the first town election in 1894, there were only seven houses. Merchants and businessmen erected about a dozen wooden buildings within the next year, and by 1895 Eveleth had 764 people, 71.5 per cent of whom were foreign born — more than half of them Finns. Named for Erwin Eveleth, an early timber cruiser and ore prospector, the settlement grew steadily but failed to match the size of neighboring Virginia.[53]

It soon became apparent, however, that the Eveleth townsite lay on top of a rich iron deposit. As a result, Adams and his partners financed the platting of a new location in 1896, and "its twelve or fifteen hundred inhabitants" began to move the village "on an average of one-fourth of a mile, up the hill, to the east." By the turn of the century more than 100 buildings had been moved to the new Eveleth to permit the opening of the Spruce Mine. Writing 20 years later, Adams confessed that he had "some disappointments" with Eveleth. Having platted Virginia earlier, he said, "made it, indeed, hard . . . to induce people to purchase lots and settle in . . . Eveleth."

Virginia and Eveleth served numerous mines in the immediate vicinity, including the successful Missabe Mountain, Shaw, Lone Jack, and Ohio. With the opening of the Mountain Iron and Biwabik properties, many prospectors expected to find ore between the two settlements. But as they moved east from Virginia and found the territory barren, Adams and others widened their search and located rich deposits at the end of a southward-pointing bend, or "horn," in the formation. At Eveleth the Adams Mining Company opened an area that by 1901 included the Adams, Cloquet-Vega, Fayal, and Spruce mines. By the end of the 1890s they had shipped over 4,000,000 tons of high-quality iron ore.[54]

In the first three years of the 1890s the major communities and many of the significant Mesabi mines were located through the efforts of hundreds of individuals. Capital had been raised, often within Minnesota, to finance exploration and begin mining operations, and the people who would make the region one of the most ethnically diverse in the United States had begun to arrive. The Merritts, Frank Hibbing, David Adams, and A. M. Chisholm led this initial exploratory phase, but they were never alone. With others, they had tapped a storehouse of wealth in the new El Dorado. Like Charlemagne Tower before them, the Mesabi entrepreneurs now faced the next step — the construction of facilities to transport the ore to Lake Superior for shipment to eastern furnaces.

|5|

THE MESABI GIANT
GOES TO MARKET

THE ORES OF THE MESABI could not be sent down the Great
Lakes to the nation's steel mills until someone financed the construction of a railroad from the mines to Lake Superior and built the necessary docking facilities there. The general terrain from the range to the
lakehead at Duluth or Superior offered an easy downgrade for heavily
loaded ore trains, suggesting that a railroad could be built economically. Since the Merritts were the first to discover Mesabi ore in marketable quantities, they also expressed an early interest in transporting
it to the lake shore.

By 1890 several railroad companies were operating in Duluth and
Superior, and the Merritts tried to convince them to build a branch
line to the mines. They especially hoped to persuade either Northern
Pacific or St. Paul & Duluth officials that carrying iron ore could be
profitable, but they also pointed out that the timber tributary to the
proposed Mesabi rail line would in itself yield sufficient traffic to justify
construction. According to Charles A. Norcross, the Merritts did not
look upon the Duluth and Iron Range as a possibility because they
regarded the methods of its parent Minnesota Iron Company as "op-
pressive to independent operators." In any case, none of the established railroads responded favorably, perhaps because they were skeptical of the new range that had failed thus far to "interest any of the big
iron operators."[1]

More than any single factor, the geologic structure of the Mesabi led
to the rapid development of its mines while at the same time retarding
acceptance of its ores. Calling the Mesabi "a poor man's range," the
Duluth Tribune of January 19, 1892, wrote: "In the history of Lake
Superior ranges . . . none has shown such an amount of ore in so short
a time. . . . The ore bodies lying almost flat with simply a shallow

covering of gravel and here and there a soft capping of soft sandstone makes exploring very cheap." Familiar only with the hard ores of the older ranges, however, the steel industry at first looked with suspicion on the powdery hematite so readily scooped up on the Mesabi. When the Merritts first traveled east in 1891 to meet with prominent steelmakers in Pittsburgh, they failed to elicit any immediate enthusiasm for the new source of ore. Lon later said that he was rudely received by Henry C. Frick, chairman of Carnegie Brothers & Company, but that he did persuade Frick to dispatch his "leading expert" to have a look at the range. According to Andrus, John and Wilbur Merritt guided the Carnegie man over the Biwabik property. Upon observing the high-grade, soft, granulated product, the expert "swore it was not iron ore." John tried to convince him to have a sample assayed in Duluth, but the easterner refused, saying that he "wouldn't make a fool of himself by lugging around a sample of dirt." As the family recalled the matter, Carnegie then declined to invest.[2]

Such rejections were offset, however, by the local encouragement expressed by newspapers at the head of the lake. The *Duluth Daily News* of April 20, 1890, insisted that "The great need in the matter of developing this mining country is railroad facilities. . . . Possibly existing roads are not in favor of new lines, but even if this be so the general interests of the country and the people are paramount, and the necessities are such that they will certainly come." Recalling King Croesus, the paper trumpeted two years later: "Gold is the semblance of power; iron is power itself. . . . When one says in cold type that . . . close to and directly tributary to Duluth and the head of Lake Superior [is] the largest deposit of high grade Bessemer iron ore known to exist in the world, he will probably attract the doubting attention of not only iron men but the public generally, yet such is the fact. No such ore formations are known on the globe as those of the Mesaba range. It is hard to write of those deposits in a way that will tell the truth and yet not be thought a gross exaggeration."[3]

Since the Merritts could not mobilize support in the East and failed to entice an established railroad to tap the range, they followed local public opinion and set out on their own. On February 11, 1891, five of the Merritt brothers and four associates incorporated the Duluth, Missabe and Northern Railway Company with a capital stock of $5,000,000. Actually these entrepreneurs obtained the charter and changed the name of the nonexistent Lake Superior and Northwestern Railway, which had been incorporated in May, 1883, by Leonidas Merritt, George Stuntz, and seven other Duluth and Superior men.

The early plans for the Lake Superior and Northwestern had been ambitious. The incorporators had originally intended to build tracks from Minnesota Point in Duluth to the Red River and to construct branch lines to the Mississippi at Brainerd, to the Canadian border via the Vermilion Range, and along the north shore of Lake Superior.[4]

The newly incorporated Duluth, Missabe and Northern formed a board of directors and elected Kelsey Chase president, Leonidas Merritt vice-president, S. R. Payne secretary, and Cassius Merritt treasurer. In August, 1891, the directors authorized the survey and location of a line of track from Stony Brook Junction (Brookston), about 26 miles west of Duluth, north to the Mountain Iron Mine, and directed the president to borrow $2,500 to do so. Two months later the *Duluth Daily News* reported an offer of $3,000,000 from an English syndicate for the Merritt mines at Mountain Iron and Biwabik and the right of way of the Duluth, Missabe and Northern. The newspaper was unable to decide whether the offer was a serious one or whether it was merely "an attempt to boom the companies' stock." In any case, nothing came of it.[5]

On January 28, 1892, after accepting the completed survey, the railroad's officers awarded a construction contract to Donald Grant of Faribault. He agreed to build before September 1 the line as surveyed, a distance of 48.5 miles to the Mountain Iron Mine, and to "furnish all the labor and material requisite for that purpose." The railroad company provided the right of way plus $900,000 in first mortgage bonds and $945,200 in fully paid common stock. In addition, both the Mountain Iron Company and the Biwabik Mountain Iron Company, which were also parties to the contract, contributed $200,000 of their fully paid common stock.[6]

With the survey completed and the construction firm under contract, the Duluth, Missabe and Northern directors next concluded a traffic agreement with Charlemagne Tower's old rival, the Duluth and Winnipeg Railroad Company. On April 14, 1892, the two firms agreed that if the DM&N built a line from the mines to the Duluth and Winnipeg tracks at Stony Brook Junction, the latter would "transport to and deliver at the navigable water of Lake Superior . . . all freight delivered to it." The two railroads established a mutually acceptable system of rates with each road furnishing "its proportion of all cars necessary to transport all freight . . . with all convenient speed . . . and without discrimination by either party." During this formative period of their operations, the Merritts viewed the 10-year agreement as especially advantageous because the 26-mile section of Duluth and

Winnipeg track from Stony Brook to Lake Superior passed through an area which would have required heavy construction costs.[7]

Incorporated on February 11, 1878, the Duluth and Winnipeg had originally set out to construct a line from Duluth northwest to a connecting link which Canadian interests planned to build south from Winnipeg. It failed to lay any track and, as we have seen, lost a land grant to Charlemagne Tower's Duluth and Iron Range Railroad. As a result the Duluth and Winnipeg was reorganized, and by 1890 it had completed 75 miles of line westward from Cloquet. Two years later, when the traffic agreement with the Merritts was signed, the firm was operating just over 100 miles of track stretching from Lake Superior to Deer River, Minnesota, but it had as yet no docks to handle ore in either Duluth or Superior. Nevertheless, section 6 of the agreement called for the Duluth and Winnipeg "to furnish and maintain at its own cost . . . all necessary terminal facilities on Lake Superior . . . including sufficient and suitable docks for the cheap and convenient handling of the iron." A usable portion of the dock was to be available by August 1, 1892.[8]

In choosing a dock site the railroad took advantage of the continuing economic rivalry between Duluth and Superior and dealt with the highest bidder. On July 22, 1891, the Duluth and Winnipeg entered into a contract with the Superior Consolidated Land Company, a Wisconsin firm formed to promote the development of Superior's west end. The latter promised to build for $250,000 a single line of track over the 12 miles from the proposed ore docks on Allouez Bay at Superior to the Minnesota state line and to bridge the St. Louis River there. For its part the Duluth and Winnipeg promised to lease track rights for five years at 3 per cent of the cost and maintenance. In return for 580 feet of lake frontage and 40 acres of land on Allouez Bay, the railroad also agreed to build an ore dock and the spur lines on the site and to ship "not less than 100,000 tons of ore" over the leased route.[9]

To carry out the construction of the dock, the Duluth and Winnipeg then created a subsidiary firm known as the Duluth and Winnipeg Terminal Company for the purpose of "building, Owning, Operating or selling railroads, docks, wharves and steamship lines." Incorporated on June 4, 1892, this subsidiary issued all of its $500,000 of capital stock to the railroad in exchange for the Superior real estate bordering Allouez Bay. The ore dock, constructed of heavy pine timbers, extended 1,000 feet into the bay with 100 pockets capable of holding 180 tons each. Contractors lined the pockets with three-inch seasoned maple planks, bolted and bound with steel straps. Each pocket had a heavy steel plate at its mouth and a 30-foot-long chute through which the ore passed from the rail cars to the lake ships. Mechanized hoists used steel cables to raise and lower the chutes. Upon completion of a portion of the dock on November 13, 1892, William H. Fisher, general manager of the Duluth and Winnipeg, told the *Superior Daily Leader*: "I can not think when Mesaba ore will be shipped except from Allouez bay. . . . We are close to deep water, with short towing or no towing, with late freezing in the fall and early thawing in the spring. There are no river or other currents to trouble us, no floodwater, no sediment. We have chosen the best place for ore shipping at the head of Lake Superior and in a year or so will do a business that will surprise the world."[10]

Nine days after signing the traffic agreement with the Duluth and Winnipeg, the Merritts established their first business connections with eastern iron manufacturers and financiers. On April 23, 1892, Peter Kimberly, a second-generation furnaceman in the western Pennsylvania community of Sharon, leased the Merritts' Biwabik property. He agreed to mine a minimum of 300,000 tons annually beginning in 1893 at a royalty of $.50 per ton. An analysis of the ore revealed a high iron content of 65 per cent and a low phosphorous average of 0.03 per cent, both essential qualities for Bessemer steel production. In return the Duluth, Missabe and Northern agreed to build a track to the mine and to carry annually any amount of ore produced. If the railroad proved unable to handle the output, Kimberly was free to void the contract and negotiate a more favorable agreement with another line. Finally Leonidas and Alfred verbally promised to reduce the freight rate from the high $.80 per ton as soon as the traffic warranted such a move. On April 24, 1892, the *Duluth Daily News* remarked hopefully: "The deal for the lease of the Biwabik mine . . . is of the utmost importance to the new Mesaba range. The men who have taken hold of the mine are all well known and practical miners and iron men, and of large capital and established business, and this fact shows most clearly the value of Mesaba ore as looked upon by men in the iron business. . . . This lease will do more than anything else to stop the mouths of those carping critics of the Mesaba, who have been crying it down."[11]

Some three months later the Merritts and their associates completed still another important transaction when they leased the Missabe Mountain Mine near Virginia to Henry W. Oliver. Born of Scottish parents in Ireland on February 25, 1840, Oliver moved with his family to Pittsburgh two years later. He quit public school at the age of 13 and became a messenger for the National Telegraph Company, working with young Andrew Carnegie. After serving in the Civil War, Oliver, with his brothers James and David, organized a small manufacturing plant that produced nuts and bolts. Their business expanded to such proportions that in 1888 they incorporated as the Oliver Iron and Steel Company. During the 1870s and 1880s Oliver had also been involved in many iron-related industries, including railroads. A partial owner of the Pittsburgh & Lake Erie, he later served as president of the Pittsburgh & Western Railway Company. Oliver knew the intricacies of the iron and steel industry and his support of the Mesabi was to be a prominent factor in its subsequent history.[12]

He saw the range for the first time in June, 1892, when he attended the Republican National Convention in Minneapolis, the first major party gathering west of the Mississippi. After the delegates nominated Benjamin Harrison for president, Oliver traveled to Duluth in his private railroad car to see for himself the "raging mining boom." From there he went via the Duluth and Iron Range Railroad to Mesaba Station and then "jolted thirty-six miles in a buckboard to the Cincinnati mine, over an unspeakable road," the Mesabi Trail. "Not far away, the Merritts were developing the Missabe Mountain," recalled John H. Hearding, veteran mining man, "and trudging over to it, Oliver saw in a flash that he had found what he was looking for, a vast body of high grade ore cheaply mined."[13]

Returning to Duluth, the affable Scotch-Irishman quickly gained the approval of Cincinnati Mine stockholders and leased the property on June 20 for 19 years, agreeing to mine a minimum of 150,000 tons annually at a royalty of $.55 per ton with an advance payment of $25,000. Then he opened negotiations for the Merritts' nearby Missabe Mountain Mine. On August 3, by the terms of his lease with the Merritts, Oliver agreed to mine a minimum of 200,000 tons in 1893 and 400,000 tons a year thereafter. He also agreed to pay the Missabe Mountain Iron Company (in which Lewis J. Merritt and his son Hulett were the heaviest individual stockholders) a high royalty of $.65 per ton in addition to an advance payment of $75,000. Assays of this ore body showed an average of about 64 per cent iron and 0.035 per cent phosphorous. Oliver and the Merritts supplemented the lease with a traffic agreement on July 1, 1892, under which the Duluth, Missabe and Northern promised to transport to Lake Superior "at reasonable rates" all the ore produced. Upon one year's notice the Pittsburgh entrepreneur could compel the railroad to handle 1,000,000 tons annually or forfeit the traffic contract. Because of Oliver's close ties to Andrew Carnegie and Henry Frick this transaction ultimately proved to have far-reaching implications for the over-all development of Mesabi iron resources, as we shall see in later chapters, for in September, 1892, Henry Oliver organized the Oliver Mining Company.[14]

While the Merritts were negotiating the Kimberly and Oliver leases, contractor Grant was clearing the railroad right of way from Stony Brook Junction north to Mountain Iron. Throughout the summer of 1892 he worked his crews hard, proceeding at a rate as high as two miles a day. In July the section gangs began laying rails weighing 60 pounds to the linear yard, the conventional weight for most railroad construction of the day. Gleefully anticipating the completion of the tracks, the *Duluth Daily News* of August 12, 1892, called for a celebration. "This will be one of the most important events in the history of Duluth and St. Louis county. Nay, more, one of the most important in the records of iron mining in this country," the paper exulted. "The Duluth, Missabe and Northern will be a Duluth road. . . . Let our people on that occasion unite and pay all honor to the men who have built the road, and at the same time show their realization of its importance to our city and county."[15]

The first train over the new tracks "consisting of mixed freight and passengers" left Duluth's Union Depot on the morning of October 6, 1892, with members of the Merritt family and other dignitaries on board. According to the *Duluth News Tribune* of that date, the company intended to run one train a day "until the road is in shape to begin hauling ore." To mark the completion of the 48.5-mile track, a special train with one ore car (number 342) made the run south from Mountain Iron on October 18 and was put on display in Duluth. The *Duluth Evening Herald* reported that the train carried at "the center of the car a young pine tree [that] waved and nodded welcome to the crowd." This car "was loaded with twenty tons of dark brownish purple soft ore . . . [which] assays a trifle better than 65 per cent iron . . . and was consigned to Leonidas Merritt."[16]

On November 1, 1892, veteran engineer A. C. Herbert brought down the first train of nine wooden ore cars carrying 22 tons each from the mine to the Allouez Bay dock. Construction crews had completed only 26 pockets, divided equally on each side of the dock, which necessitated "an unusual amount of snorting, whistling and bellowing" as the train was "pushed upon the high dock and out over Allouez bay." Cold weather brought an early closing of the lake because of ice, problems concerning track ballast, and the freezing of the ore in the rail cars and ore pockets. These factors limited the 1892 shipments from the Mountain Iron Mine to 4,245 tons, all of which was sold to Oglebay, Norton and Company of Cleveland. This was the first Mesabi Range ore to go to market.[17]

All of this activity in 1892 generated a second boom-town cycle in Duluth reminiscent of that created by the arrival of Jay Cooke's railroad in 1870. Proudly quoting the St. *Paul Pioneer Press*, the *Duluth Daily News* of February 21, 1892, proclaimed: "Recent discoveries and developments in the iron producing region of Northern Minnesota have aroused the greatest public interest, and are advertising for the first time the fact of the enormous mineral wealth of this state. New mining companies are being formed daily, and unappropriated mineral lands are sought after eagerly. Through these days of biting cold specialists and prospecting parties are busy in the northern wilds. Untold wealth lies beneath this uninviting country. . . . The boom in iron properties in Minnesota is fairly under way, and the state can now look forward to the rapid development of one of its most important natural resources." The paper listed a total of 21 mining companies "with capital stock aggregating $45,200,000" and cautioned that it would be well for investors "to look out for" fake corporations. A few weeks later on March 7 the *Duluth Daily News* warned editorially that "Every 'forty' is not an iron mine, every family cannot have one, and every man will not grow rich out of iron stocks," and urged the people of Duluth to "frown down upon gross exaggeration, wild booming or mad speculation." Then on March 20 the same ebullient newspaper reiterated its warning in verse:

> Now all who ever tackle stocks,
> From millionaire to barber,
> Are blowing in their hard earned rocks
> On the iron of Mesaba.
>
> While some will win and some will lose,
> Each wight the thought doth harbor,
> That fortune him can ne'er refuse
> All the riches of Mesaba.

But the newspapers were also quick to denounce any criticism of the Mesabi El Dorado, especially that emanating from Michigan's Upper Peninsula, an area fearful of an emerging rival. On March 14 the *Duluth Daily News* complained that "No paper has more to say against the so-called Mesaba boom than the Marquette Journal. . . . It has never seen anything good in Duluth or in the region directly tributary to this city. Marquette is a fine little city . . . but it has no more enterprise than the most lifeless New England village. While it has been sleeping away the years, Duluth has been growing and prospering, until today the one time important Lake Superior city of Marquette is scarcely heard of beside it. All this time the Journal has been croaking like an old raven." Horace V. Winchell, Minnesota assistant state geologist, also took exception to the negative judgments expressed by editors on the Upper Peninsula. Writing in the *Duluth News Tribune* of October 22, Winchell told readers: "As to quality, there are pits of ore in this new iron-clad district which cannot be excelled in any mine in the country today. The ore is so clean and rich and so soft and easily mined that it beats anything known in the world at the present time."

Many mining engineers, company executives, and competent observers of the iron trade toured the Mesabi and also reacted positively. George Booth, city clerk of Pittsburgh and a man actively concerned with that steel center's growth, visited the range in early March. He told the *Duluth Tribune* of March 7 that he had seen "nothing to equal" its rich and extensive ore beds, and predicted that Mesabi ore could be mined for half the price per ton of Vermilion deposits. Ernest Nunemacher, a Wisconsin mining speculator, traveled over the Mesabi in mid-April. He informed a *Duluth Tribune* reporter on April 13 that "The Missabe will close up every mine on the Menominee and Gogebic ranges. You can beat those districts . . . [by] about one dollar a ton, and that is a tremendous advantage. . . . I see no reason why the Missabe should not control the ore market for years to come." In September Andrew Carnegie, Henry Frick, and other prominent eastern iron and steel men toured the Vermilion and Mesabi ranges escorted by Samuel Mather, H. H. Porter, and J. H. Chandler of the Minnesota Iron Company and Jacob L. Greatsinger of the Duluth and Iron Range Railroad. The *Duluth Daily News* of September 7 rejoiced that Frick, who had rejected the Mesabi in 1891, "arranged to have a carload of Mesaba ore shipped to the Carnegie works" in Pittsburgh for analysis.[18]

During the winter of 1892–93, when nearly half a dozen furnaces received a portion of the first Mesabi ore shipment, Duluth and the iron trade anxiously awaited the results. Quoting Cleveland sources, the *Duluth News Tribune* of January 23, 1893, worried about criticism from sales agents who said "that the interests of those connected with the production of the Mesaba ore generally would have been better subserved if the entire 4,000 tons had been given to one furnace concern." By distributing the ore between several furnaces, the agents complained, "it is not known just what percentage can be used with safety."

The agents had accurately pinpointed what was to be the initial obstacle in the iron trade's acceptance of Mesabi ores. Existing blast furnaces had been constructed to use hard ores, and, as a result, the powdery or dustlike structure of the Mesabi product caused problems. At first manufacturers found it impractical to charge their furnaces with more than one-third of the soft ore. Early attempts to use a greater portion either submerged the surrounding countryside in great quantities of dust and evoked angry damage suits, or, when occasionally the ore was packed too firmly, exploded the furnace. In time furnacemen learned how to increase the percentage of Mesabi ore when they discovered that the open-hearth process, a refinement of Bessemer techniques, required less coke and that the addition of a small proportion of hard ore could prevent dangerous gas accumulation. Nevertheless the Edith Furnace Company of Pittsburgh, owned by Henry Oliver, successfully smelted 1,000 tons of the first shipment and informed him that the ores "will prove easy smelters [although] the waste . . . will be greater than in more granular ores." By 1895 the *Superior Sunday Forum* announced that Oliver was "using 66 per cent of the Mesaba ore in one of his furnaces, and the results have been very satisfactory."[19]

From the outset the city of Duluth expected to grow and prosper with the Mesabi Range and with the fortunes of the Merritts. Newspaper editors exulted that since the Mesabi "mines are owned almost exclusively by Duluth capitalists, the revenues derived from this source will eventually be invested in Duluth real estate and manufacturing enterprises." The city also expected to benefit from "the building of large saw mills to manufacture the vast bodies of pine timber tributary" to the ore-carrying railroad routes. Observing that the development of the Michigan ports of Escanaba and Marquette "subsisted entirely upon the shipments of ore, and selling of supplies" to mining companies and small towns, Duluth boosters predicted a versatile and profitable future for their city.[20]

In the fall of 1892 contractor Grant completed a 15.7-mile extension from Iron Junction northeast to the Biwabik Mine and nearby settlements. Early the following spring his crews opened the Missabe Mountain spur that extended from the main line at Wolf, a small station north of Iron Junction, 6.4 miles to Virginia to handle the production of mines surrounding that village. The Duluth, Missabe and Northern served almost all the mines then operating on the Mesabi Range, and the Merritts wanted Duluth to profit directly. On January 29, 1893, the

Duluth News Tribune expressed popular opinion when it editorialized that any postponement of "the carrying out of the plans" the Duluth, Missabe and Northern "company has made for improvements here, is a great misfortune to Duluth, the Mesaba iron range and to the whole county of St. Louis."[21]

For their part, the Merritt family possessed an almost fanatical devotion to Duluth, and their loyalty to the city where they had resided from its birth was to contribute to their downfall. On June 30, 1891, the *Duluth Tribune*, quoting an unidentified Duluth, Missabe and Northern Railroad spokesman, assured city residents: "You need have no fear of us going away from Duluth to locate our terminals. They will be in our city and every dollar spent by the road will go into Duluth. Our terminals, passenger and freight and ore docks will be here. The Mesaba road will be a Duluth road, first and last. We have been offered big inducements by the Superior people; but we have not listened to them." Even though the Merritts initially used the Duluth and Winnipeg tracks and shipped their first ore from Superior, they had always planned to connect with Duluth. They hoped to build a track from Stony Brook Junction and to erect ore docks in West Duluth near their old home section of Oneota as soon as possible.

Within this atmosphere of mutual esteem and common goals, Duluth's economic and political leaders vigorously debated how best to aid the Merritt railroad so as to benefit the city. Beginning in the summer of 1891 committees of leading citizens and the Chamber of Commerce held informal conferences with railroad officials, but were unable to agree upon a plan. On February 20, 1892, the *Duluth Tribune* announced, "This is a very favorable time for the building" of the Duluth, Missabe and Northern ore dock in Duluth, and the company was "now in a position to begin work on it." Playing upon the city's rivalry with Superior, committee members then stressed that company officials must "receive aid at once" to enable them "to go right ahead with their work and to abandon the offer made them by Superior." To counter Superior's inducements, the committee suggested that Duluth provide a financial bonus of $200,000 to entice the Duluth, Missabe and Northern to make its headquarters in the city, purchase the necessary property there, and let contracts for building yards and an ore dock. The West Duluth council quickly gave the railroad permission to use a street to reach the lake front, but arguments developed over the best means of raising the bonus money, where the docks were to be located, and whether the Duluth and Winnipeg should share in the bounty. Ultimately, the citizens' committee proposed two funding alternatives: collecting private subscriptions from city property owners, or appealing to the entire county to approve a bond issue. Mass meetings called by the Chamber of Commerce on February 23 and March 19 "were favorable to granting assistance" to both the Duluth, Missabe and Northern and the Duluth and Winnipeg but uncertain how the city could achieve this feat of having its cake and eating it too.[22]

What the *Duluth Tribune* of March 27, 1892, called "the chaos that has characterized this matter and perplexed the citizens" was neither easily nor quickly resolved. For more than a year the various factions wrangled and debated. Some city council members believed that the DM&N should be restricted in its choice of an ore dock location to a point within the city limits. Eventually they were persuaded to give the railroad a free hand after Lon Merritt was quoted in the *Duluth Daily News* of April 22, 1892, as saying, "We are bound to come into Duluth, if we can get there; but after we have spent hundreds of thousands of dollars in railroad building in order to ship ore, we are not going to lie around waiting for Duluth to open its doors to us." The issue of track rights into Duluth also delayed final action. At first the Merritts attempted to negotiate with the St. Paul & Duluth for the use of its city tracks. When this effort failed, it became clear that the Merritts would be forced to build their own line to the harbor. To do so they needed permission from the city council to cross certain streets. On April 8 the council passed the necessary ordinance over some opposition.

By the end of 1892 the Duluth, Missabe and Northern directors were fully decided to establish their lake terminal and ore dock along the Minnesota shore. "It now transpires . . . that the dock on Allouez bay is, so far as the Missabe road is concerned, merely a temporary convenience to serve until Lon Merritt's cherished hope — now pretty well assured of fruition — to locate the permanent docks at Oneota is attained," reported the *Duluth News Tribune* of November 29, 1892. But before the docks "are definitely located here," Lon told a jobbers banquet on November 30, the city would have to provide "considerable encouragement."[23]

Duluth Evening Herald reporters interviewed many port city residents concerning the highly charged bonus issue. On December 3 Mayor Charles D'Autremont was quoted as saying, "I certainly do . . . favor a reasonable amount of aid, if in return we get the . . . docks. I want these as well as the terminals." The same day Lon Merritt repeated the railroad's often expressed point of view: "Our intentions have never been changed. We are of the same opinion now that we were two years ago; we want to come to Duluth. We cannot, however, if we are to have obstacles thrown in our way by Duluth people. We have opened up the range, built the road to it, designed to particularly benefit Duluth, and now it remains for Duluth to say whether it wants the terminals or not. We ought to have the right of way into the city given to us."

Perhaps to help stimulate support, the DM&N late in December ran an excursion to the range with "nearly 400 guests of the road" on board, but not until May 26, 1893, did the people of St. Louis County approve a $250,000 bond issue by a vote of 2,481 to 533. Even though participation was light, residents of Duluth and the Mesabi Range towns cast a strong vote in favor of bonds to bring the tracks into Duluth and to help build a dock there. To many observers' surprise, Vermilion Range voters were not quite as solidly opposed as had been expected before the election.[24]

While Duluth debated in the fall of 1892, the Merritts unexpectedly learned that the Duluth and Winnipeg was once more experiencing financial strains. To make matters worse, in November, 1892, both Kimberly and Oliver gave a year's notice that during 1894 each expected to ship at least 1,000,000 tons over the Duluth, Missabe and Northern. Although the Winnipeg line had originally agreed to prorate the cost of engines and rolling stock, especially wooden ore cars, to accommodate any increased traffic from the range, and would, as promised, enlarge the dock facilities on Allouez Bay in 1893, the Merritts came to believe that the company was now unable to handle even the originally contracted shipments, let alone anticipated increases. Like Charlemagne Tower on the Vermilion, Leonidas and Alfred had anticipated two years of steady ore production to provide them with sufficient resources to finance a DM&N extension to Duluth. The financial distress of the Duluth and Winnipeg forced them, they thought, to consider building tracks and an ore dock there much earlier than they had planned.[25]

Unwilling to forfeit their railroad's ore-hauling contracts with Kimberly and Oliver and with them its control of Mesabi ore traffic, the Merritts believed themselves to be in a tight spot. Moreover, they now faced stiff competition on the range. In the spring of 1892 Duluth and Iron Range officials had disclosed plans to build extensions to various mines on the Mesabi. When the company awarded a construction bid, the *Duluth Daily News* of April 22 predicted "lively times in the new iron regions for the next six months," for the Duluth and Iron Range expected to be ready to handle Mesabi ore before contractor Grant could complete the Merritt road. The *Duluth Tribune* summed up the situation on April 30, 1892, by remarking: "It was not to be supposed that the Duluth, Missabe & Northern would be allowed undisputed possession of the range, but it was never guessed that the Duluth & Iron Range would so soon take such vigorous steps to corral the ore shipments from the new El Dorado. . . . The whole scheme looks like an aggressive move on the part of the Iron Range to secure the Missabe ore and to bring that range within its system."

During the summer the Winston Brothers Construction Company of Minneapolis laid some 16 miles of track west from Allen Junction, a small station near Colby Lake about 20 miles south of Tower on the road's main line, to tap the mines near the newly established villages of McKinley and Biwabik. Almost immediately the company added branch lines to other operating mines with which the Duluth, Missabe and Northern had failed to secure traffic agreements. As additional mining properties were developed in 1893, the Duluth and Iron Range expanded its network to mines near Virginia, and two years later it moved south to begin operations at Eveleth. The parent Minnesota Iron Company was also reportedly interested in leasing Mesabi Range lands for mining purposes. (See map, page 92.)[26]

These events prompted the Merritts to push their own line into Duluth — a decision fraught with danger for their infant empire. The brothers realized that unless they quickly raised a substantial amount of capital with which to construct a DM&N extension and an ore dock, they would be engulfed by the twin problems of competition from the D&IR and of carrying the heavy financial burden of their already completed transportation network and mining developments. Attempting to solve their dilemma, the brothers sought the advice of Alexander McDougall, an old and trusted friend who had made his home in Duluth since 1871.[27]

The Scottish-born McDougall had received patents in 1881 and 1882 for the uniquely designed steel lake freighter known as the whaleback, and it had been his whaleback 102, carrying 2,073 tons of ore, that had transported the Merritts' first Mountain Iron Mine shipment down the lakes in the fall of 1892. Closely resembling a floating cigar, the whaleback had a long, narrow, double hull with a flat bottom. The sides were rounded and the ship tapered to a snout at both ends. McDougall believed the vessels — the forerunners of modern ore boats — would offer minimal wind resistance and be extremely seaworthy. In the course of developing his fleet of ships, the retired lake captain had obtained capital from eastern financiers, and he was confident that these same men would aid the Merritts.[28]

McDougall put Lon and Alfred in touch with officials of the American Steel Barge Company, a New York corporation engaged in shipbuilding and transportation on the Great Lakes as well as on the Atlantic and Pacific coasts. The firm had purchased all the patents for McDougall's whaleback vessels for $25,000. John D. Rockefeller had invested substantially in the barge company, which was to be the first, though indirect, link between the Standard Oil king and the Merritts.[29]

On behalf of the American Steel Barge Company, Charles W. Wetmore, one of its vice-presidents, visited Duluth in November, 1892, to inspect the Merritts' operations at firsthand. This expectant capitalist is one of the most enigmatic figures in American business history. Born in Ohio in 1854 and educated by private tutors, Wetmore graduated from Harvard University in 1875 and Harvard Law School two years later. He moved to New York City and began a career as a lawyer and financier. Professional associations placed him on the board of American Steel Barge which, in turn, gave him a circuitous contact with the Rockefeller interests.[30]

After touring the Mesabi Range, Wetmore analyzed the principal mine owners and the railroad connections already established there. Reporting directly to barge company president Colgate Hoyt, he supported the Merritts' point of view, saying that "if the Duluth, Missabe and Northern can be completed to Duluth, and adequate equipment furnished, it can by its natural advantages alone control the greater part of the transportation of this region." Convinced that the Mesabi was "by far the greatest iron producing region ever discovered anywhere," Wetmore believed that it would shortly dominate the Lake Superior iron ore market. He informed President Hoyt that the Merritts were "as they frankly said 'hard up'" and needed assistance in placing bonds on the eastern market to pay for the portion of the DM&N already completed as well as for the construction of the extension into Duluth and the necessary ore docks. If the barge company could assist in placing bonds and raising the needed capital, he said, the Merritts would give American Steel Barge whalebacks the exclusive right to transport all their Mesabi ore. Wetmore emphasized that this "alliance is the *most* important matter that we have on hand. . . . It is *the great* opportunity of the present."[31]

Several weeks passed before Hoyt cautiously approved Wetmore's suggestions. The barge company president was pleased by the prospect of an alliance that would give the firm the exclusive lake transport of the Merritts' iron ore, but he reminded Wetmore that Rockefeller had "large interests" in the Minnesota Iron Company on the Vermilion Range and in several mining operations on the Gogebic. The barge company "must not do anything without his full knowledge and consent, which in any way might be considered prejudicial to his present interests," Hoyt wrote. He also made it clear that the American Steel Barge Company should not take any direct interest in the building of railroads or in the purchasing of ore mines, and that it should remain solely a ship construction and transportation firm.[32]

The Merritts' proposal to lay track into Duluth and to search for eastern financial assistance divided the leaders of the Duluth, Missabe and Northern. President Chase and contractor Grant, who was also a large stockholder, broke with the Merritts, fearing — correctly, as it turned out — that the new plans would place the whole enterprise in financial peril. Simultaneously Chase secretly opened negotiations with Donald H. Bacon of the Minnesota Iron Company to sell 51 per cent of the stock in the railroad and the mines for $7,000,000 or $7,500,000. According to Lon, the Merritts discovered this plan inadvertently when they picked up a letter dropped on the floor by Chase. Grant and several other stockholders who had invested heavily in the Duluth and Winnipeg did not wish to see that railroad's interests jeopardized by the loss of DM&N traffic. In December, 1892, the Merritts' control of the Missabe road was precarious, for family members owned only 30.8 per cent of the stock. Since they had no wish to sell to the Minnesota Iron Company, they began to seek the votes they needed to defeat the Chase proposal at an upcoming meeting of the railroad's stockholders. Both the Chase-Grant and the Merritt factions conducted lively campaigns to secure adherents. When a vote was taken on January 20, 1893, after several adjournments and parliamentary maneuvers by each group, an overwhelming majority of the shareholders opposed the sale to the Minnesota Iron Company and authorized the directors to execute contracts with Wetmore and the American Steel Barge Company by a vote of 9,173 to 1,082.[33]

Refusing to give up the struggle quietly, Donald Grant and others obtained a court injunction to stop the Merritts from transferring railroad and mining company stock to American Steel Barge. According to Wetmore, the suit alleged that "a large part of the stock . . . held by the Merritts and their immediate associates had been illegally issued under the laws of Minnesota." Shortly thereafter Chase and Grant successfully petitioned Judge Walter H. Sanborn of the United States Circuit Court in St. Paul to issue a restraining order preventing the Missabe road from ratifying its contract with Wetmore on the grounds that the document transferred the road to the Rockefeller combine for inadequate compensation.[34]

Throughout these legal maneuvers, however, the Merritt forces continued to gather additional stockholder support. Then on February 3, 1893, the Duluth, Missabe and Northern suddenly announced that the Chase-Grant faction had withdrawn the pending injunction proceedings and had sold its 5,000 shares of stock for $250,000, half to members of the Merritt family and half to representatives of the American Barge Company. An additional $400,000 worth of treasury stock was sold at par, 75 per cent to the Merritts and the remainder to Wetmore's associates. Lon and Alfred had acquired the money to buy up the Chase-Grant stock as well as 75 per cent of the new issue by borrowing $432,575 from Wetmore under the terms of five promissory notes subject to 24-hour call. Following Chase's resignation, Alfred became president of the DM&N on February 7, 1893, and Charles Wetmore joined the railroad's board of directors. The Merritt-Wetmore syndicate was now in complete control of both the railroad and the mining enterprises.[35]

As vice-president of the railroad, Lon had signed a contract with Wetmore on December 24, 1892, although the Chase-Grant legal action had delayed its official acceptance. Under its terms Wetmore agreed to purchase in installments $2,000,000 worth of railroad bonds at 80 per cent par value ($1,600,000); he was also to receive a bonus of 6,666 shares. He planned to sell the bonds, which carried an annual interest rate of 6 per cent, in the New York money markets. The contract stated that all capital received from the bond sales had to be spent for specified projects: $600,000 for additional rolling stock not then under contract, $397,614 for the extension of the main line into Duluth including necessary side tracks, $300,000 for ore docks at Duluth, and $200,000 for the approach to the docks that required viaducts over the streets and tracks of other railroads. In addition $250,000 was allotted for the construction of a branch line from Wolf 17 miles west to the village of Hibbing, an area already sufficiently developed to make its mines important feeders for the system. Finally Wetmore paid $50,000 in cash at the contract signing and agreed to pay $450,000 within 60 days. In a second contract also dated December 24, 1892, the Duluth, Missabe and Northern agreed "to give to the Barge Company exclusively" for water transportation for 15 years all Mesabi ores owned or controlled by the DM&N. American Steel Barge was also given the management of the Duluth dock, which was, however, to be built and maintained by the railway company.[36]

Confident that they now possessed sound financial backing but still facing the threatened loss of the Kimberly and Oliver traffic contracts, the Merritts moved rapidly to complete the Duluth extension and ore dock. As president of the Missabe road, Alfred directed the over-all construction activities, Cassius looked after the financial details, and Leonidas secured additional traffic agreements on the range.[37] In an interview reported in the *Duluth News Tribune* of February 8, 1893, Lon described the railroad's immediate plans: "The policy decided upon by the company is to build the Stony Brook extension to Duluth as fast as men and money can accomplish it. . . . The general policy of the road is to cheapen transportation by rail and water from the mines of the range to the furnaces of the East, to such an extent that the Mesaba mines will be formidable competitors for any known producing district."

In March, 1893, Wolf and King of Ottumwa, Iowa, the contractors engaged to build the 26-mile line from Stony Brook Junction east to Duluth, began to clear the right of way, grade the roadbed, and fill the cuts. The heaviest section of track thus far constructed in northern Minnesota was necessary in the six-mile, double-tracked portion running up and down the steep hills overlooking Duluth, where rails purchased from the Illinois Steel Company of Chicago for $180,000 were laid, weighing 80 pounds to the linear yard. The remainder of the extension utilized the more conventional 60-pound rails. Except for occasional delays caused by bridge construction, the crews pushed steadily forward. By June 29 only "about ten miles of track" were still to be laid, and over 1,000 men were employed.[38]

Construction of the ore dock on St. Louis Bay at the west end of Duluth near Oneota had begun in January, 1893, under Alfred's direction. Day laborers hired by Duluth contractor George J. Anderson had driven the first of 8,000 wooden pilings each 60 to 70 feet long. The men worked for several months on the ice, which made it easier to erect the "framing of the first story of the superstructure" of the dock. The railroad company obtained most of its timber locally but imported the longest pilings from the Pacific Coast.[39]

On July 22, 1893, the first train of Mesabi ore shipped entirely over the Duluth, Missabe and Northern tracks arrived in Duluth. According to the *Duluth News Tribune* of the following day, the contents were "dumped into one of the pockets in the new dock." Only 30 pockets had been completed. This "epoch in the history of the West end" of Duluth was witnessed by members of the Merritt family, railroad officials, employees, and "a large number of interested spectators." In a letter to her husband, Mrs. Leonidas Merritt wrote: "You ought to have been here this afternoon to help us celebrate the coming of the first ore onto the new dock. There were ten cars brought down. . . . The largest part of the Merritt family were there children and all   we had to wait a couple of hours before we saw the train   then away up the hill through the smoke and haze we could see a black line that looked like a black caterpillar creeping slowly along   soon it came onto the trestle work and then onto the dock where we all stood waiting for it . . . it moved out until the ore cars were over the pockets and then stopped to unload   nobody made a sound   not a hurrah or any thing while they came on the dock. . . .

"The engine that came down was no[.] 15 and the first ore car on the Dock no[.] 725 . . . I will send you a little of the ore from the cars that came down today so that you will realize that they have really begun to ship ore."[40]

When it was completed in the fall, the Missabe road's ore dock was the largest in the world, stretching 2,304 feet from the shore. Each of its 384 pockets held 175 tons — the contents of seven fully loaded ore cars. The ore-handling equipment matched the unsurpassed size of the dock. Automatic steam-driven Denton hoists raised and lowered the 27-foot, semicircular steel spouts. A half-mile-long trestle or viaduct connected the 52-foot-high dock to the lake shore. The viaduct crossed city streets, streetcar tracks, and the St. Paul & Duluth right of way at a 20-foot elevation.[41]

This massive dock structure was sufficient to handle Mesabi ore traffic for only a few years. In 1895–96 a second, smaller dock measuring over 1,200 feet was constructed. Its 192 pockets offered a storage capacity of 35,000 tons, 20,000 less than its companion facility. The new addition was 50 feet wide and had eight rails, giving the company a total of six tracks that allowed two trains to run abreast. In 1897 another 500-foot extension increased capacity by 15,000 tons. A third wooden ore dock, ready for use in 1900, was built 57 feet above the water to accommodate larger ore carriers. The last wooden structure, completed in 1906, was again by far the largest in the world. It rose 72.5 feet above the water, extended 2,350 feet into the bay, and could store a total of 100,000 tons.[42]

The transportation of iron ore complemented Duluth's continuing role as a major shipper of wheat and timber. The economic boom that began in the last half of the 1880s continued into the 1893 shipping season. Simultaneously the city's population increased from 46,920 in 1889 to 74,490 four years later, swelled not only by permanent residents but also by numbers of people temporarily residing in Duluth or passing through it on their way to the Mesabi or Vermilion ranges. The number of hotels rose from 27 in 1886 to 65 in 1893, the number of boardinghouses from 47 to 99. Between 1889 and 1893 the number of banks increased from 12 to 18 and loan agents from 53 to 67. In the four years following 1889 the number of lawyers attracted to the area doubled, reaching a total of 145 by 1893. The proliferation of incorporated mining companies indicated the dominant position of iron ore in Duluth's economy. In 1889 only three firms advertised their services; four years later there were 73. In the period from August 1, 1890, through July 31, 1892, for example, registered mining company capitalization in Minnesota totaled $202,000,000.[43]

Like those of the city, the fortunes of the Merritt family seemed poised for continued spectacular growth. The family controlled about 40 per cent of the stock of six Mesabi Range mining companies. They had bought out their partners in the Duluth, Missabe and Northern Railroad and had completed the construction of the Duluth extension and the dock needed to honor their traffic contracts with Kimberly and Oliver. They had what they believed to be a good contract with Wetmore to market $1,600,000 of their railroad's bonds, and they expected to ship substantial quantities of ore in 1893. Charles Wetmore had come highly recommended by the Merritts' intimate friend, Alexander McDougall, and he was associated with a powerful trust, the American Steel Barge Company. Moreover, Leonidas believed Wetmore to be a close associate of John D. Rockefeller. After all, George Welwood Murray, Rockefeller's confidential attorney, had taken an active part in formulating and executing the Merritt-Wetmore contract. In the early months of 1893 the brothers could not foresee the impending long and severe depression of the 1890s which would engulf them before the summer shipping season ended.[44]

THE MERRITT FAMILY in 1871. Standing, from left, Leonidas, Lewis J., Andrus, Alfred, and Lucien. Seated from left, Cassius, mother Hephzibah and father Lewis H., Jerome, and Napoleon. St. Louis County Historical Society Collection.

JOHN E. MERRITT, Lucien's son, was the first manager of the Biwabik Mine, opened in 1893. St. Louis County Historical Society Collection.

STUDIO PORTRAITS of Leonidas Merritt and his wife Elizabeth (left) and Alfred Merritt and his second wife Jane (right) were taken about 1905. St. Louis County Historical Society Collection.

ABOUT 1890 the Merritt brothers (standing) gathered in front of the family hotel at Oneota with their wives and children. From left are the families of Alfred, Leonidas, Lucien, Napoleon, nephew John E., Andrus, and Lewis J. St. Louis County Historical Society Collection.

MERRITT as it looked before many of its buildings were moved to nearby Biwabik. Nute Collection.

A REAL ESTATE dealer was already active in the mining camp of Hibbing in 1893. His shop stood on what became First Street in the future "Iron Ore Capital of the World."

VIRGINIA was the largest town on the Mesabi in the 19th century. In 1899 it had over 4,000 people and could boast of two-story brick buildings. Note the ore trains in the foreground. Marquette County Historical Society.

IN MOUNTAIN IRON (above) cordwood for fuel was stacked along the main street in the snowy winter of 1891. Curtains in the second-floor window of the Mountain Iron Depot (below left) probably marked the living quarters of the agent. The frame structure at right housed the Mountain Iron Bank, the first bank on the Mesabi Range, in 1893. St. Louis County Historical Society Collection.

EDMUND J. LONGYEAR invented the diamond drill which was widely used in ore exploration on the Mesabi Range.

THE YOUTHFUL Longyear (at left) drilled the first of more than 7,000 exploratory holes on the Mesabi in the summer of 1890. The site of the first venture is marked and accessible northwest of present Hoyt Lakes. Nute Collection.

THE MOUNTAIN IRON MINE as it looked about 1892 when it shipped the first ore from the Mesabi Range. Photograph by L. P. Gallagher.

ORE, removed from the Adams Mine near Eveleth in 1899 by the milling method, was taken to chutes and dropped into the cars waiting below. Photograph by Edmund Brush, St. Louis County Historical Society Collection.

HEADFRAMES of shafts nos. 1 and 2 at the Hull Mine near Hibbing in 1899. Such shafts became unnecessary after surface mining began there in 1905. The mine later became part of the Hull-Rust-Mahoning complex — the largest open pit on the Mesabi, now a National Historic Landmark. Franklin A. King Collection.

STRIPPING overburden in 1893 to reach ore at the Missabe Mountain Mine near Virginia soon after it had been leased by Henry W. Oliver. From Charles K. Leith, *Mesabi Iron-Bearing District* (1903).

STEAM SHOVELS that moved on tracks were widely used to strip overburden and load ore on the Mesabi. At the Burt Mine (left) in 1906 tracklayers prepared to move the shovel. A crew of 11 men (below) was needed to keep this shovel working in the Biwabik Mine about 1895. Library of Congress (left) and Nute Collection.

WOODEN side-dump cars used for stripping and a working steam shovel may be seen in the foreground of this view of the Mahoning Mine near Hibbing in 1899. Two strings of loaded ore cars are visible at left and right.

THE FIRST CARLOAD of ore from the Mesabi Range was placed on display in Duluth in October, 1892. It was shipped from the Merritt's Mountain Iron Mine. St. Louis County Historical Society Collection.

NOT UNTIL November, 1892, was the first ore from the Mountain Iron Mine loaded aboard Whaleback 102 at the partially completed Duluth and Winnipeg dock in Superior. Burlington Northern Collection.

CELEBRATING the completion of the new Duluth, Missabe and Northern dock at Duluth, members of the Merritt family, in the crowd at left, watched the arrival of the first ore train on July 22, 1893. Leith, *Mesabi Iron-Bearing District* (1903).

THE SHIPYARD of the American Steel Barge Company at Superior, where whalebacks were built to Alexander McDougall's unusual design between 1888 and 1898. St. Louis County Historical Society Collection.

|6|

THE PANIC
OF 1893

THE FINANCIAL CRISIS that inundated the United States beginning in 1893 marked a watershed between the first and second periods
in the history of Minnesota's iron mining industry. Before 1893 the
rapid discovery and relatively inexpensive exploitation of new ore
sources had permitted the rise of numerous small mining companies
and individual entrepreneurs. After the panic of 1893, the ensuing
severe depression years witnessed a concentration and combination of
holdings. Before the panic, pioneer miners laboriously extracted ore by
crude methods. During and after the panic more efficient machines
and techniques revolutionized the extraction and shipment of ore on
the Mesabi. After 1893 the American business world was never quite
the same. The era of the individual entrepreneur gave way to the age of
the corporation. The stock market, which in 1890 was almost entirely
composed of railroad stocks, became by the end of the decade a market
dominated by the stocks of industrial corporations. Indeed the famous
Dow industrial average was developed in 1897, and the National Association of Manufacturers was formed two years earlier. The rise of
large national corporations, the expanding markets provided by ur-
banization and the growth of cities, and the presence of a mechanism
for raising capital in the form of a market for preferred and common
industrial stocks transformed the business scene at the end of the 19th
century. A signature on a personal note from a local banker was
replaced as a means of raising business capital by the issuing of corporate stock. These developments profoundly influenced the history of
iron mining in Minnesota.[1]

Financial panics, followed by periods of depression and instability,
occurred nearly every decade throughout the 19th century. That of
1857 depopulated Duluth, while the effects of the panic cycle triggered
by the fall of Jay Cooke's banking house in 1873 were sharper and

deeper. Broadly speaking the depression periods brought about adjustments reflecting technological and administrative changes in the market economy of the United States. The downturn signaled by the failure of the Philadelphia and Reading Railway Company on February 26, 1893, was both longer and more severe than those of earlier years. With the exception of the 1930s, the 1890s are usually regarded as the most economically devastating decade in the nation's history, characterized by widespread hardship, the disastrous Pullman strike that began in 1893, and the march on Washington of Coxey's Army in 1894. Scholars agree that the effects continued for at least four years. They were exacerbated, one economist explained, by the fact that the nation "was in the process of fundamental transformation from an agricultural-industrial economy to one in which, although large-scale agriculture still was a salient feature, the outstanding characteristic was a manufacturing-industrial complex."

The collapse of the Philadelphia and Reading Railway — the first dramatic event of the panic — was caused by internal organizational problems. But because the railroad company possessed ties to several "powerful Wall Street financial houses," businessmen grew increasingly suspicious concerning "the soundness of other railways and the financial houses behind them." Other indicators of the deteriorating situation quickly emerged — credit contraction, a "dwindling favorable trade balance," and the failure of well-known corporate establishments. The gold reserve dipped below the established minimum of $100,000,000, while the prices of pig iron, wheat, and other staple commodities declined. On May 3, 1893, the stock market reached its lowest level in a decade. The fact that industrial securities bore the brunt of the crash, combined with many bank failures, led to worsening conditions. On July 26 the collapse of the Erie Railroad marked the beginning of extremely hard times for the entire railway industry. When the panic intensified, "depression forces developed more rapidly as employment fell off and price declines became more marked and more widespread." By August banks throughout the country "were forced to suspend" and "currency was at a premium."[2]

The mining industry, producing approximately 2.5 per cent of the national income during the 1890s, soon felt the widening ripples. Pig iron production registered the greatest decline thus far in its history, plunging from 10,256,000 tons in 1892 to 7,979,000 tons in 1893. Output continued falling in 1894 to 7,456,000 tons before increasing to 10,580,000 tons the following year. Except for a slight dip in 1896, pig iron production thereafter steadily increased into the first decade of the

20th century. Iron ore shipments dropped from a previously unsurpassed high of 16,297,000 tons in 1892 to 11,588,000 tons in 1893, rose slightly to 11,880,000 tons the following year, and then steadily advanced to 27,300,000 tons at the turn of the century.[3]

In 1890 the Lake Superior ranges provided 56.16 per cent of the entire United States iron ore supply. When total tonnage fell so precipitously in 1893, Lake Superior's share slipped to 52.4 per cent. In succeeding years, however, the western Great Lakes area dominated the nation's output. In 1894 its percentage jumped to 65.3. It continued to climb (except for a slight drop to 62.2 in 1896) until by 1899 it supplied 74.2 per cent of the country's iron ore — the peak production for the early period. Despite its dominant position, the lake region did not escape the impact of the depressed economy. Shipments fell from 9,080,684 tons in 1892, the highest figure to that date, to 6,075,328 tons the following year. From this nadir, production in the Lake Superior region (especially on the Mesabi) rose steadily, paused briefly only in 1896, and then continued to climb into the first decade of the 20th century.[4]

Before the 1893 shipping season opened, all appearances indicated a larger business than ever before. As early as May 1, however, many iron ore producers felt the cold winds of change when they learned that more than 2,000,000 tons of the preceding year's ore still remained unsold on Lake Erie docks. Prices for iron ore at Lake Erie ports reflected the deteriorating economic situation. Old Range number 1 Bessemer hematite, for example, dropped from $4.50 per ton in 1892 to $4.00 in 1893, and still further to between $2.65 and $2.85 in 1897. The country's worsening economic climate and fear of future competition from the Mesabi combined to create financial havoc among both large and small ore producers on the older, better established ranges.[5]

Although some companies continued to ship ore and to maintain a small labor force, most mine superintendents made a concerted effort to cut costs. Before the older ranges recovered from this economic dislocation, nearly 12,000 wage earners, approximately 80 per cent of the work force in Michigan and Wisconsin, found themselves either temporarily or permanently unemployed. Shutdowns and production curtailments were common on the Marquette, Menominee, Gogebic, and Vermilion ranges. For example, the Cleveland-Cliffs Company, one of the largest mining firms operating on Lake Superior's south shore, shut down all but one of its mines on the Gogebic, laying off more than 200 men in May, 1893, alone. On the Marquette Range, in addition to more cutbacks by Cleveland-Cliffs, the Republic Iron

Company, another major operator, suspended work at all mines for 60 days, throwing 300 men out of employment. The Chapin Mine, the Menominee's largest single producer, reduced output by 40 per cent during May and June before briefly closing down altogether, leaving nearly 500 laborers temporarily unemployed.[6]

On the Vermilion Range, the Minnesota Iron Company "had a very large amount of money tied up in ore both at the mines and at ports," according to the *Duluth News Tribune* of July 24, 1893. The firm failed to pay its usual dividend in the spring of 1893 and closed almost all of its mines, an action that affected more than 900 men — nearly 50 per cent of the company's labor force. The Chandler Mine, in which the Minnesota Iron Company owned "a majority of the stock," laid off its night shift, and the nearby Pioneer Mine at Ely halted all shipments. Even the Soudan, the oldest of the Vermilion mines, suspended in 1894. But the Minnesota Iron Company entered the depression with relatively little debt. After paying interest on a Duluth and Iron Range Railroad mortgage of $5,209,000, the parent firm reported net earnings of $1,351,721 for the fiscal year ending on April 30, 1893. By that time this strong organization owned mining property on the Mesabi as well as a fleet of eight steel ore carriers, built between 1890 and 1892, and it continued to lay D&IR tracks farther and farther west on the Mesabi in 1893 and 1894.

Unlike its competitors, however, the Mesabi Range experienced steady development during the depression years. It not only increased its tonnage, but it also became the largest single producer in the Lake Superior region. Reviewing the 1893 market conditions on January 2, 1894, the *Duluth News Tribune* concluded, "The only important development of the year in iron ore circles . . . has been the removal of the new Mesaba range from the field of speculation to that of production." From a minimal beginning of 4,245 tons in 1892, Mesabi ore shipments rose to 613,645 tons in the depression year, to 1,793,052 tons in 1894, to 2,781,587 in 1895, and to 7,809,534 tons in 1900.[7]

Even when the older ranges recovered from the economic downturn, the Mesabi not only retained its leading position among Lake Superior iron shippers, but also moved into dominance of United States ore production. In 1892 and 1893 the new El Dorado stood last among the five major lake ranges. The following year it ranked third behind the Marquette and Gogebic. Then in 1895 the Mesabi attained the number one position and steadily increased its lead on into the 20th century. In 1895, for example, it supplied 26.63 per cent of Lake Superior ores and 17.43 per cent of total United States production.

Five years later, at the turn of the century, its share of the lake market was 47.03 per cent and of the United States market 32.98 per cent.[8]

From the time the Merritts opened the Mountain Iron Mine with hand tools in 1892, claims were made for the cheapness with which ore could be extracted from the new El Dorado. The *New York Times* of December 26, 1893, reported that "Mining experts from ore-producing regions" were swarming over the range "on investigating tours. The report they telegraphed to their employers was, in effect, that the glory of the other iron ore regions had departed — that they would surely be forced into a competitive war for the control of the American market that could have but one outcome, and that was bankruptcy . . . all men who are at all acquainted with the Mesaba Range know that iron ore is mined and loaded on cars in that region for from 4 to 40 cents per ton."

The competitive edge enjoyed by the Mesabi was due to a number of factors. Among them were (1) low taxes and labor costs; (2) the nature of the ore and its generally horizontal formation; (3) the refinements in mining techniques and the development of automatic machinery. In this period mining taxes in Minnesota were not onerous, for the provisions of the first law secured by George Stone to aid Charlemagne Tower remained in force from 1881 until 1897. During those years the state collected 1 per cent per ton in lieu of all other taxes. After the law was repealed in 1897, iron mines were taxed like all other property until 1914. After that Minnesota state and local governments exacted various types of ore taxes that greatly enriched their treasuries. Nor did mining companies in this period have any problems with federal taxes. Corporation excise taxes were not levied until 1909 and corporate income taxes were not collected until 1913.[9]

Labor costs during the period from 1892 to 1901 are not readily available — especially for the first four years. But they were clearly affected by the panic of 1893 and the ensuing depression. From an average daily wage of $2.00 in 1892, the take-home pay two years later fell to $1.35. One study of Minnesota's iron ranges showed that maximum daily wages of miners employed by the Oliver company were $1.75 by 1896, rising a little each succeeding year to a total of $2.40 at the turn of the century. In 1901 the Minnesota Bureau of Labor, reporting on daily wages for underground work on the Mesabi and Vermilion, listed a low of $1.77 for mine laborers to a high of $3.41 for diamond drillers.[10]

As for mining methods on the Mesabi in the 1890s, some of the literature of the period leaves readers with the exaggerated impression that it was merely necessary to scrape away a little earth, scoop up the rich ore, and load it into railroad cars. At some times and some places on the big range, that was almost literally the case. But in general three mining methods, previously used in other ore districts, were employed — underground shafts, the method most used on the older hard-rock ranges; open-pit mining with steam shovels; and the milling system, which was a combination of the first two. Of these, underground mining was the most expensive; the open-pit, steam-shovel operation was the cheapest, and milling was in between. For example, the *New York Daily Tribune* of December 2, 1895, reported that the open-pit method in use at seven Mesabi Range mines cost from $.05 to $.10 per ton of ore removed; milling at four mines cost from $.15 to $.20 per ton; and underground methods at seven mines cost from $.35 to $.50.[11]

Previous experience with underground shafts on other ranges at first encouraged Mesabi entrepreneurs to utilize that technique. As on the Vermilion, it involved digging a shaft, cutting drifts, constructing a timbered framework and chutes, and purchasing pumps and other equipment. But because of the shallow, horizontal nature of the Mesabi deposits and the soft character of much of the ore, underground mining was frequently neither effective nor efficient. Nonetheless, the range was never without underground shafts in this period, for they could be started without the costs involved in removing masses of overburden and poor ore. Indeed 46 per cent of all the Mesabi ore shipped in 1902 was still produced by underground methods.[12]

When Minnesotans think of the Mesabi, however, they think of the huge open pits for which the range became famous. A common misconception says that such open pits originated on this range. In truth, open-pit mining had been fairly common throughout the Lake Superior basin since the 1850s. The first miners on the Marquette and Vermilion ranges, for example, had removed the ore from pits using buckets, wheelbarrows, and wagons. Even steam shovels, later so closely identified with the Mesabi, had been used on the south shore before ore was discovered on the big range. But the magnitude of the Mesabi open pits and the tremendous volume of overburden and ore dwarfed previous operations on the older ranges.[13]

Steam shovels were first developed primarily for use in railroad construction. Since they required tracks on which to operate, these shovels were nonrevolving, with an arm capable of a limited swing. In 1884 Joseph Sellwood employed a steam-powered shovel at the Colby Mine on the Gogebic near Bessemer, Michigan, but its effectiveness was limited by the physical structure of that ore body. Eight years later a shovel was used successfully to strip overburden at the Biwabik Mine soon after Peter Kimberly leased the property from the Merritts. Its introduction there was the novel idea of John T. Jones, a skillful mining engineer from the Menominee Range who directed the opening of the Biwabik, which was the first large-scale open-pit operation on the Mesabi. An eyewitness account of the use of a power shovel to load the Biwabik's first shipment of ore was offered by the *Duluth News Tribune* of June 19, 1893. "A new era in the epoch of mining . . . was recorded . . . at 8:32 A.M., on the 14th of June," ran the report, "when the monster Bucyrus steam shovel, run by Jno. Curry, plunged her strong arm into the rich ore in the Biwabik mine, raised two tons and swung around suspending it over the D&IR ore car No. 781. For a moment it stood, when the small, delicate hand of little 5-year-old Alice Jones, daughter of J. T. Jones, reached forth, pulled the cord and deposited the precious load into the pocket of the car. . . . One ore car was loaded in three minutes and 10 seconds."[14]

During the first few years of steam-shovel mining, independent contractors usually handled the stripping operations under agreements that paid them $.30 to $.50 a cubic yard, making the cost of stripping an acre about $16,000 if 40 feet of gravel had to be removed. But the contractor's success quite often depended upon such variable factors as weather conditions, labor supply, and experienced superintendents. By the end of the 19th century, the major companies were doing their own preliminary stripping, eliminating this middleman. In the early years generally accepted practice called for the removal of up to one foot of overburden for every foot of ore, if the stripping necessary did not exceed 35 to 40 feet. The cost of such work at the Biwabik was said to be $.08 a ton of ore in 1893. As Mesabi production increased, the cost factors made open pits more and more desirable. In 1896 a 65-ton shovel was in operation at the Mountain Iron Mine. Its dipper could lift 16 feet above the tracks and load 10 tons a minute, or 20 ore cars per hour. Larger and more dependable machinery enabled successful entrepreneurs to undertake deeper stripping, removing a greater ratio of glacial debris to ore. In a few years taking off about two feet of overburden for each foot of ore was the rule, using Marion or Bucyrus 90- or 120-ton shovels.[15]

"The cost of open-pit work," wrote a knowledgeable geologist, "depends primarily on the amount of overburden to be removed and the ratio of this to the size of the ore body. The average cost of loading on the car may be only 4 or 5 cents a ton. The average cost of stripping, however, to uncover a ton of ore may run from 20 to 30 cents. It is obvious that the figure would be smaller where the drift is thin or where the amount uncovered is large in proportion to the thickness of the cover, so that the cost of surface removal may be charged against a large number of tons. In general the cost of steam-shovel mining has probably averaged less than 30 cents a ton."[16]

The shovels of that day, running on railroad tracks, required a crew of 20 to 30 men to keep moving the tracks and operating the shovel. After the surface overburden was removed, the crew laid tracks to the designated part of the open-pit deposit to be mined. During the earliest seasons, the "dinkey" engines, weighing from 12 to 25 tons, and small dump cars ran on narrow-gauge tracks. These soon proved inadequate, and they were replaced in pit work by 60- to 80-pound standard-gauge tracks, automatic dump cars, and engines of greater power and weight that could negotiate grades of 2 1/2 to 3 per cent in order to get the loads out of the pits. When steeper grades developed, the shovel operators made a cut in the ore formation and loaded the ore onto rail cars brought out on a parallel track. After a series of slices had carried the bank back far enough to reduce the steep grade, a second crew began working on a lower level. This procedure was carried out on succeeding levels until the shovels had carved the entire pit into a series of terraces. Shovels and engines could then work on a terrace at any desired level. The terraces gave the mining companies access to a greater variety of ores, increased the possibilities of desirable mixing of ores, and made it feasible to procure a large output in a short time.[17]

Knowledge of the entire ore body also permitted the development of a system of grading, sampling, and shipping of the various types and grades of ore within the pit. William C. Agnew, long-time manager of the Mahoning Mine, devised a method adopted by others about 1900 that enabled him to keep ahead of the shovels on a daily basis. "It happens that Mahoning ore varies little in iron, though running to a much wider divergence in phosphorous," wrote an engineer. "As the ore is mined and loaded into cars, samples are taken of each 150-ton lot; if in 50-ton cars, of every three, if in 30-ton cars, of each five, which is about the capacity of a dock pocket. Analyses of these samples are sent to the ore-shipping dock, and so rapid is the work that often results are known before the train leaves the mine yard, always before it reaches the dock." In Agnew's "mine office is a receptacle containing a series of envelopes numbered and arranged to correspond to the pockets of the shipping dock, and as soon as made out, a copy of the assay card for each set of cars is slipped into its place. Usually trains are not broken between mine and dock, and it is an easy matter to direct where the cars shall be put and to check the pockets into which they have been dumped. When a vessel is loaded the pockets from which the cargo has been taken are reported to the mine office and by averaging the assays of cars filling these pockets a very close figure of the cargo analysis is determined. . . . By knowledge of the iron and phosphorous values in each 150-ton lot, it is easy for the mine management so to handle shipments" to increase or decrease the percentages to meet the guarantee by adding ores of higher or lower phosphorous or iron content and so avoid "the mixing of ores in cars by the shovel."[18]

Moreover, as Mesabi shipments mounted, the "construction of a special type of ship of large tonnage for the ore trade, coupled with the invention of unloading machinery of great capacity at the terminal ports" reduced the cost of transportation. Before the turn of the century a ton of ore could be "hauled one hundred miles by rail from the most distant mines . . . to a Lake Superior port, . . . loaded into the ship, . . . carried one thousand miles by water, and unloaded into cars or onto the stock pile at a Lake Erie port, at a cost of less than $1.80 per ton." One of the early tasks performed by laborers working by hand was leveling the ore in the hold of the vessel. It "was necessary for a gang of men called trimmers to place the ore in the proper way. The trimmers . . . about fifty in number" jumped "into the boat after it was loaded and level[ed] the ore in the shortest possible time. These men were paid about three cents a ton. . . . The work was very heavy" and the "crew was subject to call at all times during the day and night." As the ore industry developed larger and stronger ships, the necessity for trimming was gradually eliminated.[19]

Other mechanical improvements included cars that revolutionized "the old time tedious method of dumping small cars by hand." This primitive technique was replaced by "air actuated cylinders whose pistons pushed the dump car bottoms"; when retracted, the box would right itself. According to the *Duluth News Tribune* of May 5, 1893, "the saving in expense will be tremendous, as it will be unnecessary to have a lot of men who are putting in half their time dumping and righting the cars. . . . The cars are made to dump on two sides, but when used for stripping where the dumping grounds are all on the same side of the track but one way is essential." The cars were tested at the Biwabik Mine in 1893.

Crushers also made their appearance in this period. Ironically, in view of the complaints of furnacemen about soft, powdery Mesabi ore, they soon found it more economical. As a result ore crushers were introduced to deal with the hard ores on the Marquette, and they reached the Vermilion by 1893. The *Duluth News Tribune* of May 22, 1893, described the crushers as being "of the usual form observed in the ordinary rock breakers, immense jaws working back and forth. . . . Owing to the machine's capacity it will not be necessary to break the ore under ground, which will be one great saving in the cost of mining. In fact this item, together with the saving in work at stockpiles, is expected to more than offset the expense of crushing, besides enabling the company to offer its high grade Bessemers to the furnace men in such shape as will render their hematites the most desirable in the market. The present cost of breaking the Minnesota [Iron Company] ores at stockpile before loading is about 7 cents per ton. The cost of crushing will not exceed this, and may be less."[20]

Open-pit methods were not devoid of problems, however, and several disadvantages appeared almost immediately. The heavy cost of a steam shovel meant that it must operate almost continuously in order to be profitable. Such expensive machinery, handled by large, skillful crews, contributed nothing when down for repairs, waiting for tracks to be moved, or idling while locomotives pulled loaded cars out of the way and replaced them with empty ones. The technique was also costly in that it required the procurement of additional acreage where the stripped waste materials could be dumped. Moreover, open-pit expenses increased with the depth of the overburden, while underground mining could ignore the depth of the surface covering. Without a large annual production, expenses for stripping, rolling stock, and steam shovels increased the cost per ton for mining companies. Large, well-financed firms were best equipped to surmount these drawbacks and achieve maximum efficiency by extracting several million tons each season.[21]

A third alternative mining method known as the milling system combined several features of the steam shovel and underground techniques. Although it could not match the over-all efficiency of open-pit shovel mining, milling required less timbering than shaft mining while recovering all of the ore. After stripping revealed an ore body, laborers sank a hoisting shaft well outside the limits of the deposit. When they reached the bottom of the ore formation, the men ran a tunnel, called a drift, from the hoisting shaft under the ore. They then constructed a series of chutes or raises some 40 to 50 feet apart between the drift and the surface. Working from the top, they milled the ore down through the chutes by hand or by blasting it down with black powder. At the bottom of the raises tram cars caught the ore, and mules or men known as "trammers" hauled the cars to the main shaft. From there mechanical hoists lifted it to the surface.[22]

Milling required a much smaller initial investment because only minimal stripping exposed enough ore to begin production immediately. Moreover when the pits were exhausted, they could be utilized as a waste dump for future stripping. Against these benefits the mining companies weighed the expense of a hoisting shaft and its equipment, timbering for the drift, and construction of the chutes. The greatest drawback was the danger of flooding or of plugging the raises during severe storms.[23]

Ideal conditions for the milling system were a medium-sized ore body having surface debris of only moderate depth; the ore body had to be large enough so that it was worth while to strip but not so large that it would be more economical to use locomotives for hauling. The output per employee by milling was higher than any other method that required manual labor to move the ore. By the end of the 19th century four to six tons per day for each miner was a satisfactory yield for underground mines, while 40 tons per day, and higher under favorable circumstances, was the average at milling sites. Many officials agreed that milling ores cost at least $.25 to $.30 less per ton to produce than those mined by fully underground techniques, but $.10 to $.15 more per ton than those mined with steam shovels.[24]

At the end of the formative period of mining activity in northeastern Minnesota, the various companies extracted 47 per cent of the ore by steam shovel, 7 per cent by the milling system, and the remainder by various underground techniques. By 1903, however, the trend on the Mesabi was clearly toward the use of open pits in one form or another. An observer writing in that year confirmed the "recent tendency . . . greatly to increase the use of the open-pit steam-shovel method. . . . More of the new mines are opened in this way than formerly, and several . . . which have in the past used underground methods will produce their ore by steam shovel in the future. There is also appearing a marked change in the policy of conserving ores. In the past it has often been the practice, because of market conditions or because of desire for immediate large profits, to take out high-grade ores finding ready sale. . . . The better and for the most part the later, practice has been to determine [by drilling] well in advance, in some cases even before any mining has been done[,] the grades of ore and their distribution for the entire deposit, and in the mining to make such selection and combination of these grades as to leave the lowest surplus of undesirable ores. . . . The change is possible largely because of the new conditions of ownership whereby the control of the mines is in the hands of a few large steel interests." The amount of ore mined by steam shovel jumped from 43 per cent in 1903 to 50 per cent a year later.[25]

Many factors encouraged this development. Compared to underground mining, steam shovels offered the advantages of increased production capacity, elimination of the high cost and declining supply of bracing timber, and less outlay for lighting, pumping, ventilating, and hoisting equipment. The ore could be raised mechanically by locomotive up a grade rather than through a shaft. Miners could sort the ore more efficiently and recover all of it, while in underground mining some 10 per cent was lost. Shovels became more powerful, and improvements increased their effectiveness, reducing the cost of hauling both ore and glacial debris. Contemporary figures indicated, for example, that within 20 hours a shovel could remove 2,000 cubic yards of overburden and considerably more ore. A Bucyrus or Marion shovel could load an average of 75 wooden ore cars of 25-ton capacity in a 10-hour shift. Although it required a large initial capital outlay and a sizable operating crew, in the final analysis the outstanding features of the steam shovel were its economy and capacity to move millions of tons of soft Mesabi iron ore.[26]

But these increasingly efficient mining methods and the bright over-all outlines of the rapid development of the largest iron range the world had known do not convey the full picture of the changes set in motion by the panic of 1893 and the subsequent depression. As we have seen, the nature of the ore deposits on the "poor man's range" had attracted hundreds of mining firms to the newly opened Mesabi, where opportunities for explorers with limited capital seemed promising. Perhaps a single prosperous season would have established many of these entrepreneurs, but the hard times came too soon and lasted too long. The small, independent mining concerns most acutely affected by the deteriorating economic situation did not have the financial resources to weather the storm. The affairs of the Merritt brothers were a case in point.

The panic of 1893 played havoc with the Merritt-Wetmore plans for the Mesabi. Construction of the Duluth, Missabe and Northern Railroad extension and ore dock at Duluth got under way early in 1893, and Wetmore met the first of his contracted installments for $100,000. He had been successful in making a large sale of DM&N bonds to John D. Rockefeller, whom he approached on December 19, 1892, even before the railroad's board of directors ratified the agreement Leonidas had negotiated with him. In a transaction dated December 24, 1892 — the date of Wetmore's then unratified agreement with DM&N — Rockefeller acquired $500,000 worth of the $1,600,000 in bonds issued by the railroad with an option to purchase additional securities before January 1, 1894. In January and February, 1893, the Merritts signed five promissory notes borrowing $432,575 from Wetmore to purchase $132,544 of the Chase-Grant faction's stock in the DM&N as well as 3,000 shares of the new stock issued to secure control at that time. After that, no further money for the railroad was forthcoming immediately from Wetmore, and the brothers were forced to draw upon their own dwindling local credit to meet daily expenses. Rockefeller declined to participate in any formal association with the Merritt-Wetmore syndicate and flatly refused to invest in or loan money for Mesabi mines. On three separate occasions between January 18 and March 8, 1893, Wetmore or a close associate was reminded of this decision. The Merritts, however, believed that Rockefeller stood behind Wetmore and that he would "give us all the money we needed."[27]

Nothing daunted, Wetmore continued to expand his activities to include non-Merritt mining properties. In March, 1893, he, Colby, and other New Yorkers joined forces to close a contract with Frank Hibbing and Alexander Trimble's Lake Superior Iron Company. Under its terms the Hibbing group "in consideration of $100,000 cash and $150,000 in negotiable six per cent interest paper sell the eastern parties a half interest in the profits" of their mines near Hibbing and Chisholm, which the purchasers were to operate at their own expense. In May Lon and Wetmore were negotiating for another property — the McKinley Mine — for which they agreed to pay $150,000. "It is well known here," remarked the *Duluth News Tribune* of June 4, 1893, "that Messrs. Colby, et al., acted for a syndicate of capitalists at the head of whom is John D. Rockefeller, and which includes the stockholders to the American Steel Barge Company" and "the Duluth, Missabe & Northern railway." Whether Charles Wetmore in June, 1893, was or was not engaged in an effort to secure control of the Mesabi Range for Rockefeller, many Minnesotans, including the Merritts, believed that Wetmore had close ties to the oil financier.[28]

A month later the Merritt-Wetmore syndicate also acquired the Lone Jack and the Adams mining companies for "$428,000 short paper." The Lone Jack had been incorporated in August, 1892, by A. E. Humphreys and others; Peter Kimberly and David Adams had major interests in the Adams, which was reorganized on July 14, 1893, with Lon Merritt as president and Wetmore, Adams, and Kimberly as directors. In an interview in the *Duluth News Tribune* of July 20, Adams was quoted as saying that the purchase of the Adams Mine "places the Wetmore-Merritt syndicate with its immense holdings of properties that can be so cheaply mined, in a position to control the Bessemer ore market of the Lake Superior region. The greatest soreheads," he added, "are now compelled to admit, since they have seen the practical operation of the steam shovels, that ore can be mined very cheaply on the Mesaba." The Merritts, he continued, "are on top and there to stay. C. W. Wetmore, who is their right hand bower and associate in their railroad and mining deals, is considered one of the ablest financiers of the day and has the confidence and aid of the wealthiest men in the country." The syndicate also acquired "not less than three-fourths of the Rathbone [*Rathbun*] property" adjoining the Mountain Iron Mine.[29]

As for the Duluth, Missabe and Northern, repeated correspondence with Wetmore failed to elicit further payments. After the stock market crashed on May 3, 1893, Wetmore apparently did not immediately reveal to the Merritts his own increasingly precarious financial situation. He had obtained a personal loan from Rockefeller as early as March 31, 1893, for which he put up 150 DM&N bonds as collateral. On April 24 he secured additional personal funds from Rockefeller by pledging 5,650 DM&N shares he held as collateral on the Merritts' five promissory notes of January and February, and he transferred the notes themselves to the American Steel Barge Company. Later it was charged that the $432,575 Wetmore had loaned Alfred and Leonidas on these promissory notes had been taken by him from barge company funds without authorization. When President Hoyt of American Steel Barge returned from an extended trip to Europe in April, 1893, he immediately "demanded security" on behalf of the firm. As a result, Wetmore assigned the Merritt notes to the company as well as 2,313 shares of DM&N stock, 1,535 shares of Mountain Iron, and 5 shares of Missabe Mountain as collateral for his alleged unauthorized borrowings.[30]

Four of the five notes — all of which carried a provision that they could be called for payment upon 24-hour notice — fell due on July 20, 1893; the fifth was payable on February 2, 1894. Their transfer from Wetmore, whom the Merritts regarded as a friend, to the barge corporation constituted a hazard to the brothers, and the due dates of these notes were to loom large in their efforts to save themselves. Perhaps not fully aware of the financial shoals already on the horizon, Leonidas and Alfred, wishing to help Wetmore out of his seemingly temporary difficulties, compounded their problems. On May 3, 1893, the brothers — after the fact and evidently without the formal approval of the DM&N directors — signed a letter authorizing Wetmore to use their "notes and [DM&N] collaterals pledged as security . . . in such manner as you may deem expedient in order to reimburse yourself for the moneys which you have advanced to us thereon and the rehypothecation of such securities and notes heretofore made by you, are hereby accepted and approved."

On May 17 a worried Alfred as president of the Missabe road sent Leonidas to New York City, where he was joined by other family members from time to time during the summer. Alfred had let contracts for labor and materials to build the DM&N dock and tracks into Duluth in his own name, and the outlay for this work was running between $200,000 and $300,000 per month. The payroll alone for June, 1893, totaled $51,307. The Merritts feared that any further delay in Wetmore's installments would not only embarrass the railroad's position locally, but also cripple or destroy their personal credit. Upon Lon's arrival in New York City, he found that Wetmore had been unable to sell the remaining DM&N bonds. In the tight money market then prevailing, few investors desired to purchase the securities of what a Rockefeller associate described as "a small, distant ore railroad," running to as yet undeveloped mining properties on an untested Minnesota range.[31]

In the face of these obstacles, Wetmore told Lon that he had begun negotiations in March to create a consolidation of various Lake Superior ore interests in order to make the Mesabi more attractive to investors by attaching it to mines already shipping ore. In March and April he had been in touch with Charles Colby and Colgate Hoyt, his associates in the American Steel Barge Company, and had on April 6 incorporated the New York and Missabe Iron Company, a Wisconsin holding firm. Wetmore tried to gather into his scheme Colby's holdings on the Gogebic Range, especially the two richest properties, the Aurora and Tilden mines. With them as a foundation, he hoped to add the mines and transportation facilities of the Minnesota Iron Company on the Vermilion Range and the Merritt properties on the Mesabi. Periodic negotiations drew up cost and production estimates, but after nearly two months of conferences the merger attempt collapsed.[32]

Now in May, Wetmore urged Lon to support a second, less ambitious scheme to resolve their financial difficulties. The New Yorker believed that Lon's firsthand knowledge of the iron range and his familiarity with the traffic potential of the Duluth, Missabe and Northern would be helpful in attracting investors to the new project. From Wetmore's Wall Street office, the two men wrote personal letters to bankers encouraging them to subscribe to the railroad bonds. They found many who professed interest, but they failed to attract any actual buyers. After several frustrating weeks both men apparently became convinced that their difficulty was the newness of the project, the unproven value of Mesabi ore, and the publicly expressed skepticism of furnacemen.[33]

Although Wetmore's first attempt at consolidation had failed, he decided to try again after passage of the Duluth-St. Louis County bond issue on May 23. If several large producing mines could be combined with the Mesabi holdings, it would not only eliminate destructive competition, but also facilitate the floating of the unsold DM&N bonds. Once again Wetmore asked the south shore mining executives to join in a merger. He made a special effort to attract the support of Charles Colby, who had joined him in purchasing Frank Hibbing's mines three months before. As president of the Penokee and Gogebic Consolidated Mines and the Aurora Iron Company, Colby controlled valuable properties in Iron County, Wisconsin, and Gogebic County, Michigan. In 1892 mines on the Gogebic Range, the top Lake Superior producer that year, had shipped nearly 3,000,000 tons of ore. Concerned with the competition offered by the new Minnesota field, Colby expressed interest in some form of consolidation. A preliminary written agreement in June, 1893, anticipated that such a merger would have "the common advantage in maintaining staple prices for the product . . . and for the transportation of said product to the eastern markets" and that a process of purchase and absorption could ultimately gather all the leading Lake Superior mines into one combine. The preliminary agreement provided for a five-man committee, composed of Leonidas Merritt, Charles Wetmore, Charles Colby, his brother Joseph, and Edward W. Oglebay, to appraise the Gogebic and Mesabi properties. Lon insisted on including in the agreement the statement that "the parties hereto hereby covenant and agree that such valuations if unanimous shall be the basis of the consolidation," for he fully intended that his family should attain a controlling position in the contemplated merger.[34]

Before any additional steps could be taken, however, the appraisal committee was thwarted by organizational disputes. Ultimately Charles Colby refused to approve a merger with the Merritt-Wetmore associates. As a result, the second attempt at consolidation failed, leaving the Mesabi entrepreneurs in desperate straits. The DM&N bonds could not be sold, additional funds could not be borrowed, and it was alleged that Wetmore owed the American Steel Barge Company $622,000. Now in June and July, 1893, the Merritts' creditors as well as the DM&N dock workers and tracklayers were demanding their money. Frederick T. Gates, Rockefeller's financial adviser, later summed up the DM&N's situation when he wrote: "The railroad was trembling on the brink of receivership. Interest on the bonds was not paid. . . . Contractors were knocked down on the Merritt Railroad by their enraged men. Knives were drawn. Men actually entered the railroad offices in Duluth and demanded cash on their pay checks at the ends of drawn revolvers." [35]

|7|

A MERGER
AND A NEW PARTNER

CHARLES WETMORE'S IDEA for a merger of the ore holdings of
the Merritts and those of older south shore entrepreneurs was an idea
whose time had come. Wetmore, who might be characterized as a
promoter, was apparently cut from a similar piece of cloth as the more
successful John W. "Bet-a-Million" Gates of the same era. Having
failed in his two previous attempts at consolidation, Wetmore and
Leonidas Merritt now came into more direct contact with John D.
Rockefeller.[1]

One of the nation's wealthiest men, Rockefeller was then president
of Standard Oil of Ohio as well as a director of Standard Oil of New
York, New Jersey, Indiana, and other firms and railroad companies.
He had long since won the battles with refiners and crude oil producers
that led to the formation of what came to be known as the Standard Oil
Trust, which had already been widely investigated by the courts, a
hostile press, legislatures of various states, and the United States Congress. Two later historians wrote that the trust was regarded "by major
elements of the public" as "a vicious, venal menace . . . a dangerously
powerful monopoly," which had "achieved its monopolistic position as
a result of gross discrimination by the leading trunk-line railroad systems," espionage, ruinous price cutting, bribery, and "various associations to kill off small refiners or to force them into the combination or to
compel them to sell at ruinously low prices. . . . John D. Rockefeller
was regarded as the leading figure among the Standard Oil executives
responsible for the entire record of antisocial behavior." In 1892 the
trust was dissolved as a result of a decision handed down by the Ohio
Supreme Court, and its various elements had been reorganized into 20
separate companies. In 1893 Rockefeller was still personally active, but
after 1891 "he began, to use his own words to 'taper off,'" leaving the
actual negotiations and details of carrying on the business more and

more to various subordinates. By 1895 he retired and turned to the philanthropic activities that characterized the rest of his life.[2]

Among Rockefeller's trusted associates in 1893 was Frederick Taylor Gates, and as a first step Wetmore apparently approached him. Gates knew little or nothing about iron mining, but he had visited the Minnesota ranges in May, 1893, at Colgate Hoyt's invitation. Wetmore had been a participant in the same trip during which Gates met various members of the Merritt family for the first time. Born in 1853 in Broome County, New York, and educated at the University of Rochester and Rochester Theological Seminary, Gates after his ordination in 1880 served as pastor of the Central Baptist Church in Minneapolis for eight years. When the American Baptist Education Society was created in 1888, the young minister left the pulpit to become the organization's secretary. The society conducted a thorough study of Baptist educational interests throughout the country and concluded that it should establish an institution of higher learning. Under Gates's direction, the society raised the first million dollars for the University of Chicago. During that financial campaign he became acquainted with John D. Rockefeller, the largest single contributor to the university's endowment fund.[3]

In March, 1891, Rockefeller, whose health was poor at the time, told Gates that his philanthropic efforts were absorbing too much of his energy. He invited the minister "to come to New York and assist him in his benevolent work by taking the interviews and inquiries and reporting results" to him "for his action," Gates recalled. The minister accepted, and beginning in September, 1891, he went to work on Rockefeller's humanitarian affairs, eventually finding that he was also "compelled to turn to Rockefeller's disordered investments" as well. From 1891 until his death in February, 1929, Gates rendered loyal and varied services. In 1909 his employer wrote: "I was fortunate in making the acquaintance of Mr. Frederick T. Gates. . . . It occurred to me that Mr. Gates, who had a great store of common sense, though no especial technical information about factories and mills, might aid me. . . . Right here I may stop to give credit to Mr. Gates for possessing a combination of rare business ability, very highly developed and very honourably exercised, overshadowed by a passion to accomplish some great and far-reaching benefits to mankind."[4]

By 1893 Gates was managing a wide range of the financier's activities, and he is usually credited (or blamed) for Rockefeller's involvement in the Mesabi. According to Gates, on July 1, 1893, Wetmore and Lon went to him with what the minister recalled as "a series of financial propositions" designed to obtain Rockefeller's direct support of their Mesabi enterprises. As the family's spokesman in New York City, Lon was officially designated by the others as their representative. In a telegram, on July 11, six family members gave him "full authority to sign contracts for us . . . which you and Wetmore may make which you approve." The family also expressed confidence in Wetmore. On July 18 Andrus wrote the New Yorker: "I can tell you that we boys appreciate you and your efforts as men only can who have found in time of great need and have proved him good and true. . . . We have implicit faith and confidence not only in your integrity but in your ability to bring to a successful issue the plans that you have laid out."[5]

Over the course of the next two months, in formal conferences and informal discussions, Lon and Wetmore on the one side (joined at times by Alfred, Andrus, and Hulett) and Gates and George Welwood Murray, Rockefeller's attorney, on the other side, presented their recommendations, accepted amendments, and approved a series of agreements. Rockefeller personally took no part in these discussions, which were conducted on his behalf entirely by Gates and Murray. Under the provisions of agreements signed in July and early August, Rockefeller eventually loaned the Wetmore-Merritt syndicate over $1,000,000, at least $538,000 of which was received by the Duluth, Missabe and Northern. Some of this money was sent in cash to Duluth to pay the irate railroad and ore-dock workers and to take up the claims of contractors and suppliers who had filed mechanics' liens against the company. Alfred later admitted that it was necessary to send the money directly to the banks because "we daren't have it come to the railroad" offices, that the DM&N owed about $1,000,000 in floating debt at the time, and that suits were brought to foreclose and appoint a receiver.[6]

Throughout July Charles Wetmore's personal situation also worsened. As early as June the Merritts in Duluth must have been aware of Lon's concern about him, for Hulett wrote his uncle, "Keep Wetmore braced and don't let him worry too much." A telegram of July 17 from Leonidas in New York City to Gates in Cleveland read: "Am afraid Wetmore will break down unless immediately relieved from financial [strain], must have some money also at once to save Merritt boys collateral which means control of best properties." In a second telegram of the same date, Leonidas wired: 'I understand situation fully   if you will secure the loan I and my brothers will give Mr. R. & Wetmore one-half of all our mining interests at actual cost outside of those covered by contracts already signed   that is I accept Wetmores suggestion of to-day." The suggestion apparently referred to was the creation of the Lake Superior Consolidated Iron Mines, a New Jersey holding company not to be confused with Hibbing and Trimble's Lake Superior Iron Company mentioned in the preceding chapter. Wetmore later claimed credit for the Consolidated, writing "I do not recall any contract leading to the consolidation, whose essential terms were not first suggested by myself." The Consolidated, pooling the iron interests of Rockefeller, the Merritts, and Wetmore, was incorporated on July 21, 1893, a day after the first installment of the Merritts' promissory notes to American Steel Barge fell due. Three days later Wetmore, without the Merritts' knowledge, transferred his one-quarter interest in the Consolidated to Rockefeller. This transfer, which is preserved in the Rockefeller Archives, bears a penciled note "This must not be shown to Merritts."[7]

As the weeks went by Wetmore's personal finances had become hopeless, but not until September did Leonidas learn that the New Yorker was insolvent. Meanwhile the fortunes of the Merritt family had become deeply entangled with Wetmore's. Concerned about the New Yorker's health as well as his financial position, Lon had attempted to aid him in the belief that his problems were only temporary. Collateral in the form of DM&N stocks and bonds plus the stocks of various Merritt mining companies held in the names of several of the brothers were pledged by Wetmore to Rockefeller to obtain personal funds in a vain attempt to stay afloat in much the same way he had cavalierly transferred Merritt stocks and notes to American Steel Barge three months before. In July Lon also loaned Wetmore some cash he had borrowed from Rockefeller — a move which drew protests from Gates.[8]

Wetmore's state of mind as well as a glimpse of Wall Street speculation during the panic summer are revealed in an unfinished letter to Rockefeller drafted by Wetmore about the last week of July, 1893, and perhaps never sent. In it Wetmore wrote: "Mr Gates has explained to me that the use temporarily made of a part of the money advanced by you to the Merritts on Thursday last was regarded by him and yourself as a violation of the understanding assented to by Mr Leonidas Merritt at the time when the money was advanced. I feel that I am solely responsible for what was done and I therefore ask the privilege of laying the circumstances before you.

"As you are aware the conditions prevailing in Wall St from Monday to Thursday last were exceedingly dangerous — on Wednesday for a time a condition of actual panic existed. It was thought by many that the stock exchange would have to close unless quick relief was obtained. During that time I had to protect not only my own loans but also those of the North American Company besides doing what I could for the relief of the Merritts and the [DM&N] Railway Company. The demands upon my time were more than I could meet especially during stock exchange hours from ten until three. The market declined so rapidly that my brokers could not tell me from hour to hour where I stood. When you came to my relief with additional securities I supposed at the time that these would be sufficient but instead of this being the case I was met with a still further demand for margins. There was no time for further communication with you. My brokers insisted that there was danger of the securities which I had pledged (borrowed from you) would be thrown on the market and I had no alternative but to appeal to Mr. Merritt for temporary assistance.

"He had obtained $50,000 from you and $15,000 went for the purposes for which it was intended. The balance he did not actually need for 48 hours and he accordingly placed it at my service on my representation that I could return it to him in time to meet his requirements. This I believed I could do as soon as the brokers could give me a statement of account for with a moderate recovery in the market which occurred on Thursday I supposed that they would have sufficient securities to make good the loans leaving the cash free to be returned.

"To my surprise and dissapointment [sic] on Friday morning they informed me that the state of my account did not permit the return of the cash — they are of course borrowing in their own names the amounts which I am borrowing from them. They are looking to their own *ample* protection. They are required to keep their margins good to the *bid* price and of course require the same from me. If the market turns, the lenders do not immediately return the excess of margin but require them to wait until prices have acquired sufficient stability to make it reasonably certain that they will not relapse for say a few days. In these times lenders are especially unreasonable and brokers especially sensitive about their own credit and protection. However unjust it appears to me I have no alternative except to acquiesce.

"So great has been the hurry of the past week that it has been impossible for me to study accounts or to remember the prices at which I have received securities from you, but as far as I can tell at the moment if I had received and given securities at the same prices I should not have been obliged to deposit any cash for margins or ask Mr. Merritt for assistance. The reasons why I thought it was his duty to aid me and that you would approve if you knew the circumstance are briefly these.

"The Merritts were heavily indebted to me. From day to day for a month or more last past I have struggled successfully to save them from any default which might imperil their securities. They have placed themselves absolutely in my hands, with unquestioning confidence, otherwise I could not have accomplished what I have. The loans which I had to protect were marginal with your securities. If I permitted their loans to be in default and your securities were sacrificed though I could ultimately pay any loss that might have resulted, yet I felt you would properly blame me for not first exhausting every resource. It may be that I am mistaken but the motive that has influenced me in every step I have taken has been to protect and preserve the magnificent Combination, which has been founded, and at the same time to preserve from impairment all the resources which you have so kindly placed in my hands.

"When the money was given to Mr. Merritt he had no idea of using any of it for my benefit, altho' under the strict letter of his authority he might have done so. Yet both he and I intended that it should be used to retire additional obligations of the Merritts in the railroad and not to repay me for obligations which I had already taken up. Whatever blame there is therefore must fall upon me. In this connection I want to give you my assurance after six months of most intimate association under most trying circumstances that I have never met a man of a more honorable character or more generous and fair disposition in all things than Leonidas Merritt, and his brothers support and stand by him as one man.

"At the risk of wear[y]ing you I want to write a little further of our business relations with you.

"There are four matters about which in these times of unsettled feeling I assume that you are especially concerned.

"First. The financial condition of the Merritts

"Second. The financial condition of the Railway Company

"Third. My own and

"Fourth that of the North American Company.

"As to the first Mr Gates has all the figures and I believe has laid them before you. And in my opinion sufficient provision has been made for their wants in your contract with them.

"As to the second, Mr Gates has sent you the figures and it seems to me that it is not a difficult matter to finance the railway to a point where its earnings will take care of all expenditures.

"As to the third, I believe that I personally require no assistance except in the event of a further decline in the market to be supplied with additional securities at the same prices at which I have to deliver them — and except A.S.B. [*American Steel Barge?*].

"As to the fourth, having carried the North American Co. safely through the past week, and [*one word illegible*] its loan to the lowest [*the letter ends here.*]" [9]

Two years later Charles Colby tried to describe the effects of the panic during the summer of 1893 from a different point of view, stressing not the stock market speculation that was Wetmore's foremost concern, but rather the unavailability of funds to carry on the operations of the Penokee and Gogebic Company. That summer, Colby said, "there was a very bad condition of things financially throughout the country and banks everywhere were much more slow to accommodate their customers than they had been. Many of the large enterprises . . . were dependent largely upon the banks for money needed to carry on their work, and it was so in regard to many mining companies . . . before that time for instance, our Company could go to the banks and get a large amount of money at any time, but during the summer of 1893 it was found that where we had expected and been calling for money we were not going to get it. . . . Instead of borrowing money at the banks on paper we were obliged to make some arrangement to have money from individuals . . . we would borrow money of Mr. Rockefeller and such people as chose to unite with him in lending the money. . . .

"The financial methods were undergoing a very decided change all over the country. A good many corporations . . . were accustomed, for instance, to pay for their plant and get ready for business, and when they commenced to get an output they relied upon the banks for their practical working capital; and all those companies would seem to be in a good, solvent condition and have a good business, and they found no difficulty in getting what money they needed from the banks; for instance, our company would borrow money from several banks six or seven hundred thousand dollars and our loans ran up perhaps to a million and a half at different times, that being in the ordinary course of business. But . . . the summer of '93 — everything was changed." [10]

The borrowing procedures outlined by Colby had been followed by Wetmore and the Merritts, who then turned to John D. Rockefeller in the hope that he would bail them out of a desperate situation. Writing on August 9, 1893, Leonidas told Gates: "You are the only people to whom we can come now with any hope of success." On August 28,

1893, for the sake of convenience all the parties involved in the two months of negotiations signed a single contract that bound together all the points that had been agreed upon. The result was the first consolidation of iron mining interests on the Mesabi and Gogebic ranges under the aegis of the Lake Superior Consolidated Iron Mines. The terms of the August 28 agreement provided that the capital stock of the new company would total $30,000,000 at a par value of $100 a share. Lake Superior Consolidated obtained a full control of the New York and Missabe Iron Company (the Merritt-Wetmore holding company) as well as the Merritts' ore dock and railroad. Out of a total issue of 5,000 shares of DM&N stock, the Consolidated purchased 4,700 and converted them into 9,400 shares of Consolidated stock. In addition, the new combination retained in its treasury 6,646 shares of Consolidated stock and issued the remaining 23,354 shares in lieu of payment for majority control of ten Mesabi mining companies. It was to acquire 70 per cent of the Mountain Iron stock; 51 per cent of the Missabe Mountain, Biwabik, and Adams mines; 75 per cent of the McKinley, and 100 per cent of the Great Northern, Great Western, Lone Jack, Rathbun, and Shaw.[11]

A second part of the contract transferred all of Rockefeller's iron ore interests, consisting of three major blocks of securities, to the new combine. The first block embraced his stock in the Aurora Iron Mining Company and the Penokee and Gogebic Consolidated Mines — the same two firms headed by Colby that had been involved in Wetmore's earlier, unsuccessful merger attempts. A second block comprised $712,800 in stocks and first-mortgage bonds of the West Superior Iron and Steel Company purchased by Rockefeller from Colby and Hoyt. The third Rockefeller contribution included the stocks and bonds of the Spanish-American Iron Company of Santiago, Cuba. In the 1880s the Standard Oil entrepreneur had been persuaded to invest in this firm by Colby and Hoyt. It had since acquired 4,000 acres of iron ore land in the vicinity of Santiago, sunk shafts, and constructed railroad and ore dock facilities. By 1893 the company appeared ready to ship its first ore, but the panic delayed this effort. As a result, the Consolidated acquired Spanish-American stock and debenture bonds with a par value of $1,174,300.[12]

Other transactions, several of which did not involve stocks or bonds in the new Consolidated, were also called for in the August 28 contract. For example, Rockefeller agreed to provide $500,000 to continue work on the Duluth extension and ore dock. (According to Gates, Rockefeller's direct support eventually totaled more than $2,000,000.) In addition, he advanced a personal loan of $150,000 (taking Consolidated stock as collateral) to enable the Merritts to regain stock they had previously deposited with Duluth creditors; he agreed to negotiate to purchase the entire 1893 output of the Mountain Iron Mine; and he agreed to sell to the Consolidated his 7,001 shares of Minnesota Iron Company stock for what they cost him in return for first-mortgage Consolidated bonds at 90 per cent par.

Two conflicting points of view begin to make their appearance at this stage of the relationship between Rockefeller and the Merritts. In Gates's view, "The contract creating the Lake Superior Consolidated Iron Mines in the midst of the panic, saved and finished the railroad, opened the mines, and carried the Merritts successfully through the panic without the loss of one dollar or of one share of stock." Others, including members of the Merritt family, later portrayed a well-planned conspiracy in which Gates and Wetmore participated to wrest control of the Mesabi mines and railroad from the Merritts in order to add them to the Rockefeller empire.[13]

The opposing points of view extend to seemingly simple events and make it difficult, and sometimes impossible, to arrive at a clear-cut perception of the facts. For example, according to Gates, on July 12, 1893, Leonidas expressed a strong desire to meet Rockefeller. Gates had telephoned Wetmore's office, knowing Merritt would also be present, with news that his employer had signed the preliminary agreements. Lon, "who had never seen" Rockefeller, wished to go over to the Standard Oil Building and shake his hand. Gates later said that the two men met for "a few moments" in his presence and exchanged light conversation about the climate of Minnesota and that of other places. When Lon "began to extol" the virtues of the Mesabi Range, Rockefeller, according to Gates, "listened a moment politely" and then excused himself. The two never met again.[14]

The Merritt family version agreed that only a single meeting occurred but disagreed concerning who initiated the encounter, when it took place, and what was discussed. According to Lon, the meeting was held "about the middle or latter part of June, 1893," which would have placed it before the signing of any preliminary agreements and after the collapse of the negotiations with Colby and Hoyt. Describing the conversation before a Duluth jury in June, 1895, Leonidas said: "Mr. Gates came to the office 36 Wall Street and said that Mr. Rockefeller desired to meet me at his office 26 Broadway, and an appointment was made; he desired to meet me and talk with me about the consolidation. . . . Mr. Gates met me on the 8th floor of 26 Broadway and took me into a room on the right of the entry room and gave me a chair; he went out, and in a few minutes returned with Mr. Rockefeller. I was introduced to Mr. Rockefeller, he took a seat opposite to me, and after the usual formalities were over Mr. Gates went out. Mr. Rockefeller then talked about the weather, and about my brothers, and about the climate here and there, and this place and the other, a sort of general conversation, for ten or fifteen minutes, until Mr. Gates returned. He then said, 'Mr. Merritt, I have called you here to talk over the matter of the Consolidation of your mines.'

"Q. What did he say with reference to the matters of these mines, as to whether he had paid any attention to them?

"A. He says, 'I have looked you up, and I have looked up your brothers, and I am satisfied that you are good business men, and well capable of taking care of the practical part of the great operation. I have also had your property looked up, and I am satisfied that they can be made a great success as a business venture.'

"Q. What, if anything did he say with respect to the consolidation of the property?

"A. He said, 'you need consideration [*consolidation?*], you need to get your properties under one management, and you will find it will be a great advantage to you in many ways.' He asked me what my ideas of consolidation were; I said that really I had no definite ideas; that the matter had been suggested to me by parties which Mr. Gates mentioned, that I could see some advantages in consolidation, such as bringing the properties under one management, and save expenses of management and so forth. He went on quite elaborately into what he thought, would be the advantages of, consolidation. He said, we need to bring our properties under one management, and under one control; that they were great properties, and it would require a large amount of capital. He said I had no idea what amount of capital it would require to carry on the business, that I should have the properties in such a shape that I could get out a large amount of ore and ship it to Cleveland and have plenty of money to back us so we could compete successfully with other mining locations.

"Q. Did he make any mention in that conversation of Mr. Wetmore?

"A. He spoke of Mr. Wetmore; he said while a man like Mr. Wetmore might carry on the business which he contemplated, he would find himself handicapped from the lack of funds. He says, 'Now you know how it is. A man like Mr. Wetmore might carry it through successfully but he would be hampered from the need of money, and he would be obliged to spend a large amount of his time in renewing notes.' He cautioned me, however, he said, however, he wished me to understand that nothing he might say would be accounted as talking against Wetmore, because he liked Mr. Wetmore, and had a good deal of business with him, and he admired his grit, and thought he was a man of great ability; he simply lacked capital to carry on such an enterprise as that.

"Q. Did he say anything in that conversation with respect to the interest which he had in any mining properties?

"A. He said, if that consolidation could be brought about upon the basis which I have stated, I would be very glad to take hold with you. He says, 'I have certain properties, I have some interests in some mining properties[']; he called them by name Penokee-Gogebic and Aurora Mines, and some shares in mines in Spanish-American Iron Company. He said, 'these properties I would like to put in with you, they are prosperous, and if the matter could be arranged, I would be very glad to consolidate the matter with you, and I would be proud to be a partner of yours, that is of yourself and brothers in such a consolidation.' He said the Penokee-Gogebic was a large shipper of ore, the Penokee-Gogebic and Aurora mines were large shippers of ore, and their being in the market with large amounts of ore that were already known in the market, it would be a great advantage to the consolidation and would help us greatly. He said that the Spanish-American was a great mine situated on the Island of Cuba and was all ready to mine, they had a railroad built to it, and that the ores could be mined very cheaply, and they could be shipped as far west as the Allegheny Mountains, and compete with the best Lake Superior ores. He requested me to communicate with my brothers and associates upon that subject, and said if the matter could be arranged satisfactorily as to the value, and the basis upon which the properties should go in, he would be glad to become our partners in an enterprise of that kind.

"Q. What did he say with reference to Mr. Gates in that connection?

"A. My recollection is that he excused himself, and said he was a very busy man, that he had much to do, and he turned to Mr. Gates and says: 'Here is Mr. Gates, he will transact this business with you, and whatever he does will be satisfactory to me.' That closed the interview. We shook hands with the usual compliments, and I came home to 36 Wall Street." [15]

Nearly two decades later Lon recalled the conversation a bit differently when he appeared before a congressional committee in 1911. He said that he thought the meeting occurred "in the spring" of 1893.

Then he went on: "Mr. Rockefeller came in and shook hands with me, and he was as gentlemanly a sort of man as you have met. He had a kindly face and a brotherly sort of manner. He talked about the weather for a while, and asked me about my family. He told me that he had had occasion to look me up and to look up my brothers, and that he had satisfied himself that we were honest, straightforward sort of fellows and men of ability; that we were men who had done great work up there, and men who in their organization seemed not to have made any mistake. This was the first mistake I made in the whole business," Lon added, "that is, it was the first serious mistake.

"He said that if it could be arranged for a consolidation, or if a consolidation could be made, he would be proud to become a partner and backer of it. He said that he wanted to put in some of his own properties, and get our eggs in one basket and take care of it. He said many things which I can not repeat now. He mentioned at that time this [DM&N] railroad, and he had doubtless heard that we contemplated putting this railroad into the hands of a receiver. He said, 'Of course, you can do that; you boys can do that; it would be perfectly legitimate; it would be absolutely right; it would take you out of this trouble.' But he said after that, 'Mr. Merritt, there is a little tinge to that. You want to avoid it, and you must avoid it. Go ahead and make [t]his consolidation. I do not want to have your stock. I am past speculation, but I have money that I would gladly invest in these bonds, and I will buy your bonds.' He then said, 'It will make things easy, and, Mr. Merritt, you have no idea how much money you may have to raise; you may have to get your ores into Cleveland; you have only practically commenced.'

"Such talk as that captured me. . . . He said that whatever Mr. Gates said would be all right. Then, afterwards, I had much to do with Gates, but understood, of course, that I was dealing direct with Mr. Rockefeller in all these things. . . . Mr. Murray, as I remember, was my chief adviser; he was the attorney for the consolidated mines." According to Andrus, Leonidas left Rockefeller's office "with a great sense of relief," thoroughly convinced that "Rockefeller was sincere" and that he "would prove to be a powerful and desirable ally in the work yet to be done."[16]

Two contemporary documents support Lon's memory that Rockefeller was interested at least as early as June, although none has been found to support his recollections of the talk. The first is a letter to Gates from M. M. Clark, Lon's secretary in Duluth, sending a map that showed lands owned and controlled by the Merritts as well as traffic contracts held by the DM&N. The second is a letter from Lon to his wife Elizabeth (Lizzie) dated June 6, 1893. In it he wrote: "They all seem to have great confidence in me. Mr. J. D. Rockefeller sent me word by his private secretary [*Gates?*] that he would do nothing in relation to the Iron business until I was consulted and that if any combination should be made affecting the Iron Interests of Lake Superior I should be made the head and have the control, all of which was, of course, very flattering to me. I told him, however, that I had some duties which I owed to my family and proposed [*sic*] hereafter to take them into account in everything. They seemed to think that the mines we were at was fast getting into such shape that the heads of the departments could have quite all of the spare time they might need and such salaries as would support them very comfortably, etc." [17]

Two accounts of the meeting by Rockefeller also exist. In the earliest, he testified by deposition in a Duluth legal case in 1895. At that time he was asked, "Did you ever meet or have any conversation with Leonidas Merritt and if so, how often and when and where?" In reply Rockefeller said, "I have met Leonidas Merritt only once, and had a brief conversation with him on that occasion, the brief interview was held in the waiting room adjoining my office at 26 Broadway in the City of New York about the middle of July, 1893. . . . The interview consisted of the usual greetings and expressions incident to introductions; expressions of pleasure on the part of each at meeting the other, each remarking that he had heard of the other, etc. The conversation turned upon the weather and from that passed to a comparison between the pure air of Minnesota and the climate of the mountainous region of New York where I had recently paid a visit. The most of the time of the interview was spent on these subjects. Near the close of the interview — which was very brief, not occupying more than five or ten minutes — there was a general reference made to the deposits of ore on the Missabe Range and to the enterprises of Merritt brothers in connection therewith. This reference, however, was of the most general nature, the whole conversation being merely social. I ca[n]not recall every word that was used, but have stated the substance of all that occurred.

"[Q.] Did you or not, during that interview say to Leonidas Merritt that you had heard of him and of his brothers and relatives, and especially of him and Alfred Merritt . . . and of their ability as practical business and mining men and that you would be glad to be associated with them as partners in the enterprise which was discussed at that time, or anything to such effect?

"A. I did not.

"[Q.] Did you or not, at that time request Leonidas Merritt to make such a statement to his brothers, or to the plaintiff in this action or any other statement of a similar import?

"A. I did not.

"[Q.] Did you or not, during this interview propose that Leonidas Merritt, or the plaintiff in this action or any other member or members of the Merritt family should consolidate their iron mining interests with yours by conveying your and their interests to a corporation organized for that purpose, or by any other method, or anything to such effect?

"A. I did not.

"[Q.] Did you or not, at any other time make such a statement, request or proposition or any other of similar import to Leonidas Merritt?

"A. I never did.

"[Q.] Did you or not in July, 1893, or at any other time make to any person other than Leonidas Merritt the statement, request or proposition just mentioned or any other of a similar nature?

"A. I did not.

"[Q.] Did you or not, at any time [authorize] the defendant Gates or any other person, to make any such statement, request or proposition to any person or any of similar import?

"A. No I did not."[18]

More than a decade later, Rockefeller, who was described by contemporaries as reticent and a man of few words, wrote briefly of his reluctant acceptance of the merger: "Going into the iron-ore fields was one of those experiences in which one finds oneself rather against the will, for it was not a deliberate plan of mine to extend my cares and responsibilities. My connection with iron ores came about through some unfortunate investments in the Northwest country. . . . Most of these properties I had not even seen. . . . Among these investments were some shares in a number of ore mines and an interest in the stocks and bonds of a railroad being built to carry the ore from the mines to lake ports. We had great faith in these mines, but to work them the railroad was necessary. It had been begun, but in the panic of 1893 it and all other developments were nearly ruined. Although we were minority holders of the stock, it seemed to be 'up to us' to keep the enterprise alive through the harrowing panic days. I had to loan my personal securities to raise money, and finally we were compelled to supply a great deal of actual cash, and to get it we were obliged to go into the then greatly upset money market and buy currency at a high premium to ship west by express to pay the labourers on the railroad and to keep them alive. . . .

"We now found ourselves in control of a great amount of ore lands, from some of which the ore could be removed by a steam shovel for a few cents a ton, but we still faced a most imperfect and inadequate method of transporting the ore to market.

"When we then realized that events were shaping themselves so that to protect our investments we should be obliged to go into the business of selling in a large way, we felt that we must not stop short of doing the work as effectively as possible; and having already put in so much money, we bought all the ore land that we thought was good that was offered to us." [19]

The timing and the content of Lon's conversation with Rockefeller are important to the conflicting interpretations of events. If it occurred in June along the lines Lon described, Rockefeller initiated the meeting and took an active part in persuading the Merritts to join the consolidation by which they later claimed they were defrauded of their properties. If, on the other hand, the conversation did not occur until July as Rockefeller, Gates, and others testified, it was the Merritts and Wetmore who approached Rockefeller to form the consolidation — a move he was reluctant to make. The conflicting accounts of this event have been presented quite fully here as examples of the contradictory evidence confronting the researcher at many points in the Merritt-Rockefeller story. Some further ramifications will be described in the next chapter.

Although the two previously unsuccessful attempts at consolidation had involved many of the same mining properties and the same railroad, the third Consolidated endeavor at first seemed satisfying to all concerned. Under the second Colby-Hoyt proposal, the Merritts' control of the merger would have depended upon committee appraisals of the valuations of iron ore properties. The Rockefeller proposition, by contrast, said nothing about committee evaluations and made Lon president of the Consolidated board of directors, which also included Charles Wetmore, Frederick Gates, George Murray, and Andrus Merritt. The final contract allowed Rockefeller to appoint several officers, so he named Gates as first vice-president, Murray as general counsel, and Charles E. Scheide as secretary and treasurer. Within months he also selected William T. Scheide as general manager. Alfred Merritt retained his position as president of the Duluth, Missabe and Northern. [20]

By the terms of their agreement, both Rockefeller and the Merritts transferred property, or parts of it, to the Consolidated. In return the financier received 6 per cent first-mortgage Consolidated bonds with a par value of $4,299,000. Based solely on par valuations, this represented a paper profit of $1,250,000 over the value of the securities he had contributed. According to Gates, however, this gain was "problematical and wholly future" because the bonds "were then unsalable at any price that would show a profit." In return for their contributions, the Merritt brothers accepted stock in the Consolidated and thus seemingly controlled its operation. The contract terms based the stock valuations on the percentage of interest in Mesabi iron lands acquired by the new holding company. Gates later charged that the Merritts could "put their own valuations on their mines" and that they used "this coveted freedom with generous liberality." Their first estimate placed the cash value of their ore lands at $7,000,000. Then, said Gates, "in secret family conclave" the brothers readjusted the valuation to $10,150,000, added a new mining property (the Shaw), "shut their eyes" and blatantly "doubled the whole" figure to a total of $21,050,000. Gates contended that the Merritts bore the full responsibility for watering the Consolidated stock.[21]

The Merritts portrayed this stock manipulation as a completely open process that enjoyed Rockefeller's approval. According to their version, Leonidas, Alfred, and Andrus, meeting in the Metropole Hotel in New York City, prepared a statement of the actual and relative worth of their mining operations which on August 21, 1893, they submitted to Gates. It listed the Mountain Iron property at $3,500,000; the Rathbun and Biwabik at $2,250,000 each; the Missabe Mountain at $1,750,000; and the Shaw at $400,000. The total of these figures matches the $10,150,000 alluded to by Gates.[22]

The brothers explained the process of doubling the values by recalling that a preliminary agreement signed on July 1, 1893, as well as the final contract of August 28 entitled the Consolidated to purchase these mines in an exchange of stock at 50 per cent of their cash valuations at par. In return Rockefeller was to receive bonds at 90 per cent par value. Several days after receiving the doubled figures Gates notified the Merritts that Rockefeller seemed satisfied with their calculations. Although he later expressed great suspicion concerning the validity of these amounts, Gates explained that in "the interest of the Merritts . . . Rockefeller took bonds instead of stock." Thus the aggregate of stock assigned to each mine property did not matter to him. The Merritts could "issue as much stock to themselves as they please." The

Duluth brothers, Wetmore, and those creditors who had converted stock in the individual mining companies they held as collateral into shares of the Consolidated received holding company stock worth more than $18,000,000, two-thirds of which the Merritts retained as their portion of the divided ownership.[23]

It is difficult to develop an unbiased evaluation of the financial maneuverings involved in the creation of the Lake Superior Consolidated Iron Mines. The major factor to keep in mind is that both sides explained their positions during and after all the bitterness and character assassination that surfaced in a series of legal disputes initiated by the Merritts. The table below depicts, by perhaps the most accurate figures available, the amount of Consolidated stock issued for the properties each faction contributed to the merger. There can be no doubt that in the beginning heavily watered stock inflated the Consolidation's worth. Even with extensive, but as yet virtually untapped, ore reserves and a most advantageously located railroad line, the evidence — based on available data concerning organizational structure and on future stock price fluctuations — does not indicate a competitive valuation approaching the initial $7,000,000 much less the $11,654,000 added by the Merritts. For example, the entire capitalization of the Minnesota Iron Company, with its well-established Vermilion Range mines, its control of the Duluth and Iron Range Railroad, its ore docks at Two Harbors, and its fleet of ships, totaled only $16,500,000 in July, 1893.[24]

CONSOLIDATED STOCK BASED ON MERRITT AND ROCKEFELLER CONTRIBUTIONS[25]

Description of Property	Consolidated Stock
Merritt Associates — all stock	
Mountain Iron	$ 4,900,000
Rathbun	4,500,000
Biwabik	2,295,000
Missabe Mountain	1,785,000
McKinley	1,500,000
Great Northern	1,000,000
Great Western	1,000,000
Shaw	800,000
New York & Missabe Iron Company	750,000
Lone Jack (later removed)	124,000
	$18,654,000

Rockefeller

Aurora Mining Company — stock	$ 708,250
Penokee & Gogebic Consolidated — stock	866,469
Penokee & Gogebic Consolidated — mortgage notes	300,000
Spanish-American Iron Company — stock	1,081,300
Spanish-American Iron Company — debenture bonds	93,000
West Superior Iron & Steel Company — stock	72,800
West Superior Iron & Steel Company — first mortgage bonds	640,000
	$ 3,761,819

Why did the Merritts agree to put their railroad and mines into the Lake Superior Consolidated? Their own attorney, who examined the August 28 agreement before all the brothers signed it, warned them on August 20 that it was unfair. Lewis reported that the brothers' Duluth lawyer told him that "This contract strips us of everything. . . . The contract is the most *damnable* piece of mechanism he ever saw." Writing 40 years later, Andrus Merritt explained the family's willingness to plunge into this merger. Their financial position was untenable, and they had stretched their local credit to the limit. As national economic conditions tightened, Duluth creditors pressed for payment of overdue loans and accounts. Wetmore was ill and insolvent, and Rockefeller had impressed Leonidas with his sincerity, said Andrus. The financier had induced the family members to accept the alliance with a promise to "immediately advertise this new Consolidated Company in such a way that we would be able to sell the small amount of stock necessary to meet our outside obligations and what we owed to Rockefeller on account of Wetmore's insolvency also." The brothers believed that the oil king's connections with the eastern financial world assured a ready market for their stock at prices ranging "from forty to fifty cents on the dollar, even in those troublous times."[26]

Leonidas, too, expected the Consolidated stock to be immediately valuable. Writing to Andrus on August 7, 1893, he said, "It is utterly impossible for me to pay the money now. I simply can't do it. As soon as the consolidation is completed of the mines will have an engine in my hands to raise money with." Later Lon offered a further reason when he wrote: "Our main object in the consolidation with Rockefeller was based on his own proposition, and in part for the purpose of obtaining money through the sale of Consolidated Bonds to secure the balance of the range . . . we fully believed then, as we know now, that Rockefeller could make more money by dealing honestly with us." On another occasion Leonidas observed, "The contracts with Mr. Rockefeller were hard, but still the times were hard and if we once got the enterprize underway it could not help but be a grand success with Mr. Rockefeller's backing — so they said and so we believed."[27]

Members of the Merritt family expressed satisfaction with their new partnership and believed their financial problems had been resolved. To a well-wisher in Duluth, Lon wrote on August 30 that "our finances are now so arranged that we have [passed] the danger point and are again sailing in an open sea. The times, as you know, have been distressingly bad, but we have all stuck to it and the final outcome is more than satisfactory." In a joint letter to Rockefeller on November 22, 1893, Leonidas, Alfred, and Andrus wrote: "You have our entire confidence and we hope that we have yours. We believe that the enterprise in which we are associated is one of magnificent possibilities. We have devoted ourselves to it unreservedly and with enthusiastic confidence. . . . We deeply realize that our success depends upon the maintenance of the most cordial and confidential relations between you and ourselves. It is in the hope of establishing such a relation once and for all so firmly that it cannot be disturbed by any rumors that we write this letter. In conclusion, dear Sir, we assure you most earnestly of our hearty esteem and confidence."[28]

Lewis Merritt and his son Hulett, who did not sign this letter to Rockefeller, saw matters in quite another light. Writing to Andrus on November 21, 1893, Lewis said: "I have for a long time believed Wetmore to be a rascal of the deepest dye, and that he intended to do us up, and I now see how he has used Lon as a lump of clay and hood-winked him until he has quite if not almost cost us all our fortunes. There is only one way out of this delema [sic], and that is to go to John D. Rockefeller and make a clean breast of the whole affair and get him to advance sufficient sums for the boys to pay their debts, and to furnish sufficient capital to operate the Consolidated Company. It is . . . foolishness for the boys to attempt or think of going outside to finance this thing. It is too late to think of such a thing. . . . We have never been allowed to know anything but have been treated as fools. . . . Some one of the boys, of course, I don't know who, has put up all our collaterals . . . and in addition to that, Lon has helped himself to our collaterals and gone out and borrowed money for himself and Wetmore. . . . Everyone of you boys are at fault for not . . . insisting that Lon go direct to Rockefeller and make a deal with him. . . . For God sake go to Rockefeller straight and quick."[29]

JOHN D. ROCKEFELLER about 1893 when he began to invest in the bonds of the Merritts' railroad on the Mesabi Range. Duluth, Missabe and Iron Range Railroad Collection.

FREDERICK T. GATES represented Rockefeller in his dealings with the Merritt brothers and Charles Wetmore. Duluth, Missabe and Iron Range Railroad.

CHARLES W. WETMORE, as a young Harvard graduate in 1875. Harvard University Archives.

DAVID T. ADAMS, founder of Virginia and Eveleth, about 1910. St. Louis County Historical Society Collection.

ALEXANDER McDOUGALL introduced the Merritts to Wetmore in 1892. St. Louis County Historical Society Collection.

163

ANDREW CARNEGIE about 1910 after his retirement. The nation's largest steelmaker, he joined U.S. Steel in 1901. Carnegie Library, Pittsburgh.

HENRY W. OLIVER about 1900. He and his company were important factors in the Mesabi's development. Carnegie Library, Pittsburgh.

"JOHN D. OCTIPUS drawn in an idle moment" by Leonidas Merritt. Merritt Papers, St. Louis County Historical Society Collection.

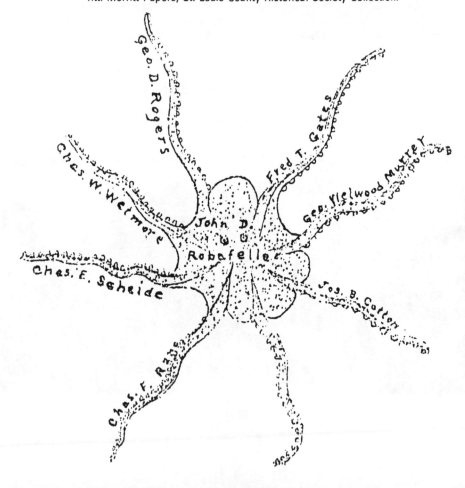

CARTOON COMMENT on the Duluth jury's award to Alfred Merritt in his lawsuit against John D. Rockefeller. From the *Minneapolis Times*, July 19, 1894.

ELBERT H. GARY, the first head of newly formed United States Steel in 1901. U.S. Steel Corporation.

FRANK HIBBING founded the Mesabi Range town that bears his name in the 1890s. St. Louis County Historical Society Collection.

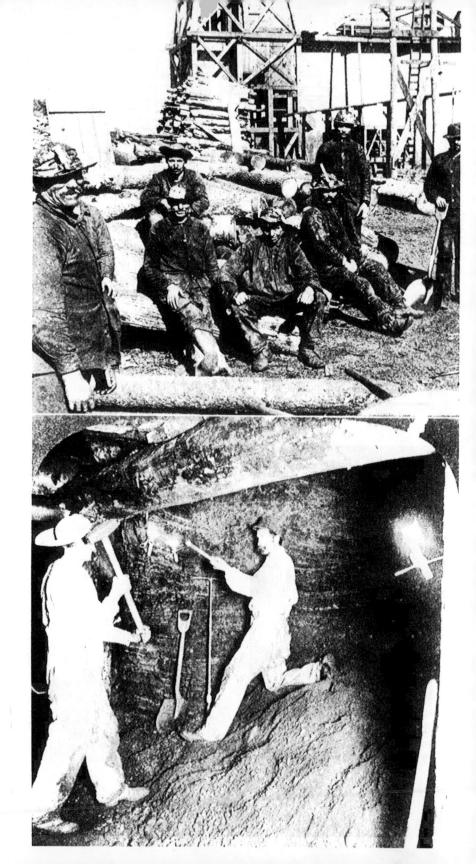

THE STREETS OF EVELETH stopped about 1910 at the edge of the productive Fayal Mine developed by the Adams Mining Company.

IRON MINERS in front of the shaft house of an underground mine near Hibbing in 1906. Library of Congress.

UNDERGROUND MINERS worked by candlelight in the Sellers Mine near Hibbing about 1906. The men were driving a hole in which to place explosives to blast down the ore.

UNLOADING wooden ore cars into dock pockets in 1906 required large crews, especially when the ore froze in late fall. Library of Congress.

STEAM from D&IR locomotives was used at Two Harbors about 1900 to thaw the frozen ore so that it could be dumped into the dock pockets in cold weather. Franklin A. King Collection.

FOUR WOODEN ORE DOCKS jutted into Duluth Harbor by 1910. Dock no. 4 (at right), the largest of the wooden ones, had been completed there by the Duluth, Missabe and Northern Railroad in 1906.

THE SUSQUEHANNA MINE near Hibbing was typical of the large open pits which dominated the Mesabi by 1910. Impressively deep, they had many terraces to reach the ore and allow reasonable grades for the trains to carry it from the pits.

ORE TRAINS, like this one totaling 135 cars, were weighed by the Duluth, Missabe and Northern at Proctor about 1920. Note the larger Santa Fe steam engines. Photograph by Gallagher's Studio, Duluth.

|8|

THE MERRITTS
VERSUS ROCKEFELLER

HAVING EXTENDED the Duluth, Missabe and Northern Railroad into Duluth and shipped the first ore from a massive new dock completed with funds supplied by John D. Rockefeller, the Merritts in the fall of 1893 set out to round up the widely scattered securities of their mining companies and railroad. Held by investors and by creditors as collateral for loans, they were slowly gathered in and replaced by stock in the new Lake Superior Consolidated Iron Mines. Although the Merritt brothers objected bitterly, these minority stockholders were allowed to enter the merger on the same conversion terms as the majority. Some of them remained unfriendly to the Merritts, especially those who had been involved in the earlier battles with Kelsey Chase and Donald Grant for control of the DM&N. The brothers, in determining the controlling percentages of the different mines, intended to issue the pooled stock only to sympathetic shareholders. They strongly asserted that any other procedure would jeopardize their controlling interest. Writing to a Faribault stockholder in September, 1893, Leonidas made it clear that "It is not the policy of this Company to invite the elements of discord or distrust into their ranks, neither are they anxious to reward either by justly deserved punishment, much less by gratuitous benefits, those who have in the past used their holdings . . . to utterly destroy the founders of their enterprise. . . . I have no words . . . to properly express my utter contempt for the men who . . . have in times past attempted to destroy this enterprise."[1]

Gates did not agree with this point of view. The former Baptist minister argued that the consolidation had no need to incur the ill will of the Duluth community. He believed that if the minority interests were alienated, the new venture's directors could find themselves "covered with litigations and injunctions or at least embarrassed by threats at every point." Then, he said, there were the best interests of the Merritt family to be considered. The "brothers have borne and do bear in Duluth an extraordinarily high reputation for fairness," Gates wrote. "Any attempt to freeze out the minority stockholders . . . would also be injurious to the high repute of our friends — the Merritts — and we wish to save them every possible embarrassment." In October Gates and Murray traveled to Duluth armed with persuasive statements from Rockefeller and successfully gained the Merritts' reluctant acquiescence to the admission of the minority shareholders on equal terms.[2]

Almost immediately the members of the family most active in the merger began to charge loudly and persistently that the Rockefeller forces were systematically undermining their position. They detected a weakening in the allegiance of formerly friendly shareholders, which they blamed on the large amounts of stock that "unfriendly forces" had been permitted to convert into Consolidated shares. They saw evidence of this antagonism in the increasing criticism of their work and of their ability to manage effectively the railroad and mining enterprises. During October, 1893, the New York office of the DM&N was closed; both Leonidas and Wetmore resigned as vice-presidents of the railroad, and a few months later the DM&N board also discontinued Wetmore's duties as general counsel, resolving "that he is no longer identified with this Co."[3]

Something of the atmosphere at the time was suggested by Leonidas in a long letter to Frank Jenkins, whom he had left in charge of the DM&N New York office. On October 20, 1893, Lon wrote from Duluth: "As soon as I arrived home, without going to the Duluth office at all, I took the special car and took a trip up to Mountain Iron, thinking I would have a little rest and enjoyment, but every where I went I found so much to do that it kind of kept my mind on business in hand so I could not forget my sorrows as I had expected to do. After Mr. Gates and Mr. Murray came here, of course we were quite busy; they seemed to have a notion that they understood how to take care of matters a little better than we did out here, and of course the old process of education had to be resorted to. . . .

"We have had many conferences, and in many cases have made much out of little. For my part I do not consider the situation at all grave. I have estimated as nearly as possible the indebtedness of the railroad, and find that it is comparatively very small. The railroad is everything that we could expect as a carrier, and will earn for itself and the consolidated mines an amount of money during the next season which will astonish the natives. . . . I am trying very hard to get Mr. Gates to see the situation in the right light; and while I am quite sure that he does see the earning capacity of the railroad and mines and appreciates it fully, yet the two or three hundred thousand dollars which we owe seems to almost overwhelm him. I presume his trouble comes somewhat from the fact that he does so well see the situation, but is afraid he cannot communicate it properly to Mr. Rockefeller. . . . The disposition therefore is to find fault with the Wetmore management, and probably vengeance will be wreaked on the New York Office. Indeed a resolution has been passed by the Board of Directors at the solicitation and instigation of Mr. Gates abolishing the railroad office at New York. . . .

"I have consented to this for many reasons; principally because I have no reason to believe that Mr. Wetmore can materially further aid us in financiel [*sic*] matters, at least not for the present, and perhaps, as the Rockefeller parties seem to have no faith in his methods, it would be better for them to deal straight with the Company here. . . . I expect to make a loan sufficient to help myself and my brothers out of their present difficulty. . . . I sincerely hope . . . that, if I ever do undertake anything in New York again, it will be when there is more than one well to get water at, and will be in such times and under such circumstances as will not oblige me to tie myself and friends up in such an uncomfortable manner as I had to do last Summer."[4]

Two weeks later, writing to Gates, Lon touched on some of his problems. "Your letter dated Oct. 26th acknowledging receipt of several telegrams received. I cared very little for the answers only that we got the money asked for to help out the pinch. The funds asked for came and we were made happy thereby for the time being. . . .

"I find myself just at this time and for the past fortnight nearly exhausted by the nervous strain which I have been forced by the circumstances to undergo for the past few months and am utterly unable to proceed with anything like the dispatch or effective execution which I could always heretofore command. I must be shortly relieved from some of this burden made doubly heavy on account of the Rockefeller interests, and the fear that he does not have as much confidence in our business ability, being a stranger to us, as he would if he was better acquainted with all the circumstances surrounding us here. . . .

"You have at all times manifested a kindly and helpful interest in the personal welfare and financial success of the Merritt family, which I need not tell you we all very much appreciate, and for which I promise you, feeling absolutely certain that we are not only willing but fully responsible and perfectly able to repay you in a substantial manner. While it was supposed that we were in the power of the New York people and were likely to be swallowed by them our personal credit was impaired to a great extent, and in some cases utterly destroyed. It is now understood in its right light and perfect confidence is restored. We are taking care of everything without trouble, and this without any collaterals. We will, however, have to begin soon to take care of some of our liabilities, especially those incurred by indorsements. . . . Mr. [Thomas H.] Phillips, our bank cashier, who will take them to New York and try and get from your auditor our transactions with Mr. Wetmore, and then see if there is any one in the world who is willing to lend a hundred thousand on five or six million assets.

"The days are past when we want to rate ourselves as paupers and it is my purpose from this time forth to demand of the world at large the recognition and respect which is due to us as individuals and the respect which is due to the Consolidated Iron Mines. . . .

"I have had several letters and telegrams from Mr. Wetmore asking me to come down there and undertake some financial scheme with him. I shall have Mr. Phillips look over whatever plan he has very carefully, and if he has got something good for himself and us, we will avail ourselves of it.

"I very much wish to see Mr. Wetmore get out of his present difficulty, and do not feel like blaming him for the manner in which he has carried on the financial end of the business in New York, because I believe, under the circumstances, he did the best he knew how, or the best probably that could be done.

"I do not see any reason now why the Consolidated Iron Mines cannot go ahead and put the railroad in such shape that everything will be easy, and hope that you will have some plan arranged looking to that end."[5]

The last paragraph of this letter probably referred to Lon's ambitious plans to buy up more mining lands on the Mesabi very cheaply during the hard times then prevailing in order to ensure additional traffic for the DM&N. As for the amount of money owed by the brothers, Gates wrote: "I knew the Merritts were heavily in debt, but Leonidas had carefully concealed from me the full extent of their obligations, if, indeed, he knew himself." Gates estimated that their debts by July, 1893, had totaled about $2,000,000, before they borrowed any money from Rockefeller.[6]

Further misunderstandings between the family members and the

Rockefeller men developed over operating methods. As early as July 31, 1893, Gates had proposed an audit of the affairs of Wetmore and the Merritts in order to ascertain how the new Consolidated stood. At that time he thought optimistically that it would "take only two or three days," but by December, 1893, the work was still in progress. John Merritt, president of the Mountain Iron Company, also experienced difficulties. He asserted that the New York office had ordered expansion of his mining operations but neglected to provide sufficient funds. In addition, a branch being built from the DM&N main line into Hibbing encountered delays in the shipment of rails, labor difficulties, and overextended credit. A family spokesman claimed in 1912 that these, and other similar problems, demonstrated conclusively the existence of a "scheme to despoil, legally, the Merritts of their possessions" in such a way that "it would appear as the legitimate consequences of incapacity." Merritt family leaders soon came to believe that Rockefeller had thrown his full energy and financial resources into gaining control of Minnesota's iron ore.[7]

A more realistic assessment suggests that the Merritts misconstrued the true nature of Rockefeller's activities. According to Gates, Rockefeller's financial records showed that by the fall of 1893 more than $5,900,000 in cash had been loaned to 58 different individuals and companies. To provide this financial support the New York entrepreneur, with greater access to funds than most people during the depression, had borrowed between $3,000,000 and $4,000,000. Declining Andrus Merritt's request of September 30, 1893, for a loan of $100,000, Gates summarized Rockefeller's situation in precise terms: "It is not true that his resources for this [Consolidated] enterprise are unlimited. This is only a portion of what he carries. Other enterprises with which he has been long associated and other men who have long been his friends have been constantly coming to him during these hard times for assistance . . . with more urgency than ever before. I have to-day on my desk urgent, imperative appeals to save old friends from ruin amounting to many hundreds of thousands of dollars. I have incurred the enmity of important business enterprises with which Mr. Rockefeller is connected because I have had to decline to assist them within the last few days. Each one of them supposes that it could not make any very great difference to Mr. Rockefeller were he to help them out, forgetful that his request is one of many which make an aggregate absolutely impossible in these times for any man or combination of men to carry."[8]

The Rockefeller spokesman also suggested that Leonidas Merritt and his family were "the unconscious victims" of several illusions which may have contributed to their difficulties. Gates believed that since the initial discovery at Mountain Iron in 1890, Lon had remained convinced that this new iron ore field would immediately dominate the country's market; that the Merritt interests controlled the most valuable Mesabi properties; and that the Consolidated stock would be quickly marketable at good prices. Gates extrapolated from these premises that the Merritts expected to mine their ore cheaply, control its transportation, sell large quantities at high prices, "speedily pay their debts," and enjoy "fabulous riches." In reply Lon pointed out that Henry Oliver was one of the first "to become imbued with the same crazy illusion," and that in his company were men such as Kimberly, Franklin Rockefeller, James J. Hill, and Andrew Carnegie. "The Merritts were not," he said, "the only dreamers."[9]

Unfortunately neither the structure of Mesabi Range ores nor the country's depressed economy allowed the Merritts to realize their dream as rapidly as necessary for their financial well being. Existing furnaces, constructed to handle hard, rocklike ores, lacked the technology to utilize immediately large quantities of loose, powdery ore from the Mesabi. It took several years for iron ore consumers either to design and build new furnaces or to dismantle and remodel their old plants.[10]

Meanwhile the panic and succeeding depression continued to have a devastating effect on the nation's iron and steel industry. The price of all classifications of Lake Superior ores, for example, began falling in 1893, continued down the following year, recovered slightly through 1895 and 1896, and then dropped even lower in 1897. Although Mesabi shipments increased throughout the decade, "it was very hard to find a market for ore," and "when it was found prices were low in the extreme . . . any company operating on a narrow margin, or trusting to large immediate returns to make good a deficiency in its treasury, was bound to fail," wrote an early scholar of the industry. The panic of 1893–97 "tied up the country's financial resources and brought general business to a standstill, the iron trade . . . being the severest sufferer. The price of ore, which at the beginning stood at the lowest figure then known, went down continuously, with only one short season of recovery, until the product of the mines brought little more than half the low price prevailing in 1892. Coincident with this unprecedented depression occurred the development of the Mesabi range, which was able to produce ore at figures far below any the old ranges had yet reached. The hard times and monetary stringency quickly bankrupted most of the small concerns just starting."

Within this economic atmosphere the Merritt brothers sought unsuccessfully to sell their Lake Superior Consolidated stock at a profit. Securities in nearly all facets of the economy fell to disastrous levels during 1893 and 1894, and the Merritts were unable to market their watered stock at even 10 per cent of par value. Late in 1893 their financial position and that of Wetmore hit bottom as endorsements piled up against the brothers and creditors demanded payment on long overdue loans.[11]

A revealing memorandum written by Frederick Gates to Rockefeller on December 11, 1893, described the developing strains in the Merritt-Wetmore-Rockefeller relationship. At a meeting of the Lake Superior Consolidated Mines board of directors on December 8, "Lon Merritt complained bitterly of the delay in getting out the Consolidated stock," wrote Gates. "He said that when he came to New York three or four weeks ago he could borrow money on the Consolidated stock at 50 but the delay in getting out the stock and the uncertainty as to the railroad being financed and the uncertainty as to the policy of the Consolidated Company in opening up new mines etc[.] had created such distrust that he was now unable to borrow money upon this stock. He remarked that not one thing had been done by the Consolidated directors looking to giving the stockholders any assurances as to the value of their property; that we owed it to the stockholders of this concern to take such measures as would enable them to use their stock the same as all other forms of property.

"I remarked that so far as I was concerned I had done everything in my power to push the work of getting out the Consolidated stock; that as to financing the railroad it had been impossible for [many?] to arrive at any conclusions regarding them in the absence of any authoritative statements as to the condition of the road; that the instant that I learned in Duluth that Mr. Rockefeller would have to come to the relief of the railroad to save it from its creditors I had telegraphed for experts to look into its accounts and although these experts had been working for weeks both in Duluth and New York they are unable even yet to give a final and exact statement of the railroad, so inextricably tangled are its accounts; that we had to take things as we found them and we had found them both with reference to the Consolidated and the railroad in such shape that more rapid progress was utterly impossible.

"Mr. Merritt then remarked that he had been here and could testify that everybody had worked as hard as they could. He was not blaming anybody but as a matter of fact he would be obliged to return to Duluth and take advantage of the bankrupt law.

"On Saturday (Dec. 9) Mr. Wetmore came in and modestly inquired if there was any hope of a loan to save the Merritts. I told him I saw no hope. I urged Mr. Wetmore in their interest, however, to bring in his balance and his 300 bonds; that the equities in these bonds would probably amount to $60,000 and this would very materially reduce his debt. He replied that these bonds are up as collateral; that they are sandwiched in with various loans with other collateral; that the interest on these bonds he was dependent upon in order to pay the interest on his various loans; that if he were to take them out of his collateral and surrender them he would be absolutely and immediately ruined; that he had no means of supporting and no means of paying this interest except from these coupons.

"I remarked then that if the railroad was not financed by Mr. Rockefeller he would necessarily be ruined. He said that that was true. Furthermore said I, the Merritt brothers would also be ruined in substantially the same manner. He replied that this was true. Then further said I, the Rail Road and the Consolidation will also be ruined. Exactly he replied. Then I said, if Mr. Rockefeller does not finance the railroad then you and the Merritts' and the Rail Road and the whole consolidation will be absolutely ruined, and this comes to Mr. Rockefeller after he has put in $100,000 in the Rail Road more than you assured us would be necessary to save it. He admitted that to be true. And now one step further said I, if Mr. Rockefeller contemplates financing the railroad and looks about him to avail himself of such legitimate profits of his investments as may properly come to him, you reply that you must have, yourself, a considerable portion of these profits or else be ruined any how.

"This again he admitted with apparent sadness but with no attempt to justify himself, further than this: He said that he had worked very hard for this consolidation and had staked everything he had upon it; that he had saved Mr. Rockefeller's Steel bonds and his Penokee & Gogebic investment; that he had got together this consolidation at very slight cost; that he felt that he was entitled to more recognition and more gratitude than he was receiving; that latterly he had not even been consulted on any of the matters connected with the consolidation as to its future policy. He desired to give himself up to the best interests of the business but he was not asked to do anything in connection therewith.

"I asked him to specify where he had been overlooked. He cited the case of the Lake Superior Mines transaction. I explained to him that when they came to us he was at the end of his rope; it was utterly impossible for him to do anything; however sanguine he might be, we knew perfectly that unless Mr. Rockefeller took that up and financed it through himself it would be lost; he then cited the present case of the railroad. To this I replied that he was at liberty of course to do anything he could to save the railroad end; if he could get somebody to finance the railroad he was of course at liberty to do so. The facts are however that Mr. Rockefeller would have to do this or it would not be done. Of that fact he himself was perfectly well aware, and if Mr. Rockefeller did it he would himself have to name the terms on which he would do it; that when these terms were named an opportunity would be given for everybody to come in on the same or better terms. He would then have an opportunity to exert all the financial influence he possessed.

"On his departure Lon Merritt came in. He had evidently broken completely with Wetmore. He had heard that Mr. Wetmore desired to have two or three years time. He hoped that Mr. Rockefeller would give him so much time. He did not want the securities of the Merritt brothers tied up here for all that period. He wanted Mr. J. D. to take in all of Wetmore['s] securities at the price, apply them on the loans as far as they would go and release a corresponding amount of their securities so that they could use them in tiding over their affairs." [12]

As the Merritts broke their ties with Wetmore and grew more suspicious of Rockefeller and Gates in December, 1893, American Steel Barge demanded "immediate payment" of the five promissory notes "with interest at six per cent." Writing to Leonidas on December 20, President Colgate Hoyt notified him "that unless these notes are paid with interest on or before the expiration of ten days from date, that this Company will at its pleasure, sell out the collateral pledged to the payment of these notes, at public auction or private sale," apply the proceeds on the notes, and hold Lon and Alfred "responsible for the difference." In an earlier exchange of letters American Steel Barge had refused the extension "on your notes as requested in your communication through Mr. Gates of September 28th." On October 5, 1893, Hoyt had informed the Merritts that the financial condition of American Steel Barge was "such as to make it absolutely impossible to consent to this extension. We are willing and anxious to oblige you in every reasonable way," Hoyt wrote, "but are so sorely in need of the money that is due that we cannot grant this request." [13]

With creditors pressing them on all sides, the brothers began to seek a buyer for their share of the Lake Superior Consolidated. In an attempt to sell the family's interest, Alfred Merritt began negotiations with

James J. Hill, the Great Northern Railroad entrepreneur whose head-quarters were in St. Paul. Near the end of December, Hill was informed that "the Merritts of Duluth Missabe & Northern Ry find it necessary to sell their interests in that road and the consolidated mines" and that "they would prefer to sell to you on account of past differences" with the Duluth and Iron Range and the Canadian Pacific railroads. The asking price was $5,000,000 for "the entire railway, docks, and terminals . . . and all the Merritt mines." Hill, who professed no firsthand knowledge of the Mesabi mines but was interested in the DM&N, apparently made inquiries of H. H. Porter of the Minnesota Iron Company and Jay C. Morse, president of the Illinois Steel Company. Porter's reply was succinct: "I have been carefully investigating the property and methods of the parties [the Merritts] . . . and I am afraid they are so mixed up that no prudent man would now attempt to unravel them." Morse concurred, writing Hill "We have been approached from two directions within the last week in regard to the purchase . . . but on investigation decided the properties are in such bad shape that it would be unsafe to touch them at any price."[14]

According to the Merritts, Hill agreed to purchase all their Consolidated stock for $7,000,000, one-seventh at the contract signing and the remainder within 30 days. The only obstacle was the pending action of Congress to remove tariff duties on imported iron, a bill that passed the House on February 1, 1894. Hill believed that the placement of iron ore on the free list would allow foreign sources to demoralize the industry and retard Mesabi development. As a result of the bill's passage by the House, he is said to have refused to conclude his deal with the Merritts.

Greatly chagrined, the brothers placed the blame on Rockefeller. Andrus later raised the specter of conspiracy, accusing him of manipulating the tariff action. He claimed that Gates and counselor Murray "took occasion to say to Lon and me, with smiles of great satisfaction, that the tariff would be taken off of iron, that Rockefeller was spending $40,000 to accomplish it." In fact, however, the tariff was not removed. The Senate refused to concur in the House bill, and when finally passed by both chambers on August 28, 1894, the Wilson-Gorman Tariff lowered, but did not remove, the rate on iron ore from $.75 to $.40 per ton.

With Hill out of the picture and their impatient creditors unpaid, the Merritts once again turned to Rockefeller. To Gates's professed surprise, on December 23, 1893, some family members offered to sell their portion of the Consolidated merger at $40 per share. Although they considered this price far below the real value of the mining companies and the railroad, they thought it was the most they could expect to get under the circumstances. The family asked for a total payment of $600,000 and cancellation of a $150,000 loan and interest borrowed from Rockefeller on July 15, 1893 — the first installment of which was due on January 15, 1894.[15]

Seeking to retire all their public and private indebtedness, the brothers offered to sell their stock "for no other reason than to relieve our present financial condition, and for the best interests of the Consolidated Mines. It is clear," they wrote Gates, "that with the present financial stringency and the disturbed iron trade, by reason of proposed tariff legislation, and our own pressing needs, which have been extended from time to time in the past, and with the mines not as yet on a paying basis, we cannot hope to hold and save intact our Consolidated stock for any considerable period without being obliged either to sacrifice the same or to imperil our all under legal complications and bankruptcy proceedings. We have given so much of our past to the mining and railway problem and have had such high hope of their ultimate success as business propositions, that it is with much regret that we find it necessary in the line of our interest to submit to your favorable consideration a proposition to sell, and thereby sever our past confidential and close business relations. We beg leave to say that we appreciate heartily your earnest work in behalf of and faithful attention to the true and substantial interests of the Consolidated Mines and our own interests therein as well as personal, and to suggest that we desire always your personal esteem and confidence.

"If this proposition should meet with favor and the sale thereby consum[m]ated, as we hope, we beg to acknowledge the fair treatment we have always received from Mr. Rockefeller and yourself and to express the entire satisfaction with and hearty approval of the work accomplished by the Consolidation and the results of his and your personal efforts in its and our behalf.

"We submit this feeling that the best interests will be enhanced by the sale, and as a practical solution of our difficulties *which had their inception before our relations with you began* [author's italics]."

Six days later Gates told Lon by telephone that "Mr. Rockefeller's mind in the matter seemed to be about this: He had already put in about two and one-half millions of cash into this enterprise and was now being called upon to put indefinitely large amounts more into a second [DM&N] mortgage and would no doubt be called upon in addition thereto to contribute largely in financing the Consolidated Mines. In view of these large probable requirements in cash and the very discouraging outlook for mining stocks Mr. Rockefeller did not dare to put more cash into iron than is absolutely required to keep this consolidation together. . . .

"Mr. Merritt replied saying that he supposed that Mr. Rockefeller would have no objection to their selling their stocks to others. I told him certainly not, since the stocks had been offered to Mr. Rockefeller. That however much he might regret to see them sell out he could raise no objection. Mr. Merritt replied that they may as well face the music now. I told him I was sorry. He said he on the other hand was glad; that he knew they would have to come to it sooner or later and they might as well face it now and end the matter as not." [16]

Despite all the friendly rhetoric, Rockefeller rejected the proposal to buy at $40 per share. Apparently unable to find other buyers, the brothers in January, 1894, made a second appeal to Rockefeller to purchase their Consolidated stock at $10 a share payable within one year. The financier accepted this more realistic overture, based on the still depressed securities market, and signed two agreements dated February 1 and 21, 1894. They called for the sale of "any Lake Superior Consolidated shares up to a par value of $1,200,000 and all purchased at 10% par value, purchase price to be paid in cash as fast as possible." All the certificates sold were to be delivered to Rockefeller's New York City office within ten days. In addition the Merritts specifically consented to use the funds they received "to pay the outstanding claims" against them. Rockefeller purchased 90,000 shares of Consolidated stock, half of it on February 1 and the other half 20 days later. The Merritts received $900,000 and an option allowing them one year within which to repurchase 55,000 of those shares at $10 each plus 6 per cent interest. [17]

This transaction constituted the end of the partnership between the developers of the Mesabi and the oil king. Gates did not understand why the brothers "were compelled to sell just at that time." [18] Rockefeller included the option clause, Gates said, to give the Duluth family time "for working out their salvation," and the option would have been extended if the brothers had asked for renewals. Indeed Lewis Merritt and his oldest son, Hulett, did exactly that in an action which caused the first major split in family solidarity. Hulett asked for and received renewals that eventually allowed him and his father to sell their Consolidated stock to the newly formed United States Steel Corporation in

1901 at a substantial profit. Gates pointed out that Alfred and Leonidas could have exercised the same option and been more than $9,000,000 richer.[19]

None of the other Merritts took advantage of the option, nor did they ask that it be extended. Instead they severed all ties with Rockefeller and the Lake Superior Consolidated Iron Mines. Lon resigned as president and as a director, and Alfred left the presidency of the DM&N on February 6, 1894. Then they turned to the courts in an effort to obtain funds and regain their former holdings.[20]

To represent them in the series of lawsuits they brought in 1894 and 1895 the Merritts employed two attorneys new to Duluth. When the brothers established the DM&N and organized their mining companies, they had received competent legal assistance from Joseph B. Cotton. In February, 1894, Cotton asked the Merritts to release him from any further obligations to them because he had been offered a five-year contract to serve as Rockefeller's local counsel in Duluth. The brothers consented, although a family spokesman later claimed, "They were too dazed as yet to realize the importance to Rockefeller of the acquisition of their trusted attorney, familiar with all their affairs and the history of all the Missabe transactions."[21]

The new lawyers who handled all the suits brought by the family in 1894 and 1895 were Anak A. Harris and his son Henry. They were retained by Andrus upon the recommendation of the Merritts' friend, United States Circuit Court Judge John F. Dillon. Duluth citizens knew little about the two men except that they had recently moved to the city from Fort Scott, Kansas, and were said to be "both honorable and efficient." Members of the family and Gates later painted a different picture of the middle-aged lawyer and his son, incorrectly asserting that the elder Harris had arrived in Minnesota seeking to start a new career for the fourth time in as many states.[22]

In May, 1894, Leonidas instituted the first litigation to come to trial when he sued the Duluth, Missabe and Northern Railroad for payment of his salary as vice-president from October, 1891, to October, 1893, charging that he had served as the managing agent of the company's New York City office and negotiated loans to finance construction of the Duluth extension and ore dock. Even though a contracted salary had never existed, Lon sought $40,000 as the sum his services "were fairly and reasonably worth," as well as reimbursement for office rent of $1,050, clerk's salary of $1,400, stenographer's fees of $607.50, telephone $72.50, hotel $2,250, and railroad fare (trips to New York) $500, making a total of $45,880 plus 7 per cent interest. The trial began on September 12, 1894, in the St. Louis County District Court in Duluth, with Judge Josiah D. Ensign presiding. On September 22 the jury issued a verdict in Lon's favor, awarding him $12,530 for his services to the railroad, plus interest and court costs, bringing the total compensation to $13,303.[23]

In separate proceedings in June, 1894, Leonidas also brought suit against two mining companies he had earlier helped incorporate. In a trial held in the District Court in Duluth, he asked $31,479 for his services as president of the Biwabik Mountain Iron Company from December, 1890, to February, 1894. Again no salary had been contracted, but Lon considered this sum to be "reasonable compensation" for his work. On November 16, 1894, after a five-day trial before Judge Charles L. Lewis, the jury awarded Leonidas $513.10 plus $31.90 in court costs.[24]

Immediately following the Biwabik case, Judge Lewis was scheduled to hear a second suit brought by Leonidas against the Missabe Mountain Iron Company for $11,096 in payment for his services as president from February, 1892, to March, 1893. The Missabe Mountain case, however, never came to trial, for Judge Lewis ruled in the Biwabik case that under Minnesota law "a director or executive officer of a corporation is, in his relationship, a trustee and he cannot recover for services except by express contract made by resolution of the board of directors." Leonidas, through his attorney, then averred that he lacked the proper witnesses to continue against the Missabe Mountain Company, and Judge Lewis dismissed the case on November 15, 1894, "without prejudice" to either party. As is evident, Leonidas gained little recompense from these small suits against his former companies.[25]

In a far more complicated action with bigger stakes, Alfred and Leonidas sued the American Steel Barge Company in an effort to recover for the securities pledged to Wetmore as collateral on the five promissory notes totaling $432,575. This legal action began on April 11, 1894, in Judge Lewis' District Court, but upon petition of the barge company, the case was removed on April 18 to the United States Circuit Court, also meeting in Duluth, on the grounds that the litigants resided in two separate states.[26]

In their complaint, the Merritts declared that both parties had agreed that the notes were nonnegotiable and that the stock could not be repledged, sold, or disposed of in any fashion. Wetmore, as vice-president and acting managing officer of American Steel Barge, had no right to transfer them to the company to cover a personal debt. Nor was

American Steel Barge at liberty to convert the securities in the fall of 1893 into trustee certificates representing 11,331.3 shares of Lake Superior Consolidated capital stock. Alfred and Leonidas alleged that, since Wetmore was "wholly insolvent," the barge company must return the Consolidated shares or pay them $500,000.[27]

Barge company spokesmen readily admitted that the firm had accepted the Merritt securities from Wetmore as payment for the latter's indebtedness and had converted them into Consolidated stock. Company officials denied, however, that the Merritts had been unaware of these transactions. They introduced letters from Alfred and Leonidas to Wetmore and to American Steel Barge President Hoyt in which the brothers authorized the transfer by Wetmore and requested the conversion into Consolidated stock.[28]

While this case was pending in Duluth, the barge company on May 31, 1894, filed a countersuit against Leonidas and Alfred in the New York Supreme Court. In it, American Steel Barge alleged that the Duluth brothers were "insolvent and unable to pay any of the . . . promissory notes." The company asked that it be given "right and title to said notes and collateral and its lawful lien . . . be established and declared." The New York court appointed Franklin Bien as referee in this dispute. On September 11 he reported findings favorable to the barge company, and on October 1 the court ruled against the Duluth men. Agreeing with Bien's report, Judge Abraham R. Lawrence decided that the Merritts owed the barge company the original sum of $432,575 plus accrued interest of $41,122.62, making a total of $473,697.62. The company then sold the disputed Consolidated stock at public auction, accepted the proceeds as partial payment toward the debt, and held the Merritts responsible for the balance.[29]

After news of this verdict reached the federal Circuit Court in Duluth, Judge Rensselaer R. Nelson also ruled in favor of the barge company on October 16, 1895. He held that the decision of the New York court, rendered first, barred further action. The Minnesota court, with original jurisdiction, would have retained the case if the Merritts had sued for the return of Minnesota stocks. Instead they brought legal action to recover the value of Consolidated stock, and thus allowed American Steel Barge quite properly to make use of the eastern court, because "the certificates representing the shares of stock in controversy were situated in the city of New York."[30]

While these cases were under way, the Merritts also brought suit against John D. Rockefeller. At first the brothers also planned to include Frederick Gates in this legal action, but they soon dismissed the idea. Although Lon and Alfred looked upon the former Baptist minister as their greatest enemy, they realistically perceived Rockefeller to be a more rewarding legal target. Alfred was selected by the family to file the test case, and he did so on October 29, 1894, in the District Court at Duluth, seeking to recover $1,226,400 in damages. That exact sum had been arrived at by estimating Alfred's share of the merger at $1,533,000 and then subtracting the $306,600 he had received in Consolidated stock. After Alfred won his case, other members of the family planned to sue Rockefeller for at least $2,000,000 more.[31]

Alfred based his litigation on the assertion that he had been defrauded because his property had been taken at unjustly low valuations through misrepresentation on Rockefeller's part. His complaint pointed specifically to the stocks and bonds of the Penokee and Gogebic Consolidated Mines and the Spanish-American Iron Company, which Rockefeller had contributed to the merger. The financier had knowingly inflated their value, Alfred charged, and had falsely assured the Merritts that the companies were solvent and prosperous, that they owned valuable mines, and that their presence in the merger would enhance the value of Consolidated stock. Had these "statements, representations, and promises" not been made, Alfred averred, "he would not have entered into the . . . agreement" of August 28, 1893. Moreover, they were "wholly false and untrue . . . and well known to be so at the times" when they were made.[32]

During the interim between the filing of Alfred's complaint in October, 1894, and the opening of the trial in June, 1895, both sides agreed to transfer the proceedings to the United States Circuit Court in Duluth. To avoid future charges of prejudice on the part of either a local or an eastern jurist, John A. Riner, federal district judge for Wyoming, was assigned to hear the case. The trial lasted from June 5 to 13, and spectators occupied every seat in the hot courtroom. "The . . . skirmish line in the great legal battle being waged against John D. Rockefeller has been withdrawn, and yesterday morning the big Merritt gun boomed forth its note of announcement that the heavy cannonading had begun," trumpeted the *Duluth News Tribune* of June 6, 1895. Local newspapers printed lengthy descriptions of the testimony and described the legal tactics employed during the trial. The 12-man jury was drawn from throughout Minnesota; four were from Duluth, one was from a small Mesabi Range community, and the others came from such widely separated parts of the state as Warren in the west and Stillwater in the east. They represented the diverse occupations of farmer, druggist, jeweler, river pilot, merchant-lawyer, justice of the peace-village council president, clerk, and liquor dealer. George Murray, with the aid of the Merritts' previous attorney Joseph Cotton, plus Cushman K. Davis, a former state governor, and John M. Shaw handled the case for Rockefeller. Anak and Henry Harris, assisted by Jed L. Washburn and Otway W. Baldwin, represented Alfred Merritt.[33]

Answering Alfred's charges, Rockefeller spokesmen denied that any deception or fraud had been perpetrated. Expert witnesses testified that the Penokee and Gogebic firm owned ore lands, especially the Tilden and Aurora mines, containing some of the richest deposits on the Gogebic Range. They admitted that the companies had gone into receivership when they experienced temporary financial difficulties caused by the panic, but they pointed out that when he became aware of these problems, Rockefeller had offered to redeem the securities these firms represented in the merger, cancel a portion of its bonds, and thus reduce the Consolidated's debt to $1,500,000 — hardly the work of a man intent on "cunningly betraying his business associates."[34]

A substantial portion of the Rockefeller defense involved establishing the fact that the properties he contributed were indeed valuable. (The Aurora and Tilden mines in subsequent years became large shippers of ore.) This, of necessity, led to a considerable body of testimony concerning the initial stages of the creation of the Consolidated in July and August, 1893. Charles Wetmore refused to appear in person and testified by deposition. He gave his age as 40, his residence as Oyster Bay, Long Island, and his occupation as lawyer. In response to written questions, he insisted that the plan to create the Consolidated "originated with myself and Messrs. Leonidas and Alfred Merritt," and that neither Rockefeller nor Gates had anything to do with its conception. Conflicting testimony from both sides revolved about the exact date of the meeting between Rockefeller and Leonidas. The latter claimed that the conversation took place in June, 1893, involved a long discussion of the Mesabi, and induced the Merritts to participate in the preliminary agreements leading to the merger. Rockefeller witnesses produced evidence that the two men met on July 12 at Lon's request after several contracts had been signed and merely exchanged light conversation, largely about the weather.[35]

One of Rockefeller's strongest lines of defense was the fact that the Merritts had approached him in the first instance to purchase their railroad and mining stock. It was not he who initiated the financial relationship. Gates testified, "The proposition came from them; the terms were named by them." As for the value of the mining companies Rockefeller contributed to the merger, Gates said, "Mr. Wetmore was then, and had been for some years the secretary and attorney of those companies. The office of those companies was at 36 Wall Street. Mr. Wetmore's office was the same office, same building, same number," and there was nothing to prevent Leonidas from examining the books to learn all he wished to know about the financial condition of the firms Rockefeller contributed to the Consolidated.[36]

On June 13, after more than five hours of deliberation, the jury awarded Alfred $940,000 of the $1,226,400 he claimed as damages. The following day the *Duluth News Tribune* speculated that it took three ballots to reach a decision in Alfred's favor, and that the jury spent most of the five hours debating the amount of damages. The newspaper reported that the suggested sums ran from Alfred's demands down to $15,000 before a compromise resulted.

"The verdict was a thumper, wa'nt it?" Murray commented to Gates on June 15, 1895. The judge's "ruling on the measure of damages was most peculiar and without doubt was the cause, so far as law goes, of the heavy verdict. It gave full play for the tremendous local prejudice concerning which I do not need to inform you. . . . It seems to me that Judge Riner committed reversible error most certainly, with respect to the measure of damages. . . . We may now either (a) make a motion . . . for a new trial, or, (b) sue out a writ of error which is equivalent to an appeal from the judgment."[37]

In contemplating their next move, the Rockefeller forces realistically took note of what Murray referred to as "the tremendous local prejudice." They knew that the Merritts owed substantial sums of money to many Duluth residents and, as Gates put it, that "public sentiment was not averse to paying their debts, and to the circulation of some Rockefeller money in Duluth."[38] In this, they were entirely correct. Regional sentiment favored the Merritts and was generally antagonistic toward the eastern industrial establishment. For example, on September 22, 1895, the *Superior Sunday Forum* reflected this local feeling when it said that "the Standard Oil octopus . . . would be able to swallow and digest the Merritts." Rockefeller was portrayed as looking at Duluth "with the cold eyes and steady pulse of capitalists and monopolists." The writer concluded: "Rockefeller is a financial cannibal who eats men every day."

Nor did the favorable verdict completely satisfy the Merritt family. Leonidas had testified that while he was in New York City every scrap of paper and memoranda pertaining to his financial affairs had been stolen from his hotel room. In addition he later declared that someone entered his Oneota home, taking his letters to his wife containing detailed accounts of his activities. He alleged that certain questions asked by Rockefeller's attorneys "could have been put in no other way except from the contents of those private papers and letters," although he admitted under cross-examination that he had frequently used the letter books in Wetmore's office for his own correspondence. Andrus expressed disappointment over the amount awarded. He asserted that two of the jurors "from the southern part of the state had held out for a verdict in favor of" Rockefeller. "Finally the balance of the jury, who wanted to give $3,000,000 to Alfred Merritt compromised at the lower figure." Andrus also reported "a rumor on the street" that these two southern Minnesotans had been offered $20,000 each to create a hung jury but backed down under pressure to reach a decision. He wrote that the family was told "the next morning that a hundred men were behind the court house the whole night the jury was out with a rope, prepared to hang the men who would hang the jury." One difficulty with this story is that the jury began its deliberations soon after noon and returned its verdict by 6:00 o'clock that evening. Thus no one would have had occasion to stay at the courthouse throughout the night.[39]

Each side believed the other to be engaged in some form of conspiracy. As soon as Lon finished his courtroom testimony, Hulett handed Gates a penciled note intended as a helpful warning: "You are falsely charged with having made misrepresentations by people who from a malicious motive have combined together with a corrupt intent and by improperly meaning to defraud justice, to extort money from you and to injure," wrote Hulett. "You are in the midst of Bribery, Corruption and a great and deep laid Conspiracy." After the verdict Andrus accused Anak Harris and his son of either such gross incompetence that they completely "bungled the complaint, or so cleverly managed it" that Rockefeller was hurt as little as possible. Gates went so far as to blame all the Merritt difficulties after 1893 upon the "malign influence" of Harris, who, he said, "poisoned their minds with the hopes of vast booty."[40]

Although the following incident was not mentioned in these trials, Gates had some basis for his suspicions. During the summer of 1894, while the suits against the constituent firms and the American Steel Barge Company were in progress but before Alfred's complaint was filed, Harris appeared in New York City and had a talk with George D. Rogers, Rockefeller's private secretary since 1877. The lawyer claimed to possess enough influence with the Merritts to avoid any direct legal action against the New York financier. On three successive days from July 10 through 12, he tried to induce Rockefeller to purchase all the remaining Merritt Consolidated stock at a price well above its market value. "You folks ought to buy them out," he told Rogers, "and let them go and end all this trouble. . . . I do not want to get up any trouble, or friction, or anything between Mr. Rockefeller and the Merritts." Rogers eventually refused to accept any offer from Harris and told the Duluth lawyer that Rockefeller wanted him "to deal directly with Joseph B. Cotton, his attorney."[41]

Gates drew a direct connection between the refusal of Harris and the Merritts' legal action against Rockefeller. "This suit was trumped up by Harris, who did not hesitate to boast that he had 'discovered' it and that he expected to make [a] reputation out of it," wrote Gates.[42] Indeed the lawyer did boast to the *Duluth Evening Herald* of June 14, 1895, that when he "took hold" of the case, "the Merritts had no papers, no letters, no records and practically nothing but their memories upon which to make up a case." Harris was also quoted as saying that a "block of Minnesota Iron company stock worth $400,000 belonging to Rockefeller was attached and this was a trump card for it made Rockefeller answer in this state, he having property here. Had it not been for this he might have transferred it to New York."

The available evidence suggests, however, that Harris acted on his own in New York City and that he did not initiate any of the Merritt suits. Gates's belief that he sought to enhance his legal reputation by actively participating in highly publicized matters remains wholly plausible. Nor was Harris beyond improving his own personal finances. In conversations with Rogers he offered to sell Rockefeller not only his influence but also some stock the Merritts had given him for services rendered. "I have some influence with them and they will do what I say," he told Rogers. The stock sale, Harris said, was "a separate matter" from any Merritt legal activities, but its purchase "would have to be preliminary to any . . . settlement. . . . This stock is mine and if you buy it I do not think there will be any more trouble," Rogers quoted him as saying.[43]

The Merritts reserved their most virulent attacks for Gates and Rockefeller. They pointed to Rockefeller's absence from the Duluth trial, where he had testified by deposition rather than in person. Gates had appeared personally but that did not exempt him. James C. Merritt, Andrus' 18-year-old son, reported that he followed the minister, who was alleged to fear arrest for perjury after he testified, "as he dodged through back alleys and across the railroad yards" to a hiding place behind a pile of ties. From there, James said, Gates jumped unnoticed on a train in order to flee the state.[44]

After weighing several alternatives suggested by counselor Murray and wanting to avoid a new trial, Rockefeller took the case to the United States Eighth Circuit Court of Appeals. This three-man tribunal included Judges Walter H. Sanborn of St. Paul, Amos M. Thayer of St. Louis, and Henry C. Caldwell of Little Rock. They accepted the case on a writ of error that the lower court had "refused to permit" Rockefeller to show the actual value of Consolidated stock, and that Judge Riner had improperly instructed the jury on how to determine the amount of Alfred's loss. Riner had ruled that the damages consisted of the difference between the value of Alfred's stock placed in the merger and the actual market value of the Consolidated securities he received in return.[45]

Meeting in St. Louis on November 9, 1896, the Appeals Court judged this an improper interpretation. Instead it held that "The true measure of the damages suffered by one who is fraudulently induced to make a contract of sale, purchase, or exchange of property is the difference between the actual value of that which he parts with and the actual value of that which he receives under the contract. It is the loss which he has sustained, and not the profits which he might have made by the transaction." The judges then went on to ask what amount would fully compensate Alfred for his loss. And they answered that Alfred "had less than one-seventh" of the stock in the Consolidated, "and his share of full compensation for this injury could not possibly have been more than one-seventh of $1,619,851 and interest, or $231,407.28 and interest. But he recovered a verdict for $940,000. . . . In our opinion these damages far exceeded the just measure of full compensation for this injury. . . . They must have been composed more largely of bright anticipations of prices which the owners of the stocks hoped to realize from them than of the difference between the actual values of the stocks which were exchanged. In other words, they were speculative rather than compensatory damages." The Appeals Court then reversed the lower court's decision, charged Alfred $1,040.35 in costs, and remanded the case back to the District Court in Duluth "with directions to grant a new trial."

The Merritts saw this decision, said Andrus, as the final "staggering blow" and decided against additional litigation. Leonidas and Alfred said that ill health and lack of funds prevented them from further pressing their claims, but it is obvious that in the light of the Appeals Court's reasoning they could not hope to recover the large sums they desired. They bypassed Anak Harris and retained Jed Washburn, who had assisted in the Rockefeller case, to handle their affairs. Early in January, 1897, he and Andrus traveled to New York City and negotiated an out-of-court settlement. Within a few days both sides reached a compromise whereby Rockefeller paid his former business associates a total of $525,000.[46]

For their part 20 members of the Merritt family, including all the brothers except Lewis J., signed a statement retracting all charges of fraud against Rockefeller. In it they declared: "Certain matters of difference have existed between the undersigned and Mr. John D. Rockefeller, and a certain litigation has been pending between the undersigned Alfred Merritt and Mr. Rockefeller, in which litigation it was claimed that certain misrepresentations were made by Mr. Rockefeller and those acting for him concerning certain properties sold by him to Lake Superior Consolidated Iron Mines. It is hereby declared that from recent independent investigations made by us or under our direction we have become satisfied that no misrepresentation was made or fraud committed by Mr. Rockefeller, or by his agents or attorneys for him, upon the sale by him of any property to us or any of us, or to Lake Superior Consolidated Iron Mines, or upon the purchase by him from one or more of us of any stocks or interests in any mining or railway company or companies, or upon the pledge by us or either of us to him of stocks and securities belonging to one or more of us; and we hereby withdraw all such charges and claims and exonerate Mr. Rockefeller and his agents and attorneys therefrom."

On January 22, 1897, lawyers for both Rockefeller and the Merritts filed the statement with the clerk of the United States Circuit Court in Duluth. The court then "released and discharged" the judgment for costs and "stipulated that this action be forever discontinued and dismissed without costs to either party as against the other."

Although reluctant to exonerate Rockefeller, Leonidas later explained why he convinced family members to retract their charges and accept the settlement. "I was willing to do almost anything to get out of the trouble," he said. "I had lost, deservedly, I think, my confidence in the courts. I had lost . . . my confidence in those I had naturally supposed to have been my friends. I had been stabbed right and left by those I helped most." Essentially, however, Lon blamed himself for the family's misfortunes, and he signed the retraction "because I wanted to relieve my family from their destitution and absolute poverty."[47]

On the surface it seems puzzling that Rockefeller should have agreed to a half-million-dollar payment. His reasons become clear, however, when it is understood that past experience had provided him with little faith in local juries, and that a new trial would have been held in Duluth. He knew that protracted litigation would be costly in time and money. After all, the settlement protected him against future lawsuits growing out of the whole affair. Another important consideration was that Rockefeller now controlled large iron ore deposits and a railroad in northeastern Minnesota. He needed the good will of Minnesotans to exploit these resources successfully. Further judicial battles could do additional damage to his reputation in Duluth and on the ranges.[48]

The Merritts applied the $525,000 acquired in the settlement to relieve their financial situation. Before the panic many speculators with potentially rich ore leases had lacked the financial resources to develop them. The Merritt brothers had cosigned numerous loans for these men in order to secure additional traffic contracts for their DM&N Railroad. During the depression, however, most of these small operators had fallen into bankruptcy, and the endorsers of their notes were held responsible for payment. In 1897 the family's own lawyer estimated that Alfred and Leonidas alone had endorsed paper "approximating $1,000,000" for which "there is a large row of judgments." Rockefeller's money provided funds for the Merritts to meet some of these long overdue obligations. Thus more than half of it went for the purpose "of paying for dead horses," as Andrus put it, $25,000 was given to Napoleon and Thomas A. Merritt, and Alfred divided the remainder with Leonidas and Andrus, the brothers most active in the long-drawn-out process.[49]

Two acts remained in the Merritt drama. The first was yet another lawsuit brought in 1897 by Alfred and Leonidas against their former attorneys, Anak Harris and his son Henry. The brothers charged that the Harrises refused "to deliver unless they receive $25,000" papers and letters of the Merritts, as well as 954 shares of capital stock in the Security Land and Exploration Company. The plaintiffs claimed that the stock was given to the attorneys as security for fees due them in both the American Steel Barge and Rockefeller lawsuits in which the Harrises represented the Merritts and that the lawyers had threatened to make public private family matters. Filed in the St. Louis County District Court on February 23, 1897, the case was not heard until early

June; the jury that had to decide which party was responsible for "deliberate falsification" failed to agree, and the case was continued over to the September term. Finally on October 10, 1897, the case was decided in favor of the Merritts; the Harrises were ordered to turn the stock and other papers back to the family. When the lawyers failed to comply promptly with the decision, they were cited for contempt of court, and it was not until March 5, 1898, that the judgment was satisfied.[50]

The last act in the Merritts versus Rockefeller was played on a national stage in 1911. After subsiding for nearly 15 years, the conflict resurfaced "like Banquo's ghost," said Andrus, in November, 1911, when Alfred and Leonidas testified before a House of Representatives committee investigating the United States Steel Corporation. This nine-man committee, headed by Kentucky Democrat Augustus O. Stanley, had been instructed to examine the organization and operation of the industrial giant for possible antitrust violations. In separate appearances, the two brothers publicly attacked Rockefeller from a new angle, but one that nevertheless re-emphasized the conspiracy theory. Alfred and Leonidas told the committee of a $420,000 call loan from Rockefeller acquired about the time the Consolidated was formed. Although neither could remember a specific date, both contended that the head of Standard Oil "in the fall of 1893" demanded either payment on that loan within 24 hours or the sacrifice of their Consolidated stock deposited as collateral. Under questioning Alfred admitted that such a loan could be called at any time and that Rockefeller had not legally mistreated them. He said, however, that the family had trusted Gates to allow all the time necessary to repay the total amount. The Duluth men appeared especially angered, because, by their method of valuation, the collateral stock they lost had a minimum value of $100 per share. Yet in the face of incisive questioning, both brothers admitted that no one would have purchased Consolidated stock on the open market for even $8.00 a share at the time. For this they also blamed Rockefeller, saying that he controlled the "money trust" and had intentionally deflated the stock market during the panic years.[51]

Neither the Merritt family nor Rockefeller's associates were willing to let the affair die. Gates and Rockefeller did not appear before the Stanley Committee, and they were not subpoenaed to do so. But immediately after the Merritt brothers' testimony received wide publicity, Gates published a pamphlet entitled *The Truth About Mr. Rockefeller and the Merritts*. In it he publicly defended Rockefeller and accused the Merritts of falsehood, distorted statements, and reckless conclusions. He emphatically denied the entire story of a loan on 24-hour call. "The Merritts falsely swore before the Stanley Committee that Mr. Rockefeller compelled them to sell by calling a loan of over $400,000 and giving them twenty-four hours in which to pay," Gates wrote. "In this, there is not one syllable of truth. The Merritts owed Mr. Rockefeller at that time $150,000 only, and five-sixths of that had a long time yet to run. Mr. Rockefeller never, at any time, in any way, called a loan on them, for any sum whatsoever. The story of a sudden and cruel call by Mr. Rockefeller is a calculated falsehood, to conceal the real reasons for the sudden sale." [52]

Leonidas countered by preparing a rambling manuscript entitled "Leonidas Merritt's Answer to the Gates Pamphlet," which he never published. In it, he declared that "not one line, sentence or paragraph from title page to final cover, no word of Gates therein written, bears even the semblance of truth. . . . From beginning to end the publication is a palpable web of falsehood, misconstrued statements, vile innuendo and reckless conclusions." [53]

A careful examination of the preserved documents of Rockefeller and the Merritts does not substantiate the brothers' story. If such a loan existed, there is no record of it. Both Lon and Alfred told the Stanley Committee that they never possessed a copy of the loan agreement; Alfred confessed that he had never even read the note. Congressman Stanley attempted to subpoena the document from Rockefeller but discovered no evidence of its existence. The only known loan agreement solely between Rockefeller and the Merritts was the $150,000 personal note signed by the brothers as part of the final Consolidated contract; it had, as Gates stated, long-term repayment provisions. [54]

Nevertheless the story has persisted throughout all the years since 1911, and it is still believed by Merritt descendants. It is unlikely, unless new records come to light, that the matter can be further unraveled. The present writer looked carefully and exhaustively for any contemporary documents that would support the Merritts' claims. Although it is possible that such evidence exists in the private papers still held by the family, he could find no mention by anyone in the 1890s nor any documents in the publicly preserved Merritt Papers or the Rockefeller Archives concerning such a loan. Neither was it mentioned in any of the testimony or the supporting exhibits in the numerous legal actions involving the Merritts between 1894 and 1897. The only 24-hour Merritt call loans on record in the entire tangled matter are the five promissory notes given by Alfred and Leonidas to Wetmore, transferred by Wetmore to American Steel Barge, and called by that firm early in 1894.

What of the conspiracy theory? In the intervening years the downfall of the Merritt family at the hands of John D. Rockefeller has assumed a firm place in the oral legends of the development of the Mesabi, and it is still current on the range. No contemporary documentary proof was found by the present writer to support the Merritts' contention that Rockefeller set out deliberately to defraud them of their mines and railroad with the help of Charles Wetmore and the American Steel Barge Company. That assertion rests upon the testimony of various family spokesmen later in the 20th century. The earliest possible reference to a conspiracy theory uncovered in the research for this book occurred obscurely in a letter to Alfred from James Merritt, Andrus' teen-age son, apparently written on January 8, 1894. He reported a confidential conversation purportedly held in Titusville, Pennsylvania, between Leonidas and William Scheide, the retired Standard Oil official who had been Rockefeller's choice for general manager of the Consolidated, a post he held for only a few months before resigning. James wrote: "Mr. Scheide gave us his advice and oppinion [*sic*]. The substance of which Lon knew but had lacked confirmation. He advised that we simply do nothing," that rather than "be brought . . . before a Minnesota jury," Rockefeller "will trade with us . . . he dare not allow a scandal such as the courts would make of this." The family later maintained that Scheide told Lon of Rockefeller's plans to ruin them and warned the brothers not to trust Cotton, who was then their attorney.[55]

This letter is in the Rockefeller Archives. Gates did not learn of it until 1895. He believed that it offered a possible explanation of the Merritts' conduct which he had previously found puzzling. Gates sent the letter to Rockefeller with an explanatory memorandum dated May 23, 1895, in which he wrote: "Scheide had been introduced to the Merritts as an old time employe of yours, in whom you had the utmost confidence and who was your personal choice [for general manager of the Consolidated]. By many years of highly successful service at a large salary, he had acquired competence and retired. In looking over the whole field of your acquaintance, you had drawn this man forth from his retirement to assume this position of intimacy and trust. He was, inferentially, a man to whom you would reveal your inmost thoughts and disclose your most secret plans. He had spent two months with us in New York. They [*the Merritts*] naturally assumed that he was on the inside and when he made the statements, and gave the counsel, declared in this letter of January 8th, they believed he had come forth from your own bosom and had acquired knowledge of plans to ruin them. If the statements of this letter were made to them and the advice given them, it ought not to be thought strange that they should have believed it and acted on it ever since. I now believe that fact to be, that they are honestly acting upon a false view of your character and motives and your conduct in the transactions with them, based on the Titusville interview."

The second contemporary reference to the Merritts' belief in a conspiracy on the part of Rockefeller was found in a letter from Cotton to Gates on March 8, 1894. Cotton, who had just given up his work as the Merritts' lawyer and signed a contract to represent Rockefeller in Duluth, wrote in part: "The Merritts claim that Mr. Wetmore was the agent of Mr. Rockefeller and that there was collusion between C.W.W. and Mr. R and that as the result, they, the Merritts, were squeezed and defrauded out of their property, and that the Barge Co. was also in the deal. I understand now that Col. Harris the travelling lawyer from Kansas, is to be the attorney here and will be the main spoke in the wheel. . . . I do believe that the Merritts, headed by Leonidas and Alfred, and John E., are up to some mischief, and that they mean to do all the harm they can. . . . Knowing them as well as I do, I know they will tell their attorney only such facts as are in their favor, and will purposely conceal any thing which reflects on their ruinous way of doing business."[56]

Assuming for the moment that Wetmore and American Steel Barge were the instruments of the Merritts' fate, what light can be cast upon their maneuvers? The known activities of American Steel Barge are limited to a legitimate interest in transporting ore down the lakes in the firm's ships and to the five promissory notes of Leonidas and Alfred transferred to the company by Wetmore. After the court ruled upon the ownership of the collateral accompanying these notes, American Steel Barge did not retain the stock, selling it at public auction in partial payment of the $432,544 debt it claimed.

As for Charles Wetmore, accounts in the Merritt Papers show that from March through August, 1893, he borrowed over $834,100 from Rockefeller on ten separate occasions. Of this only $49,400 is listed in the form of demand loans. In addition to collateral in Merritt and non-Merritt companies shown as belonging to Wetmore, the accounts include 4,252 11/12 shares of stock owned by various family members and pledged against $405,760 of the loans listed by specific dates. Of these, 3,619 11/12 were shares in the various Merritt mining companies and 633 were DM&N stocks. The ownership of the Merritt collateral was divided as follows: 2,371 8/12 shares (including all of the DM&N) to Leonidas; 1,735 5/12 to Lewis J. and his son Hulett; 145 10/12 to Alfred. Wetmore himself put up a total of 17,770 DM&N stocks and bonds against these loans. Whether they truly belonged to him is, of course, a question.[57]

A second record in the Merritt Papers, dated November 1, 1893, is headed "Leonidas Merritt In acct with C. W. Wetmore." It lists loans made by Wetmore to the Merritts between February 7 and October 10, 1893, totaling (with interest) $592,735.38, of which $109,602.64 had been returned to Wetmore between July 20 and October 17, leaving a balance owed by the Merritts of $483,132.74. Against these loans Alfred and Leonidas had pledged 5,650 89/100 shares of DM&N stock.

A third account also present in the Merritt Papers shows the Merritt collateral pledged to Rockefeller on three loans totaling $150,000 obtained on July 21, 26, and August 15, 1893. Also shown is the stock put up for $29,450 loaned to Wetmore and Leonidas Merritt on July 31 and for an unspecified loan to the Mountain Iron Company obtained from Rockefeller on August 28, 1893. For all of these borrowings, the Merritt collateral totaled 13,776 shares in their various mining companies and the railroad. The family ownership of this collateral breaks down as follows:

Lewis J. and Hulett	6,213 shares
Leonidas	2,369 "
Alfred	2,265 "
John	1,364 "
Cassius and Andrus	1,011 "
Eliza M. Anna Matilda T. Thomas A. Jennie S. Eugene T. Napoleon B.	554 "
Total	13,776 shares

The companies and amounts represented are as follows:

	Missabe Mountain	Mountain Iron	Biwabik	DM&N	Minn. & Northern Townsite	
To Am. S. Barge	5	1,535		3,050		
By C. W. Wetmore				17,770		
By Merritts on C. W. W. loans	500	197	2,725	633	197	11/12
On Merritt loans	4,946	2,074	3,522	3,234		
TOTAL	5,451	3,806	6,247	24,687	197	11/12

It would thus appear that the stocks listed to Lewis and his son exceeded the number of shares pledged by the others, and that the latter's contribution consisted largely of shares in the DM&N, while those of Lewis were almost exclusively mining company securities (predominantly Missabe Mountain and Biwabik). On the basis of these lists, which cannot be verified, it seems clear that what was lost to Rockefeller on known loans to the Merritts and Wetmore was the Duluth, Missabe and Northern Railroad, and that Lewis and his son (who it will be recalled exercised their option to repurchase from Rockefeller) owned the largest single block of Merritt mining stocks pledged to the eastern financier. Thus even when the collateral held by American Steel Barge is taken into account, the question of how the Merritts lost their presumed control of the mines they had discovered remains unanswered.

Lewis Merritt, writing in 1897, placed the blame upon the brothers themselves, charging that the family did not have the stock to fulfill the terms of their August 28, 1893, agreement with Rockefeller. Lewis said that he told the others to "Look at the Stock Books and see for your selves and you will find the Merritts dont own 1/3 let a lone 51% of any mine on the range. I said every time the stock has boomed any you have sold off Stock and fir[e]d it into Some thing out side untill with what you have thrown to Suckers and men for their Influence you have Lost Control. . . . They still Insisted that I was crazy but don[e] as I told them to do look up Stock Records and came back to me and said you are right."[58]

If a conspiracy existed, the shadowy Charles Wetmore must have occupied a strategic corner in the tangled web, although he repeatedly denied having done so. His activities are impossible to trace more fully in the existing records. At least one Merritt spokesman laid much of the family's financial plight at his door. Charles Norcross, who had the co-operation of Andrus in preparing a manuscript on the subject in 1912, asserted that the mysterious loan called by Rockefeller represented Wetmore's "funded obligations" to him in the form of "demand notes, subject to call."[59] As we have seen, this statement is at least open to question. It is certain, however, that the notes held by American Steel Barge were called, and it is not clear what, if any, action was taken by Wetmore to recover the $483,132 supposedly still owing to him on November 1, 1893.

A somewhat different, but wholly speculative, explanation for the call loan and the conspiracy theory could be constructed around the character and personalities of the Merritts themselves. They had brought the Mesabi to market against great odds. They had built a railroad, well located and still in use, and with Rockefeller's help they had completed it to Duluth. Their devotion to that city was well known. Indeed their decision to build into Duluth was one underlying cause of their downfall. They were proud men, looked up to in their home state, respected and well liked. If Lewis Merritt's explanation is indeed correct, could the call loan story and the conspiracy charges have been an understandable rationalization for the brothers' earlier loss of the mines, for their inability to take advantage of the option Rockefeller offered, for their lack of understanding of the seriousness of the financial morass in which they found themselves in a period of low business ethics on Wall Street, and for their staggering losses to local creditors and friends whose notes they had cheerfully endorsed to badly overextend their resources? Perhaps in the end Alfred himself offered the best explanation we will find when he said, "I was a kind of an innocent sort of a duck."[60]

Historian Allan Nevins described the Merritt family as "inexperienced, ill-advised, and greedy." In its totality this is a rather harsh indictment. The Merritts may have lacked either the experience or the proper advisers to deal in eastern financial circles in a time of panic and depression, but a balanced interpretation of the evidence suggests no unduly avaricious motivations. Representative Stanley, chairman of the House committee that conducted the United States Steel hearings, described the Merritt brothers in truly sympathetic fashion when he said: "the people of Minnesota regarded these men, in a way, as we regard Boone in Kentucky, and as they regard Houston in Texas, with gratitude, with reverence." Somewhere between these two extreme appraisals no doubt lies the truth.[61]

In 1929 Hansen Evesmith, the administrator of Cassius Merritt's estate and a man who knew the family well, wrote that the "panic of 1893 capsized the Merritt enterprises as it did many other worthy developments. The business world was strewn with wrecks. They were caught crossing the stream, when the deluge of the panic overtook them. Had they not been gathered up by the Rockefeller interest, it would have been the Great Northern interest or some other financially strong interest." [62]

Notwithstanding their conflict with eastern capital, the members of the Merritt family served northeastern Minnesota to the best of their ability. Chiefly responsible for discovering one of the world's richest sources of iron ore, they risked everything to exploit the natural wealth of the region to profit themselves and northern Minnesota. Hansen Evesmith's appraisal seems sound. As entrepreneurs in a state only 35 years old in 1893, the family lacked the financial resources to survive the greatest depression the country had experienced to that time. Nor was the young state of Minnesota then sufficiently developed to supply the very large amounts of capital the Mesabi required. Family spokesmen to the contrary, in the final analysis the Merritts were not conquered by a conspiracy of eastern capital or a money trust. Instead they suffered defeat and financial loss because of inexperience, poor advice, and inability to deal in the treacherous industrial world of the 1890s.

In a thoughtful editorial printed about the time the Merritts began their day in court, the *Duluth News Tribune* of May 16, 1895, commented: "No one will underestimate the debt of Duluth to the Merritts. It is due to them and their associates in great measure that the Mesaba range had its development. Their faith gave St. Louis county a new iron range and their loyalty brought the railroad and the ore dock to this city. If their loss was the result of false statements in which they trusted too much to others and if it can be proven that Mr. Rockefeller was party to the transactions Duluth will be pleased to see the truth prevail.

"But if on the contrary it is true as claimed, that the Merritts went into the various deals with Mr. Wetmore with wide open eyes, that Mr. Rockefeller entered later into the transactions . . . in a legitimate business way . . . and that the Merritts' loss of the control of these properties and their financial misfortunes were due to private debts and speculations other than those of the road and mines a new phase of the question will be presented.

"If this should be the finding. . . , though it would in no way lessen either respect or sympathy for the Merritts, it should mean an end to prevalent criticism of Mr. Rockefeller. Today Mr. Rockefeller controls one of Duluth's most important branches of commerce. . . . When litigation is ended . . . Duluth might do well to remember that Mr. Rockefeller is here, that his influence may easily be diverted, and that the loss would be serious."

Never again were the Merritts active on the Mesabi. Instead Leonidas, Alfred, and Andrus organized the American Exploration Company and spearheaded numerous expeditions to search for copper and silver rather than iron. During the first two decades of the 20th century they wandered in Idaho, Oregon, New Mexico, Canada's western provinces, and Mexico's northern states. Yet all three retained their permanent residences in Duluth. Lewis and his family moved to California, where he and his son became wealthy after selling their Consolidated stock to United States Steel. Cassius died in 1894, and both Leonidas and Alfred died in 1926. Andrus then moved to Kansas, Nevada, and finally California, where he died in 1939. All four men are buried in Duluth. For his part, as the 19th century gave way to the 20th, John D. Rockefeller, possessing capital and administrative genius, played a continuing role in the economic development of Minnesota's iron mining industry.[63]

|9|

OF KINGS, PRINCES, AND INDUSTRIAL EMPIRES

THE CREATION of the Lake Superior Consolidated Iron Mines, with its Mesabi mining properties and the Duluth, Missabe and Northern Railroad, was only the first of a series of moves toward combination in the iron mining and steelmaking industries. Catching the winds of change as early as November 12, 1893, the *Duluth News Tribune* remarked that "the large and promising lands, old and new, in the mining region continue to drift into the hands of big corporations that can furnish the money to keep them up in times of depression. . . . Stock in these mines will be the stock of rich men in the future."

In the last half of the 1890s four major firms with large capital resources quickly acquired most of the valuable ore deposits necessary for the manufacture of steel. At the heads of these companies were four more or less unenthusiastic iron kings whose names were well known in other phases of American industry — John D. Rockefeller, Andrew Carnegie, Elbert H. Gary, and James J. Hill. Among their princes and prime ministers were Frederick Gates, the three Henrys — Frick, Oliver, and Porter — and James N. and Louis W. Hill. By the end of the 19th century this group of ten men emerged as the leading entrepreneurs on Minnesota's Mesabi and Vermilion ranges as well as in the older south shore mining areas of Michigan and Wisconsin. But the following year saw another giant — J. Pierpont Morgan, the controlling force in American banking — take a hand, an unbeatable hand as it turned out, that led to the formation of the United States Steel Corporation and still further consolidation on the Lake Superior ranges.

Leading the march toward a new era in the history of iron mining was a professedly reluctant John D. Rockefeller. Writing in 1909, he described how the panic and depression had involved him in the iron business: "We had invested many millions [in the Merritt enterprises], and no one wanted to go in with us to buy stock. On the contrary, everybody else seemed to want to sell. The stock was offered to us in alarming quantities — substantially all of the capital stock of the companies came without any solicitation on our part . . . and we paid for it in cash."[1]

The operating head of Rockefeller's iron ore venture was Frederick Gates. In February, 1894, he replaced Leonidas Merritt as president of the Lake Superior Consolidated, and on October 16, 1894, he also assumed the presidency of the Duluth, Missabe and Northern Railroad. Gates seems to have been somewhat more successful at railroading than he was at mining. Mesabi veteran John Hearding, observing him in action on the range, commented that Gates "could dig money but not iron ore." Leonidas Merritt maintained that the Consolidated failed to acquire the best mining properties. "Their experts sent on the ground condemned practically the whole Hibbing area," he wrote, "and also threw out many of the leases we had obtained there and elsewhere. . . . Under the management of Gates, the holdings of the Consolidated . . . shrunk in comparative value." Although both men exaggerated, there is some truth to these charges. The Lake Superior Consolidated missed the greatest open-pit mine of them all — the Mahoning at Hibbing — and it failed to gather in such other large ore shippers as the Fayal and Stevenson. On the other hand, it acquired leases on the Burt, Day, Hull-Rust, and Sellers mines near Hibbing, the Duluth near Biwabik, the Ohio near Virginia, the Pillsbury at Chisholm, and the productive Spruce at Eveleth.[2]

In 1894 Gates placed the Mesabi mining companies under the immediate direction of William J. Olcott, an experienced mining superintendent from Ironwood, Michigan, who, said the *Duluth News Tribune*, "has represented Mr. Rockefeller on the Gogebic, but will now come to Duluth." The treasurer of the subsidiary mining firms during that transition year was Napoleon Merritt. He was replaced by Charles Scheide in 1895. As for the Duluth, Missabe and Northern Railroad, its directors struggled throughout 1894 to get the company's house in order, untangle the confusion in its relations with Charles Wetmore, and settle various claims and lawsuits, all the while continuing to build with money procured from the sale of second mortgage bonds to Rockefeller. Issued in 1894, these bonds had a par value of $1,900,000, on which the road realized $1,140,000 in cash. Until October the DM&N had no designated president, but the work went forward under the direction of vice-presidents Anthony D. Allibone and Alexander McDougall, the whaleback inventor who had been re-

sponsible for introducing the Merritts to Wetmore. Napoleon and
Hulett Merritt were also among the DM&N directors elected in February, 1894, and Donald M. Philbin, a lifelong railroad man, became
general manager. During that year, the road extended branches to the
Minnewas, Franklin, Adams, and Rathbun mines; in 1895 it reached
the Sellers and Pillsbury, while also completing the construction of a
second ore dock at Duluth.[3]

On February 5, 1895, after the Merritts failed to take up the repurchase option offered them by Rockefeller, the Consolidated's mining and railroad subsidiaries held their annual meetings. The reorganized boards of directors revised the company bylaws to provide
greater uniformity and increased efficiency. Gates retained the presidency of both the Consolidated and the DM&N; Hulett and Napoleon
Merritt and McDougall were again among the seven elected railroad
directors; and Lewis J. Merritt, who had not joined the family in its
fight with Rockefeller, was named to the Consolidated's board. Other
members of the Merritt family were replaced by Rockefeller men.[4]

Under the Merritts' direction in 1892–93 the railroad's gross earnings had totaled $90,363; its operating expenses were $237,406, leaving
a deficit of $147,043. In addition the firm spent $4,687,439 for equipment and the construction of 54.3 miles of track. In June, 1893, when
its fiscal year ended, its mileage totaled 85.2. All freight carried
amounted to 53,534 tons, very little of which was ore. In the transition
fiscal year of 1893–94 when the road was partially under Rockefeller
management, gross earnings were $743,805, operating expenses
$433,461, producing a surplus of $310,344 that allowed the DM&N to
retire completely the earlier deficit and show a profit of $162,984. In
this fiscal period the firm spent an additional $3,296,898 for equipment
and for the construction of the first Duluth dock and 47.9 miles of
track. Reflecting the opening of additional mines on the range, the
DM&N carried 972,395 tons of freight, of which 909,499, or 93.5 per
cent, were listed as iron ore.[5]

This rapid turnabout did not continue into the next two years when
the depression had its greatest impact. Although gross earnings rose in
fiscal 1894–95 to $1,378,268, as the tonnage carried nearly doubled to
1,802,257 (of which 1,704,679 tons represented ore), the company's
net surplus dropped to $102,625 in June, 1895, and the following year
fell to $9,163 on 7,330,822 tons of freight (of which 2,226,168 were
ore). During these years, the road actually functioned with a deficit of
expenses over income in 1894–95 of $60,359 and in 1895–96 of $93,461.
While interest paid on bonds and other notes remained fairly stable,

taxes increased from $14,159 to $27,548. Despite its declining surplus, the company spent $397,161 for construction and equipment in 1894–95 and $656,353 in 1895–96, when it built the second ore dock at Duluth. By June, 1896, it owned 136.5 miles of track. These statistics demonstrate how capital resources enabled Rockefeller to expand the transportation system on the Mesabi during a harsh depression by foregoing immediate profits — an enviable state of affairs beyond the reach of small entrepreneurs like the Merritts.[6]

Although he operated from a position of great economic strength, Rockefeller never at any time controlled all of the Mesabi. By the spring of 1894 the New Yorker was facing stiff competition from another wealthy eastern entrepreneur — Andrew Carnegie. The Scotsman had visited the range in the fall of 1893 and returned home unimpressed. The *Duluth News Tribune* of October 1, 1893, reported, "There was no special significance in the trip. Mr. Carnegie is not and never has been a mining man. You couldn't sell him mining shares for songs." Carnegie, who had emigrated from Scotland as a boy of 13 in 1848, got his start on the path to riches by working for the Pennsylvania Railroad and saving his money. During the Civil War he picked up investments in various iron companies making products needed by the railroads, and by the end of the war he was managing four such suppliers. Concentrating his attention on iron manufacture, railroading, bridge building, and the marketing of bonds through J. Pierpont Morgan's private bank, Carnegie gradually branched out. He traveled to England, where he met Henry Bessemer, and, impressed by his process, returned to organize a company in 1872 to manufacture steel rather than iron rails. To do so he built his first steel plant near Pittsburgh and became the operator of the first modern Bessemer furnace erected in the United States. Continuing to expand, Andrew Carnegie by 1880 dominated the nation's steel industry. In that year he bought the American rights to the Thomas-Gilchrist open-hearth patents and introduced the new process in his mill at Homestead, Pennsylvania, in 1888. He supplied funds in 1890 for the first of the many free public libraries for which he became famous and continued to increase the steelmaking capacity of Carnegie Brothers & Company, dismantling Bessemer converters and replacing them with open-hearth furnaces throughout the depression.[7]

The catalyst who catapulted Carnegie into a major role in iron mining was Henry Oliver. It will be recalled that this persuasive Pittsburgh industrialist had gained his initial foothold in northeastern

Minnesota when he leased the Cincinnati and Missabe Mountain mines from the Merritts and others in 1892. Despite a chronic shortage of funds, Oliver on August 5, 1892, incorporated the Standard Ore Company, capitalized at $1,500,000, and turned over to it the Cincinnati operation. Then on September 29, he organized the Oliver Mining Company, capitalized at $1,200,000, and assigned the Missabe Mountain lease to it. The economic downturn caused by the panic forced Oliver's steel firm (but not his mining companies) into receivership for two years — a temporary setback that scarcely slowed the engaging Scotch-Irishman's progress.[8]

Notwithstanding the intensity of the depression and the prevailing prejudice of furnacemen against Mesabi ores, the Oliver Mining Company successfully developed the Missabe Mountain property without delay. Under the direction of mining captain Edward Florada, the mine shipped 123,040 tons of ore in 1893 and 505,946 the following year. In March, 1893, the Missabe Mountain Iron Company (in which Lewis and Hulett Merritt held large interests) declared a stock dividend — the first to be paid from income by any Mesabi Range firm.[9]

At the end of the 1894 shipping season Oliver leased another property, the Lone Jack Mine near Virginia, which Rockefeller had previously acquired as part of the Lake Superior Consolidated. By the terms of this lease Oliver agreed to mine a minimum of 100,000 tons per year for a royalty of $.25 per ton, well below the previously accepted rate of $.50 to $.65. Once again Oliver's mining operations were successful; the Lone Jack supplied 389,338 tons in 1895 and increased its shipments by 75 per cent to 681,957 tons the following year.[10]

From the beginning of his involvement on the Mesabi, Oliver realized that he could not rely solely on his own inadequate funds. But he was the first eastern steel man to grasp the important economic advantage to be gained by controlling his own supply of raw materials. He also knew that developing iron mines, if past experience were any guide, required large outlays of capital. After obtaining his first leases in 1892, Oliver began to look about for financial assistance. Thus he approached Andrew Carnegie, a boyhood friend who had helped him get his first job as a 13-year-old messenger boy in the busy Atlantic and Ohio Telegraph Company's office in Pittsburgh and who now controlled the nation's largest steel-producing firm.

The two men were as unlike in temperament as fire and water, and Carnegie professed little respect for Oliver, whom he regarded as a speculator. Thus Oliver's approach to the doughty Scotsman was indi-

rect. He went first to the offices of Henry Frick, chairman of Carnegie
Steel and a fellow Pennsylvania industrialist. Perceiving the advantages of coke over coal for steelmaking, Frick had entered the business
world at the age of 22 in 1871, when he acquired coal lands and
pioneered the building of coke ovens at Connellsville, Pennsylvania,
with borrowed funds. Ten years later, after Frick had made his first
million dollars, Carnegie began to buy stock in the H. C. Frick Coke
Company. As coke became more and more important in open-hearth
steelmaking in the 1880s, he increased his holdings. On July 1, 1892,
he transferred them to the Carnegie Steel Company, Limited, a
merger of all the Carnegie interests, of which Frick was the operating
head.[11]

When Oliver arrived at Frick's office in August, 1892, the chairman
was recovering from Alexander Berkman's attempt to assassinate him
on July 23 during the bloody strike at Carnegie's Homestead Mill, the
first great open-hearth plant in the United States. Despite his concern
with the labor turmoil threatening him, Frick expressed great interest
in Oliver's Mesabi investments. When the Frick and Carnegie enterprises had been combined, Carnegie Steel had emerged with large
supplies of coke and limestone as well as impressive steelmaking capacity, but it did not control the ore supplies necessary to produce its basic
product. The acquisition of those supplies in the 1890s, wrote Joseph
L. Wall, changed Carnegie's "company from a highly successful business into an industrial empire." Both the Oliver and Carnegie steel
firms needed guaranteed supplies of ore and by working in combination could probably obtain lower transportation rates. Moreover, the
development of the Thomas-Gilchrist process had made open-hearth
furnaces practical from the cost point of view, and it was no longer
necessary to seek only the high-grade, low-phosphorous ore required
by Bessemer converters. The ores of the Mesabi thus became doubly
appealing.[12]

Oliver offered Carnegie Steel a half interest in his mining company
in return for a $500,000 mortgage loan to develop his mines. Con-
vinced that such a transaction presented a great opportunity for future
profit, Frick notified Carnegie, expecting enthusiastic approval. The
chairman stressed that the steel firm could obtain a mining company
and a portion of the ore supplies it needed merely by granting a loan.
No speculative investment was required. Carnegie refused, holding
firmly to his often-expressed theory that "pioneering don't pay." From
his retreat in Scotland, he wrote Frick that "Oliver's ore bargain is just

like him — nothing in it. If there is any department of business which offers no inducement, it is ore. It never has been very profitable, and the Massaba is not the last great deposit that Lake Superior is to reveal."[13]

As chairman of Carnegie Steel, Frick possessed adequate authority to complete the negotiations with Oliver, but he persisted for many months in his efforts to obtain the approval of the company's founder. Carnegie continued to resist, expressing dissatisfaction with both the proposal and Oliver's ability to manage mining operations, "good fellow though he is." Carnegie thought it best to let others assume the risks of the ore business, while his company devoted itself to perfecting mechanical equipment and an efficient organization. He "was not enthusiastic," wrote Wall, "about taking advantage of the greatest opportunity of his entire business career."[14]

The visible presence of Rockefeller on the Mesabi by early 1894 initiated a slow and reluctant retreat from this negative position, not because Carnegie was convinced of the merits of Oliver's proposition, but because of the competition Rockefeller could offer. In March, 1894, the Scotsman wrote Frick: "Oliver hasn't much of a bargain in his Mesabi, as I see it, but in view of threatened combination it is good policy to take the half as independent of its intrinsic value; it gives us a wedge that can be driven in somewhere to our advantage. . . . In less strong hands, the Oliver would be squeezed. Remember Reckafellow [*sic*] & Porter [*of the Minnesota Iron Company*] will own the R.R. and that's like owning the pipe lines — *Producers* will not have much of a show.

"Taking half with Oliver means we have all the risk, must furnish all the capital . . . besides, Oliver isn't a good manager. . . . It's a pity we have to go in at all, still I cannot but recognize we are right in flanking the combination as far as possible."[15]

Over a decade later in testimony before the Stanley Committee investigating United States Steel, Carnegie tempered his attitude toward Oliver. Where he had earlier mistrusted his "mercurial and overly expansive" boyhood friend, the steel tycoon now described him as "one of the brightest men Pittsburgh ever could boast of, and he saw far ahead, and went up to that region and loaded himself with ore leases that he purchased for a song." The profits Carnegie's company reaped from its alliance with Oliver doubtless mellowed his earlier attitude, but it is noteworthy that the Scotsman's autobiography contains not one word about Henry Oliver.[16]

At last in the spring of 1894 the Oliver Mining Company formally transferred 6,000 shares of its stock to Carnegie Steel. With new financial support and the right to supply ore for both the Carnegie and Oliver furnaces, the mining firm survived the depression years without major setbacks. At the same time that Rockefeller's Consolidated was securing additional ore leases, Oliver, with Carnegie Steel's backing, rapidly emerged as a major competitor. Both groups were establishing fiefdoms in northeastern Minnesota, and both possessed the financial strength to operate in the depressed iron ore market.[17]

Newspapers and iron-trade observers prepared to enjoy the battle of "the ruler of oil against the ruler of steel." Throughout the nation's financial centers rumors spread that Rockefeller was actively trying to purchase land between Cleveland and Duluth for the construction of a giant new steel mill. Fully aware of the negotiating value of such stories, Rockefeller and Gates did not publicly disclaim them. In actuality, however, Carnegie was not seriously worried, and Rockefeller had no intention of entering the production phase of the steel industry. Rather he seems to have been more interested in the water transportation of ore — an interest that had made him an early investor in the American Steel Barge Company. Rockefeller reportedly increased his holdings in the barge firm in 1894 and later organized the Bessemer Steamship Company as an outgrowth of his investments. The *New York Times* of December 9, 1895, announced that the financier had ordered "eight monster steamships" and ten ore carriers, "two for Pickands, Mather & Co., who are Rockefeller's brokers, and eight for Rockefeller," adding "it is hardly likely that a thoroughly organized trust to handle the product of all the leading producers and crush out small competitors is more than a few years away. If conditions the next four years follow their usual course, the beginning of the next century is likely to witness such a combination already consummated."[18]

It did not take nearly that long to effect a mutually satisfactory merger. A year later, at the end of the 1896 shipping season, news of just such an arrangement between Carnegie, Oliver, and Rockefeller shocked the iron ore trade. It was Oliver who initiated the negotiations with Gates that led to the agreement. Although Carnegie had earlier expressed doubts about the ability of Gates and Rockefeller to operate successfully in mining, he put them aside and addressed the New Yorker directly when the merger discussions seemed about to break down. "Our people have been conferring with your Mr. Gates upon an alliance which would give us all the ores we can use from your properties," Carnegie wrote. "The differences between the two seems to have been so great as to cause a failure of the negotiations. They came to see me today and explained these differences, which do not seem to me too irreconciliable [*sic*], if both parties realized as I do, the mutual advantage of such an alliance, and were prepared to meet each other halfway.

"When Mr. Gates submits the matter to you, as I suppose he will, and you concur in this, I believe you and I could fix it in a few minutes, and I shall be very glad to go and see you if you think it worth while to take the matter up. It is a big operation, and needs to be looked at in a broader light." [19]

The terms of the alliance, formalized on December 9, 1896, demonstrated that both sides knew the value of compromise. [20] As part of the Carnegie firm, the Oliver Mining Company leased all the iron ore properties held by Rockefeller's Lake Superior Consolidated at a low royalty of $.25 a ton. Carnegie-Oliver agreed to ship a minimum of 600,000 tons annually from the Rockefeller leases and another 600,000 tons from the original Oliver properties. Together these included three of the four most productive operating mines up to that time — the Mountain Iron, the initial Merritt holding, which had been in operation the longest and had shipped the greatest ore tonnage; the Lone Jack, an Oliver lease in production only two years, which was a close second; and the Missabe Mountain, which then ranked fourth on the range after the Biwabik. [21]

The agreement further provided that all the ore was to travel from the mines to Lake Superior via the Duluth, Missabe and Northern at an initial freight rate of $.80 per ton, a charge that remained constant through the 1911 shipping season. Rockefeller's Bessemer Steamship Company would then carry the ore to Lake Erie ports at a rate determined by the open market. In 1898 and 1899 this amounted to $.60 cents per ton. Finally, by the terms of this 50-year contract, Carnegie-Oliver promised not to buy or lease any additional Mesabi property without Rockefeller's permission. For his part, the head of Standard Oil agreed not to manufacture steel.

At first glance the alliance seemed to favor Carnegie, but a closer analysis reveals that each participant received substantial benefits. Rockefeller gained the most desirable iron ore customer in the nation at a price that, although low, provided an excellent return on his investment. He also acquired guaranteed cargoes for his railroad and his lake steamers. Carnegie-Oliver secured a relatively cheap ore supply independent of market deviations and eliminated a potentially dangerous competitor in the steel industry. Putting aside his former doubts about Rockefeller's ability to manage mines, Carnegie now described him as "the coming man in ore." He told the Carnegie Steel board of directors that, since Rockefeller's "ownership in ore lands will be very large, I believe it will be more to our advantage to mine ore in his territory, paying him a royalty, than to attempt to purchase ore property for ourselves." On December 23, 1896, a *New York Times* editorial predicted, "We shall be surprised if this transaction shall not perceptibly promote the establishment of American supremacy in the iron industry of the world." And that is exactly what it eventually did.[22]

News of the Carnegie-Rockefeller consolidation demoralized the remaining independent producers. They seemed especially threatened by the competitive advantages the new alliance possessed in transportation rates, in control of its own ore market, and in the minimal royalty payments required. By 1898 the *Duluth News Tribune* quoted one mining man who said the big operators had "decided the time is now ripe to gather in all available properties. . . . freeze out has been the game for the purely independent miner." Rail freight rates charged by the two Minnesota lines varied from road to road. Between 1892 and 1911 the Duluth, Missabe and Northern charged $.80 per ton from all Mesabi mines to Duluth, while its major competitor, the Duluth and Iron Range, charged $1.00 per ton during this period for the haul from Ely to Two Harbors. Lake freight rates showed greater variations. The levy from Duluth, Superior, and Two Harbors to Lake Erie ports fell from $1.25 per ton in 1892 to $.80 in 1895. It rose to $1.05 the following year, only to drop to a new low of $.70 in 1897. Thereafter, for the last two years of the decade, the rate settled at $.60 per ton.[23]

Ore prices for both of Minnesota's ranges testified to the effects of the Rockefeller-Carnegie combination and substantiated the worst fears of the independents. In 1896 Mesabi Bessemer ore, the highest grade, brought an average of $3.50 per ton at Lake Erie ports. The following season, the first to be affected by the merger, the average price fell to $2.25 per ton, where it remained until 1899 when it rose slightly to $2.40. Although the Carnegie-Rockefeller consolidation did not at first own any Vermilion property, the Lake Erie price for top grade Vermilion Bessemer fell from an average of $4.25 per ton in 1896 to $2.92 the following year. It rose slightly to $3.13 in 1898 and increased another $.22 a ton during the 1899 shipping season.[24]

The Carnegie Steel Company took full advantage of its competitive supremacy to acquire additional mining land on the Vermilion, Gogebic, Marquette, and Menominee ranges. "The very success of the Carnegie-Oliver-Rockefeller negotiations had an effect upon the ore properties of the northern ranges which neither Carnegie nor Oliver had anticipated but which was to extend their control over the ore market. . . ," wrote historian Joseph Wall. "Shareholders were eager to sell out before the Carnegie-Rockefeller combination should reduce prices to a ruinous level." Aided by the deflated panic conditions, Oliver set out to purchase shares of ailing mining companies at ridiculously low prices. During the summer of 1897 he secured options from hundreds of individual stockholders either to purchase or lease iron lands. Yet Carnegie hesitated to pursue with speed the chance to acquire them. Fully aware of his attitude, Oliver on July 27, 1897, sent a lengthy letter through Frick to the Carnegie Steel board, hoping to convince the directors of the opportunities the company could seize with relative ease. Once more demonstrating unusual foresight, Oliver wrote: "I propose at a risk of using our credit to the extent of $500,000, or possibly one million dollars, to effect a saving, in which our competitors will not share, of four to six million dollars per annum. . . . On the Gogebic Range, the mines I have selected comprise over 80% of developed ore or 'ore in sight.'. . . . The possession of a large body of ore in the Gogebic Range will strengthen our position, in holding the Rockefeller people down to low freight rates from the Mesaba Range. . . .

"I am not ignoring the strong position we hold on the Mesaba Range. . . . More Mesaba ores can be used in our mixtures, but it is not a wise policy to quickly exhaust the rich quarry we have on the Mesaba. . . . Although we are mining it at present for less than five cents per ton for labor, we must look to the future. . . . We should rather prolong the period of cheap steam shovel mining, take in the other Range properties . . . and, by working one Range against the other, keep down costs of freights. . . . Now let us take advantage of our action before a season of good times gives [rival] ore producers strength and opportunity to get together by combination."[25]

Frick transmitted this letter to Carnegie in Scotland and the company's founder softened his objections so far as to allow board members to make the final decision. The nine directors unanimously authorized Oliver to secure all available options. They bowed, however, to Carnegie's insistence that Oliver should lease rather than purchase as many properties as possible, a process that would require a smaller outlay of capital.[26]

Oliver used this approval to obtain two of the most valuable south shore mines. Rockefeller had previously acquired the lease of the Tilden Mine on the Gogebic Range near Bessemer, Michigan, from

Thomas D. Davis, head of a regional mining group known as the Keweenaw Association. Oliver easily arranged to sublease this property (which had figured in Alfred Merritt's lawsuit) at a royalty of $.25 per ton based on a minimum annual shipment of 400,000 tons. Opened in 1891, the Tilden had quickly become the second largest producer on that range. Production reached 418,188 tons in 1895, and it shipped a total of 1,551,249 tons through 1897.[27]

Acquiring the Gogebic's largest producer proved to be a more difficult task. The Norrie Mine near Ironwood, Michigan, had first shipped ore in 1885; three years later it surpassed all the other mines on that range. By the end of 1897 it had shipped a total of 6,880,122 tons, with a high of 985,216 in 1892. Although it, too, was controlled by the Keweenaw Association, the Norrie had more than 300 individual stockholders. Faced with great expense to replace obsolete equipment during a declining ore market, the Norrie directors were prepared to deal with Oliver. Difficulties arose, however, when he discovered that the shareholders wanted to sell rather than lease the mine.[28]

For several months during the summer and fall of 1897, the leaders of Carnegie Steel argued among themselves about Carnegie's previous insistence on leasing ore property. Charles M. Schwab, the firm's dynamic young president, favored buying the mine; Frick did not. On September 25, 1897, only five days before the option expired, Oliver sent a personal appeal to Carnegie in Scotland. "I am distressed at indications here that Norrie options expiring on Monday are to be refused," he cabled. "It would be a terrible mistake. . . . I could not possibly secure these options again at fifty per cent advance. . . . I will guarantee . . . to return in profits every dollar we invest in two years. Do not allow my hard summer's work to go for naught."

Carnegie relented. He cabled the directors, meeting in emergency session: "Always approve unanimous action of board after full expression of views. Sure leasing the true policy but if board decides this exception all right." The board voted unanimously to buy all of the Norrie stock on September 27, three days under the wire. The purchase made Carnegie Steel "self-sufficient"; never again "would it have to buy a ton of Bessemer ore on the open market," wrote Wall. "Oliver's victory was complete, but never did a man have to work as hard to present a fortune to others."[29]

Carnegie was not unappreciative. In a Christmas letter to Oliver, he wrote: "I don't flatter you when I say that I do not believe there is a man in the United States who could have done what you have in regard to the Norrie. . . . Your genius as a negotiator is unequalled." But he could not resist adding, "What I sometimes think Providence has re-cently been kind to you in, is providing a partner or two with less brilliant, but equally as needful qualities. You just need to be held in a little, now and then — you know you do."

At the same time that Oliver was steadily accumulating iron lands in Michigan, he also extended the competition to the Vermilion, where the Duluth and Iron Range Railroad offered the only transportation link between the mines and Lake Superior. By January, 1898, Oliver leased Section 26 on the Vermilion, and two months later he acquired the Zenith Mine. When that mine's stockholders accepted his of-fer, the *Duluth News Tribune* of March 26, 1898, quoted one of them as saying: "We sold out for enough to pay our debts. . . . Low prices for ore and the exorbitant transportation charges by the range roads have made it impossible to operate the mine," which had shipped a total of only 29,379 tons in 1892 and 1893 before closing for two sea-sons. In July Oliver finally gained control of the Pioneer Mine at Ely from a syndicate that included Wisconsin Senator John C. Spooner. (In 1897 his purchase offer had been refused by Pioneer stockholders.) Although this mine had begun operations in 1889, it had never shipped more than 12,000 tons annually before 1893 and 1894, when it shut down completely. The Pioneer experienced its greatest season in 1897, shipping 207,123 tons to Lake Erie ports. In the next two years Oliver also leased the nearby Savoy and Sibley mines.[30]

The irrepressible Oliver, in violation of Carnegie Steel's contract with Rockefeller, also leased and bought several mines on the Mesabi Range. On April 7, 1898, the *New York Daily Tribune* reported that he was negotiating for the Franklin, Duluth, and Pillsbury mines. He successfully leased the Stephens and purchased the Rouchleau out-right. When Gates discovered this fact in September, 1899, he pro-tested the violation vigorously. After months of dickering, the matter was adjusted by turning the Rouchleau and Stephens over to Rockefel-ler's Lake Superior Consolidated, in return for leases to Oliver of the Stephens, Ohio, and Shaw mines. Not the least nonplused, Oliver said he was "delighted that our plan of campaign has worked out to such a favorable result."[31]

Although Carnegie had resisted outright purchase of the Norrie, once he experienced the benefits of possession, he wanted to own rather than lease the Pioneer Mine on the Vermilion. Since the agree-ment included an option to purchase, Carnegie dispatched Oliver to reopen the negotiations. Senator Spooner convinced his partners that a sale offered immediate profits, whereas continuation of the lease could leave them without income should Carnegie-Oliver decide to cease operating the mine. "Probably there is no danger that they would throw up the lease," he wrote, "but I learned a long time ago not to put much faith in Princes." On July 19, 1898, the Oliver Mining Company purchased a majority of the Pioneer's stock in a transaction that *Iron Age* pronounced "the most brilliant of those carried through" by Carnegie-Oliver.[32]

When news of this purchase reached eastern iron ore markets, observers expressed surprise that no specific arrangements had been made to transport the ore from the Vermilion Range. *Iron Age* wondered if Carnegie-Oliver was "relying upon the magnitude, quality and cheap rate of extraction of their ore bodies" to force the Duluth and Iron Range to offer a mutually acceptable freight rate. Or perhaps, the journal theorized, Carnegie-Oliver had "so large a tonnage to offer that the construction of an independent road would not meet serious obstacles so far as financing is concerned."[33]

The Oliver Iron Mining Company fanned the flames of speculation by failing to deny reports of the development of a new iron port at Grand Marais, on the north shore of Lake Superior about 90 miles northeast of Two Harbors, and construction of a rail line between that harbor and Ely. On September 5, 1899, seven Carnegie Steel leaders, including Oliver and Chester A. Congdon, his Duluth attorney, incorporated the Virginia and Ely Railroad Company. The organizers expressed a desire to build a rail link that could be used by all Carnegie properties on both the Vermilion and Mesabi ranges even though such a plan would again violate their contracted alliance with Rockefeller. "The country from Grand Marais to Ely is exceedingly rough. . . ," commented *Iron Age*, "and no one would think seriously of constructing a road if he could get such rates over existing lines as might prove reasonable. . . . A new road, therefore, might be a veritable bull in the china shop once it started to be vigorously competitive." In reality Oliver never seriously contemplated laying new tracks through the rugged uplands of St. Louis, Lake, and Cook counties, but the rumors helped keep rail rates on both ranges at a steady, acceptable level.[34]

The significance of these moves was not overlooked by the Minnesota Iron Company, the Duluth and Iron Range's parent firm, for by the late 1890s the Carnegie-Oliver-Rockefeller combination controlled not only the Duluth, Missabe and Northern but also "through lease or ownership 34 working mines, including 10 mines of the Marquette range, 7 on the Gogebic, 5 on the Menominee, 4 on the Vermillion [*sic*], and 8 on the Mesabi" — a respectable proportion of the iron-bearing properties on the five Lake Superior ranges. It will be recalled that the Minnesota Iron Company, a syndicate of Chicago and New York industrialists headed by Henry Porter, had purchased Charlemagne Tower's holdings and the Duluth and Iron Range Railroad in 1887. Beginning in 1892 the well-established firm had challenged the Merritts' control of the Mesabi by extending its rail lines to working mines there whose operators had not signed contracts with the rival Duluth, Missabe and Northern. Throughout the depression years the Duluth and Iron Range continued to build, and, like Rockefeller and Carnegie-Oliver, Minnesota Iron bought or leased additional ore properties at greatly depressed prices. By 1897 the Duluth and Iron Range not only dominated the Vermilion, it was also operating 59 miles of track on the Mesabi, and its parent firm controlled the successful Auburn, Canton, and Norman mines there.[35]

Although the Minnesota Iron Company did not have large holdings of iron-bearing lands on either Minnesota range, the firm had remained strong through the years of varying ore prices. Despite the growing competition, it possessed all the ingredients necessary to operate successfully — mines, railroad, harbor, and steamers. Its rail lines extended into Duluth, and it had exclusive use of a second lake port in the community of Two Harbors. By the mid-1890s four wooden ore docks capable of storing a total of 87,351 tons (an increase of 53,351 tons since 1885) stretched out into Lake Superior there. In addition the firm operated a fleet of nine steamers and seven steel barges in 1898, making it capable of transporting ore from its mines to customers on the lower Great Lakes. Henry Porter realized, however, that continued internal integration would not enhance Minnesota Iron's ability to compete, and he set out to merge with the Illinois Steel Company, Carnegie Steel's foremost rival.[36]

Incorporated in 1889, the Illinois firm brought together a North Chicago rolling mill, blast furnaces and an iron mill in Milwaukee, and steelworks at South Chicago and Joliet. Within five years it had become the principal midwestern steelmaker, with invested capital of $33,000,000 and annual shipments of over 1,000,000 tons of pig iron and 563,000 tons of finished products. Its production capacity was exceeded nationally only by Carnegie Steel. As both companies expanded their operations and their output, the competitive battles between them intensified. When Carnegie-Oliver enlarged its ore holdings and signed a traffic contract with Rockefeller, Illinois Steel purchased its raw materials from the Minnesota Iron Company.[37]

Elbert H. Gary, counsel for the Illinois steelmaker, convinced the board of directors that further consolidation would enhance its competitive position. He suggested that Illinois Steel buy the Minnesota Iron Company and thus acquire a guaranteed source of high-grade iron ore as well as a transportation network. Gary, a 52-year-old corporation lawyer, who was to become a power in the American steel world, was a native of Wheaton, Illinois. Since his graduation from the Union College of Law in Chicago in 1868, he had built a lucrative practice in the Windy City. He had also served as mayor of Wheaton and as a county judge from 1882 to 1890. His interest in steel developed as a result of an appointment to the board of directors of Illinois Steel. Porter, who served as a director of both Minnesota Iron and Illinois Steel, facilitated the merger process, and New York banker J. Pierpont Morgan agreed to finance the combination. Gary and Robert B. Bacon of Morgan's staff then formed a committee with authority to create a merger and to purchase other companies that would further strengthen the consolidation.[38]

The result of their labors was the formation of the Federal Steel Company in September, 1898. Capitalized at $200,000,000, the new consolidation included all the holdings of the Illinois Steel Company and the Minnesota Iron Company, as well as the Duluth and Iron Range and the Elgin, Joliet & Eastern Railway, a 180-mile belt line around Chicago connecting the Illinois firm's mills. Federal Steel also purchased additional plants at Lorain, Ohio, and Johnstown, Pennsylvania, which made it a factor in most phases of the iron and steel industry — ore, coal, coke, transportation, and various finished steel products.[39]

Gary became the president and moving spirit of Federal Steel. Under his direction the new aggregation took full advantage of its independent and integrated structure. The 1899 annual report showed that the company possessed 80 per cent of the coke needed to meet its furnace requirements, and that the mining division could provide just under 3,000,000 tons of iron ore annually transported by its own railroad and lake steamers. A statement of management policy affirmed that "It has never been the intention or desire of the company to secure a monopoly of any line of business. The plan is to own and control sufficient iron ore, coal, coke and limestone, and other raw products, to supply all of the mills of the constituent companies; to own and control adequate facilities for transportation, both on land and water; to manufacture and deliver finished steel, and to do all with the greatest economy."[40]

Thus by the time the depression lifted in 1898 most of the iron ore resources of northeastern Minnesota belonged to three large corporations: Rockefeller's Lake Superior Consolidated, the Oliver Iron Mining Company, and the Minnesota Iron Company. All three were linked to either Carnegie or Federal, the nation's leading steel manufacturers.

Before all these participants had time to get used to the new arrangements, a fourth major entrepreneur arrived on the Mesabi. He was James J. Hill, the Empire Builder of railroad fame. Born in Canada in 1856, Hill made his way to St. Paul when he was a few months shy of his 18th birthday. Starting as a shipping clerk for local agents of the Dubuque Packet Company, he successfully engaged in steamboating, fur trading, warehousing, commission work, and selling coal. He launched his epic career in railroading in 1878, with Canadian financial backing and money he had made in the Northwestern Fuel Company, by purchasing the bonds of the bankrupt St. Paul & Pacific Railroad. For the next two decades he showed little interest in Minnesota's growing iron-mining industry, turning down an opportunity to buy the Merritt properties in 1893. Concentrating upon rail transportation, especially the shipment of wheat and timber, Hill completed the Great Northern Railway Company's network across the northern Great Plains to the Pacific by 1893. The insistence of his oldest sons, James N. and Louis W. Hill, and their interest in two railroads — the Duluth and Winnipeg and the Duluth, Mississippi River and Northern — propelled James J. Hill into iron ore.[41]

In 1892 and 1893 Hill attempted to purchase the Duluth and Winnipeg, a move strenuously opposed by that road's controlling interests. It will be recalled that the Merritts terminated their contract with the Winnipeg line and built their own tracks into Duluth in 1893. During the panic the Duluth and Winnipeg fell on hard times. Then owned by the Canadian Pacific Railway Company, the Winnipeg firm went into receivership on October 13, 1894. Twenty-three months later, as part of a reorganization plan, the Duluth, Superior and Western Railway Company was incorporated; it acquired all the property and equipment of the bankrupt Winnipeg firm.[42]

Hill never lost interest in adding to his railroad network the approximately 100 miles of Duluth and Winnipeg track operating west of Duluth. While the reorganization progressed, Hill, through the Chase National Bank of New York and Kuhn, Loeb & Company, approached Sir William C. Van Horne, president of the Canadian Pacific. On June 2, 1897, the Great Northern paid $2,500,627.40, a sum that reimbursed the Canadian Pacific for its investment plus 4 per cent interest, in return for its holdings in the Duluth and Winnipeg, its successors, and affiliated companies. For the next few months, Hill continued to gather in additional outstanding securities. When all these sales were authorized by the Duluth, Superior and Western board of directors on June 21, 1898, the Great Northern owned a total of $3,385,400 in preferred and common stock and first mortgage bonds. This acquisition included an ore dock, elevators, and warehouses in Superior, as well as an estimated 10,000 to 12,000 acres of Mesabi land held by five subordinate mining companies. Without intending to do so, Hill at his sons' prodding was about to become a major participant in Minnesota's iron ore industry.[43]

Early in 1899 the St. Paul financier, again at his sons' request, concluded a second and far more significant railroad purchase that also included iron-bearing property. Located in the heart of the Mesabi Range, these tracts would in both value and extent form the backbone of Hill's ore holdings. He bought them from Ammi W. Wright and Charles H. Davis, who had been among the Michigan lumbermen active in acquiring pinelands in northern Minnesota in the 1880s. On March 21, 1892, they had incorporated the Duluth, Mississippi River and Northern Railroad, the first logging road in Itasca County. This standard-gauge line extended 46 miles from the Mississippi River at the mouth of Swan River near Jacobson to Dewey Lake, 12 miles north of Hibbing. In that same year Wright and Davis sold the timber on 45,000 acres of their pineland holdings to the Weyerhaeuser Timber Company, which then signed a traffic contract to ship it over the Duluth, Mississippi River and Northern. After that, the logging railroad handled over 1,000,000 square feet of logs annually. In 1895 it also began to carry iron ore.[44]

In so doing it tapped a bonanza, for the great Mahoning Mine at Hibbing opened that year. Operated by the independent Mahoning Ore & Steel Company, this large and extremely profitable property had escaped both Rockefeller and Carnegie-Oliver. The company was organized in 1893 by a group of Ohio men who had leased the land from Wright and Davis; its mine quickly became a leading Mesabi producer, shipping more than 500,000 tons in its third season. Wright and Davis built spur lines to serve the mine and invested their profits in timberlands near Lake Charles, Louisiana, and in the Pacific Northwest. By the end of 1898 the lumbermen had become so involved in these new locations that they announced a willingness to sell their Minnesota railroad and pinelands.[45]

At first Hill was not interested, but his sons were. James N. Hill, who was then president of the Eastern Railway of Minnesota, persuaded his father to visit the Mahoning in 1898. Firsthand inspection convinced the Empire Builder of the property's value, and he consented to negotiate with the Michigan timber barons. After returning to St. Paul, however, he learned that Rockefeller's Lake Superior Consolidated hoped to acquire this mine and railroad. Moving swiftly, Hill met with Wright and Davis in Chicago before Rockefeller became aware of his intentions. On May 1, 1899, the stockholders and directors of the Duluth, Mississippi River and Northern held separate meetings in the St. Paul offices of the Great Northern. The stockholders resolved that the company should "convey to the Eastern Railway Company of Minnesota, the title and ownership, in perpetuity, of . . . the said railway, together with all its equipment and appurtenances." Thus Hill bought the railroad, the Mahoning Mine fee, and, with Weyerhaeuser's consent, the pinelands and timber traffic contract for $4,050,000. The negotiators set a monetary valuation on each portion of the sale: railroad from Swan River to Hibbing $1,500,000; logging railroad from Hibbing north to Dewey Lake $300,000; 45,000 acres of land $2,250,000. Hill later said, "when I made that purchase the Mahoning Mine was in plain sight. There was no risk about it. . . . The Mahoning Mine was worth all I paid for it."[46]

Over 40 years later, Louis Hill, who succeeded his father as president of the Great Northern in 1907, wrote of the events that suddenly and spectacularly launched his family on the Mesabi. "During the 1890's when the Mesabi Range development got underway, my father . . . had little interest in iron ore or in the development of railroad freight tonnage from that source," Louis began. "He was reluctant to spend any substantial amount of the [Great Northern] railroad's money in the assemblage of ore properties through fee acquisitions and leases. On the other hand, I had a confident belief that iron ore lands would prove a valuable source of traffic to the railroad. I went to Duluth in 1895 and lived at the Spalding Hotel there for some three years. During that period, I secured for the railroad interests the bulk of the properties which later became the Great Northern Iron Ore Properties. . . ."[47]

"During those years in the middle '90s . . . I came to appreciate the important [sic] of getting the railroad established on the Iron Range so as to insure the freight revenues from the hauling of iron ore. At that time, the Canadian Pacific interests owned the Duluth and Winnipeg branch, and, recognizing the importance of this mileage as a means of access to the Range, I strongly urged my father to arrange the purchase of that branch line for the Eastern Railway of Minnesota. I had first mentioned it to him here in St. Paul, but thereafter he went on a trip to England, and I cabled him several times on the subject. The result was that he arranged over in England the purchase of this branch from the Canadian Pacific people, and, on account of its connection with the Wright-Davis Logging Railroad at Swan River, it provided access to Hibbing and the iron mines of that district. The Duluth and Winnipeg branch ran from Duluth to Deer River, and later on the gap from Deer River to Fosston was built by the Great Northern interests, thus completing the east-west line from Duluth to Grand Forks. . . . Incidentally, I was personally responsible for, and personally negotiated, the purchase of all of the acreage on Allouez Bay [in Superior] upon which today stand all of the elaborate ore dock facilities of the Great Northern Railroad."

Because of increasing iron ore traffic, the Great Northern leased on June 1, 1903, the track of the Duluth, Superior and Western Terminal Company. This firm, part of the original Duluth and Winnipeg organization, operated slightly more than five miles between Saunders and Superior, Wisconsin, where it connected with the ore docks of the Allouez Bay Dock Company. That same year a third Superior ore dock was completed providing a total storage capacity of 168,000 tons. On August 1, 1908, the Hill railroad took the next logical step and purchased outright the terminal company railroad (main track, side track, and yard) for $772,161.89.[48]

Louis Hill's 1946 reminiscences continued: "When my father, at my insistence, negotiated the purchase of the Duluth and Winnipeg branch from the Canadian Pacific interests, the railroad received in the bargain a minority stock interest in the North Star Iron Company of West Virginia. This corporation had been formed by the original pioneer Merritt family . . . as a holding company for mining and timber lands which they had acquired and which they considered to have great potential values. At that time, the controlling stock interest in North Star Iron Company of West Virginia was still held in the Merritt family. . . . After our purchase of the Duluth and Winnipeg branch, various members of the Merritt family and others, being financially hard up, sought to liquidate their remaining holdings in North Star Iron Company. . . . This resulted in our gradual acquisition for the Great Northern interests of approximately 91% of the stock of North Star Iron Company of West Virginia, which corporation later on in 1906 became one of the proprietary companies of Great Northern

Iron Ore Properties. The mines of North Star Iron . . . have been very important producers and shippers, and altogether have produced [*by 1946*] approximately 18,500,000 tons of iron ore, all of which has been hauled over the Great Northern Railroad. It can be frankly and honestly stated that the iron ore revenues of the Great Northern Railroad have served to distinguish it from so-called 'marginal roads' all through the depression period [of the 1930s] and since," Louis concluded.

With these two substantial transactions, James J. Hill arrived on the Mesabi Range, a development enthusiastically applauded by the Duluth business community. In strong contrast to Rockefeller's generally unfriendly local press, this home-grown millionaire received effusive praise. Even before Hill entered iron mining, the *Duluth News Tribune* had said on New Year's Day, 1893: "Duluth is fortunate in many respects but in none more so than in being the main lake port of such a line of road as the Great Northern, headed by such a man as James J. Hill. No man has done more to build up the transportation interests of the United States, none have earned a better name or have erected a better monument to his business sagacity and indominable [*sic*] energy than Mr. Hill. He is the friend of every business interest and of every city, town and village along his great road. A master of the subject of transportation, he studies every locality which his roads penetrate, and his chief aim seems to [be to] give them the service and to study their interests that they may be prosperous, and so bring prosperity to his road. . . . Surely Mr. Hill and his road deserve the success which has attended them."

Thus it is not surprising that the same newspaper on January 28, 1899, welcomed Hill's involvement in the Mesabi. "It has been our opinion," commented the editor, "that the farther Mr. James J. Hill projected himself into the affairs of St. Louis county, the better it would be for the county in general and Duluth in particular. Hence the news of the deal between the Great Northern and the Wright-Davis concern . . . will be generally regarded, we believe, as a good thing for the county. Mr. Hill is progress personified. He is a large man. He does business on a large scale. He hitches his wagon to the star of empire. We shall watch with much interest the developing of his plans so far as this county is concerned; as we believe that they are closely linked with Duluth's 'manifest destiny.'"

On its front page, the *Tribune* announced on the same day: "An interesting feature of this deal is the checkmating of the Rockefeller interests. It is understood that John D. Rockefeller has been negotiating to buy the Wright & Davis property for some time, but some weeks ago the negotiations were dropped as Mr. Rockefeller apparently expected to get the property at a lower figure than the owners cared to accept. Mr. Rockefeller never knew that Mr. Hill had any designs on the property or it is thought he would not have tried to drive a sharp bargain. But he thought he would win and he lost. It is known that Mr. Hill never buys railroads to make playthings of and it is believed that his invasion in to the iron mining region of this state means a great deal more than can now be seen by outsiders. It is thought that it means a great deal for independent mine owners. . . . Mr. Hill, it is generally conceded, is able to become a formidable competitor of other ore carrying interests," the paper concluded gleefully.

Hill acquired the Wright and Davis railroad and mineral lands with his personal funds. Then in May, 1899, he transferred the railroad to the Great Northern system at cost, but he retained the ore lands in his own name for another year and a half. The Great Northern's charter did not permit it to "acquire, own, operate or lease ore lands." As a result in October, 1899, to control the ore lands and the Mahoning, Longyear, and Bennett mines, Hill incorporated the Lake Superior Company, Limited. This "limited partnership association" issued 100 shares of capital stock, 75 to James J. and 15 to James N. Hill. Its management was placed with the Great Northern board of directors. The Empire Builder later explained: "I did not want any personal interest in anything that was mixed up in the company's affairs, and I turned those lands over to be held for the benefit of the shareholders of the Great Northern Railway. . . . I always had a rule that if I could make money for myself in a transaction connected with the company I could make it for the shareholders."[49]

Spurred especially by the interest of Louis Hill, Lake Superior Limited officials acquired more land on the Mesabi at such a rapid rate that by 1906 the company held, by fee or lease, a total of 65,091 acres. "Large tracts were bought from private owners who had never carried on exploration and had little faith in their properties," claimed *Iron Age*. "These were bought if necessary on the principle of the small boys who trade in jackknives — 'sight unseen' — but they were bought cheaply and if a single mine could be found on them the cost price was a mere bagatelle." Hill also expanded the operation of the railroad by financially aiding independent mine purchasers in return for traffic contracts. He obtained further cargo by allowing miners to explore, test pit, and drill on his property on a royalty basis in return for their use of the Great Northern to transport all the ore they extracted.[50]

With these activities, James J. Hill laid the foundation for the considerable influence he would exert on the Mesabi Range in the 20th century, at times thoroughly unsettling his large competitors in the process. When he resigned as chairman of the Great Northern board on July 1, 1912, Hill reviewed his contributions, saying, "Most men who have really lived have had, in some shape, their great adventure. This railway is mine. I feel that a labor and a service so called into being, touching at so many points the lives of so many millions with its ability to serve the country, and its firmly established credit and reputation, will be the best evidence of its permanent value and that it no longer depends upon the life or labor of any single individual."[51]

As the 20th century dawned, the formative period of Mesabi development was drawing to a close. With the end of the long depression, confidence and optimism reappeared, spurred in part by pride in the success of the United States in the brief Spanish-American War of 1898. President William McKinley was in the White House, and business prospects brightened. The return of prosperity not only stimulated ore production, it also accelerated the movement toward consolidation in all segments of the economy. The market for industrial stocks soared. Pools gave way to trusts which in turn were replaced by holding companies as the first great merger movement in the country's history reached its peak between 1898 and 1902. Scarcely any product from crackers to tobacco was overlooked. But the largest merger of all resulted in the creation of the United States Steel Corporation, the country's first billion-dollar company, in 1901. Its formation was the work of yet another king of the American business scene — J. Pierpont Morgan.[52]

Unlike Carnegie and many of the other entrepreneurs we have discussed, Morgan was neither an immigrant nor a self-made man. Born in 1837 into a well-established Hartford, Connecticut, family, Pierpont was familiar from his youth with Europe's famous sights and spas. A patrician, he was partially educated abroad and grew up in England, where his father was a successful private banker. The young man early showed signs of mathematical genius. As an adult he was a dedicated member of the Episcopal church, an internationally known yachtsman, and a prodigious art collector. As a partner in the New York City private banking firm of Drexel, Morgan & Company (which became J. P. Morgan & Company in 1895), he replaced Jay Cooke as the nation's leading banker in the 1870s. A large man, six feet tall, with considerable magnetism and a forceful personality, Morgan in the 1890s exercised many of the functions later carried out by such government agencies as the Federal Reserve System, the Interstate Commerce Commission, and the Federal Trade Commission. In 1900 he was already famous for his stabilization of the government's gold reserves in 1895 and for his reorganization of the nation's railroad industry. Then he turned his attention to iron and steel.[53]

In 1900 the three primary suppliers of crude steel were the Carnegie Company led by Andrew Carnegie with young Charles Schwab as president, the Federal Steel Company of Chicago headed by Elbert Gary, and the National Steel Company, organized in 1899 by a merger of Ohio and Pennsylvania plants, under the direction of William H. Moore. As we have seen, Morgan was not uninterested in the steel industry, having financed the formation of Federal Steel in 1898. Although Carnegie Steel was the largest and strongest of these firms, Federal Steel — at least on paper — possessed greater independence because it controlled its own lake shipping, while Carnegie relied upon its 1896 traffic agreement with Rockefeller's Bessemer Steamship Company. Bessemer, however, could handle only half of Carnegie's Mesabi ore production, forcing the steel firm to utilize numerous independent lake steamers. Dissatisfied with this lack of control, Henry Oliver persuaded Carnegie to organize its own lake fleet in 1899 by purchasing six existing ore carriers from the Lake Superior Iron Company, a Marquette Range firm which dated back to 1853, and ordering the construction of five more ships. Carnegie then formed the Pittsburgh Steamship Company, capitalized at $4,000,000, and rapidly acquired the fourth largest fleet on the Great Lakes.[54]

The market for iron and steel suffered a temporary decline in 1900, and finished-steel producers like Federal and National were forced to cut prices. Seeking ways to reduce costs and revive their declining profit margins, these firms planned to expand their primary steel production in order to limit or eliminate their dependence upon Carnegie. Such other firms as John Gates's American Steel and Wire Company and National Tube also laid plans to build furnaces to produce their own crude steel. These plans threatened Carnegie's empire, and he set out to deal with his rivals. As a first step he announced he would construct a large and efficient steel-tube plant at Conneaut, Ohio, on Lake Erie, and his future intention was to manufacture tin plate, sheet steel, and other finished items as well as to build his own railroad to transport them.[55]

Carnegie's announcement attracted Morgan's attention, because the Scotsman's plans would spell ruin for the National Tube Company, in which Morgan was an investor, and would disrupt the recent peace he had arranged to reduce the ruinous competition among American railroads. According to historian Gabriel Kolko, the "working, informal détente" between the major steel "empires collapsed amid threatening diversification and price competition that promised to drive the over-capitalized steel mergers to bankruptcy and ruin." Fortunately, Morgan's desire to prevent a disastrous competitive struggle coincided with Carnegie's inclination to retire in order to devote full time to philanthropy. Through the efforts of Charles Schwab, who was to become the first president of United States Steel, Carnegie was persuaded to sell his holdings to Morgan for $492,000,000.[56]

Morgan then turned his piercing gaze upon Rockefeller's Lake Superior Consolidated mines and transportation network. After some weeks of negotiation, in which Henry Frick served as the go-between, Morgan agreed to buy Rockefeller's mines and the Duluth, Missabe and Northern Railroad for $80,000,000 in common and preferred stock in a new consolidation. In addition, Rockefeller received $8,500,000 in cash for the Bessemer Steamship fleet. Writing in 1909, the oil king recalled that "A [Morgan] representative . . . came to see us about selling the land, the ore, and the fleet of ships. The business was going on smoothly, and we had no pressing need to sell, but as the organizer of the new company [*Morgan*] felt that our mines and railroads and ships were a necessary part of the scheme, we told him we would be pleased to facilitate the completion of the great undertaking. . . . After some negotiation, they made an offer which we accepted. . . . The price paid was, we felt, very moderate considering the present and prospective value of the property."[57]

The end product of Morgan's efforts was the organization of the United States Steel Corporation, incorporated on February 25, 1901, under the laws of New Jersey. The new firm, with Elbert Gary as chairman of the board of directors, offered to exchange its securities for control of the companies involved, a proposition to which 98 per cent of the stockholders agreed. The participants included the Carnegie Company, Federal Steel, National Steel, American Bridge, American Sheet Steel, American Steel Hoop, American Steel and Wire, American Tin Plate, National Tube, and Rockefeller's Lake Superior Consolidated Iron Mines. Also absorbed were the operating divisions of the parent companies, including the Oliver Mining Division of Carnegie, the two ore-carrying railroads, and the Minnesota Iron Company of Federal Steel.[58]

The constituents' property included 78 blast furnaces and rolling mills, vast holdings of iron ore, coal, and limestone lands, a 112-vessel lake fleet, and over 1,000 miles of railroad. Controlling three-fifths of the nation's steel business, the new corporation estimated its annual cumulative capacity at roughly 7,400,000 tons of finished products. Led by Morgan, potential competitors had co-operated to create the greatest steel manufacturing company the world had yet seen. Among the large holders of Minnesota ore lands, only James J. Hill had been overlooked.

The press reacted to the new corporation with almost universal skepticism or outright hostility. Editors in the United States and Europe predicted that the merger would throttle individual initiative and benefit only a small clique of capitalists. A *New York Times* editorial on February 28, 1901, suggested a very specific motivation for the corporation's conception. "The new consolidation, however formidable, is a long way from controlling the iron and steel trades of the United States," the paper observed. "So far as appears, this is not its purpose. Those who know the facts understand that it has been formed primarily to eliminate MR. CARNEGIE from the trade. His competitors are tired of dancing to the music of his bagpipes, and could make no plans for their own protection until his vast capital and masterful intelligence were devoted to philanthropy rather than to business."[59]

Two articles in *Cosmopolitan* magazine provided contrasting opinions and one of the few favorable assessments of the corporation. Noted University of Wisconsin economist Richard T. Ely's essay ranged from caution to outright opposition. Especially concerned with the merger's effect on labor, Ely warned that "the wage-earner feels that, isolated and alone, he is a pigmy, a nothing, when his individual interests are pitted against amalgamated hundreds of millions." Charles S. Gleed, a well-known financial expert, portrayed the giant consolidation favorably and lauded its creators. He believed that "The character of the men in the directory is a prophecy of the success of the company. The members are men of rare strength in both money and experience."[60]

In Minnesota press reaction ran the gamut from neutral to negative. The *Minneapolis Journal's* front page on February 26, 1901, depicted an octopus labeled "United States Steal Corporation" with its arms entwined around a grain elevator, a steamship, a locomotive, and other symbols of industry. The headline of an accompanying article read, "Morgan Gets A Nice Bit . . . Insiders Get Cream," and an editorial in the same issue questioned the soundness of the merger's capitalization. Northeastern Minnesota newspapers were uniformly cautious, printing wire-service reports and very little editorial comment. On March 30, when news of the sale of the Lake Superior Consolidated Iron Mines was released, the *Hibbing News* noted carefully: "The Consolidated has been good to Hibbing, and if a change is made it is hoped the new owners will do as well for the town." The *Iron Trade Journal* of Two Harbors, the only outlying paper to express a strong opinion, quipped sarcastically on February 28 that "things look exceptionally bright for the laboring man — to starve."

Duluth editors were emphatic in their opposition. On February 14, the *Duluth News Tribune* cautioned that "the independent miner will soon be a thing of the past, and ore properties will have to be sold at such prices as a common interest among the iron and steel concerns shall dictate." The paper continued to sound this warning on March 13, commenting that "From the hour that Morgan reigns supreme in the steel world, we may look for the cessation of exploring, the shutting down of mines, the discharge of men on the ranges and by the ore-carrying railroads." Duluth editors especially feared exclusive control of the railroads. "The essential of monopoly is the prostitution to its service of the franchises the people have granted to railroad, express, telegraph and other companies," thundered the *News Tribune*. "Discrimination in rates is the prop of trust expansion. . . . Railroad rates are the knife with which monopoly cuts the throat of competition."[61]

An unsigned letter to the *Duluth News Tribune* on February 18 argued the opposite case. The writer believed that the merger "will not produce any unfavorable changes in northern Minnesota, nor anywhere else. . . . The demand for ore will undoubtedly be greater instead of less." Most important to St. Louis County's laboring population "the effect will be in favor of the wage earner, and will not be unfavorable to the owner of iron properties. . . . There will be a new demand for ore deposits. Expansion will be the fundamental principle of the new company and that has never meant lower wages or lower prices for raw material." The writer also touched a nationalistic note when he said the merger produced "the hardest blow to the iron industry of foreign countries that has ever been struck by American capital and industry."[62]

The *Duluth Evening Herald* of February 21 printed an anonymous refutation: "If the combination that is reported under way is anything like other combinations in the world of commerce, it will not help Duluth a bit. . . . Duluth gets now about as little benefit out of it as it possibly could. With the single exception of Andrew Carnegie, who has come down handsomely for the Duluth library, this city owes nothing to any of the magnates that control her chief industry. Somebody has put the situation pretty well when he said that the iron ore business

"HIS HANDS FULL," a cartoon depicting J. P. Morgan's activities in forming the United States Steel Corporation. *Minneapolis Journal*, February 26, 1901.

may be compared to a bullet, the lake navigation system to the gun, and the part of the business Duluth gets to the dirt that sticks in the barrel when the gun is discharged. I think that if the combine goes through the shooting will be pretty clean, and that even less dirt than now will stick in the barrel."

Reactions to the merger were not confined to the editorial pages of the state's newspapers. The Minnesota legislature was also interested in promoting and protecting local industry. On March 21, after its sponsor "flourished the trust bugaboo before the senate," that chamber passed a bill "preventing and restraining operations of pools, trusts, and conspiracies," originally aimed at the lumber interests. The act declared any agreement to set prices illegal and gave district courts jurisdiction to prevent combinations. A second bill passed by the Senate the same day made the 1899 antitrust statute retroactive, but this action was not approved by the House and never became law. The *New York Times* of March 24, 1901, reported that more drastic measures were being urged to provide that "the Attorney General be instructed to bring suits in the name of the State that shall not be dismissed or compromised except upon the express authority of the Legislature. Also that any corporation organized in Minnesota (such as Minnesota Iron) that permits a majority of its stock to be acquired by a corporation organized elsewhere be declared forfeited." The paper called attention to the protectionist legislation already enacted and warned: "Unless some step is taken for the immediate upbuilding of iron-making industries in the State the United States Steel Corporation is likely to find itself hampered in every direction by unfriendly legislation. The Minnesota Legislature is largely a community of farmers who are firm in the opinion that their own markets will be greatly improved by a local manufacturing centre."[63]

These concerns about monopoly control are understandable in the light of the new corporation's impressive statistics. At the end of the 1901 shipping season United States Steel possessed a majority, or near majority, control of the output of the producing mines on four of the five Lake Superior ranges. The lone exception was the Marquette, the oldest of the ranges and until 1895 the leading shipper. Through 1901 its mining operations had extracted a total of 63,079,395 tons; United States Steel controlled 37.8 per cent of the Marquette mines then producing. By contrast, the consolidation owned 50.0 per cent of the Gogebic and 53.0 per cent of the Menominee output. In St. Louis County, where it held 58.6 per cent of the Vermilion and 71.8 per cent of the Mesabi output, the steel firm enjoyed even greater dominance.[64]

By the time United States Steel came into being in 1901 the Mesabi Range had emerged as the leading producer, shipping 46 per cent of the output of all ranges. The combined production of the four largest Mesabi mines — Mountain Iron, Fayal, Adams, and Mahoning — exceeded that of the entire Menominee Range, its closest rival. Shipments from the Fayal Mine alone almost equaled the total Vermilion Range tonnage. On the strength of the Mesabi's enormous output, Minnesota in 1901 surpassed Michigan to become the nation's leading ore producer, and the formative period in the development of the eastern and central portions of that giant range came to a close.[65]

|10|

THE
WESTERN MESABI
AND THE CUYUNA

THE DECADES between 1890 and 1910 had witnessed the transformation of the Mesabi from a frontier wilderness to the world's leading producer of high-quality iron ore. Creative entrepreneurs — the Merritt brothers, Rockefeller, Oliver, and others — had played their innovative roles, gained and lost fortunes, and set the stage for a future oriented more toward corporations and less toward individuals in both the Mesabi and Vermilion ore districts. Although the entrepreneur made himself felt on the two ranges primarily before 1900, he was never entirely supplanted by the large corporations of the 20th century. He immediately reappeared, for example, in the two Minnesota ore districts still to be developed — the western Mesabi Range and the suspected, but as yet undiscovered, Cuyuna Range.

Long known to contain deposits of iron ore, the western Mesabi had been largely bypassed in the early years of production on the richer central and eastern sections of the over 100-mile-long range. Extending roughly from Nashwauk-Keewatin near the eastern border of Itasca County about 25 miles southwest to Grand Rapids, the western area had been visited by explorers Zebulon Pike in 1810, James Allen and Henry Schoolcraft in 1832, and Joseph Nicollet in 1836. Thirty years later state geologist Henry Eames had noted the presence of iron ore near the Pokegama and Prairie River falls. Following the Civil War, James Whitehead, who had learned "in some way" that there was ore near Prairie River, interested Justus Ramsey, a brother of Minnesota Governor Alexander Ramsey, in a scheme to refine it. The plan proved impractical because the closest furnace for smelting was at Galena, Illinois, but Justus acquired title to some lands near Grand Rapids.[1]

The 19th-century development of the western Mesabi was, however, desultory at best, largely because of the lower iron content and sandy composition of its ore. Nevertheless, before the panic of 1893 choked off exploration, a number of entrepreneurs active on other parts of the range became interested in the western district. Among them were David Adams, who in 1886 discovered but did not develop the ore body that was to become the Canisteo Mine near present Coleraine; the Merritt brothers, whose explorations included sections of central and eastern Itasca County; and Edmund Longyear, who took his diamond drill west in 1891, acquired land, and later helped develop townsites there. After the initial onslaught of the panic, the persistent pioneer John Mallmann opened the Arcturus Mine near Marble in 1895. Once again he was unlucky. According to historian Donald E. Boese, the Arcturus was the first to ship ore from the western Mesabi. Ten tons were sent to Grand Rapids in 1897; from there the ore traveled by rail to Elbert Gary's Illinois Steel Company, which pronounced it "unsuited to their blast furnaces." The mine was then closed.[2]

As the 20th century dawned, James J. Hill and others began to take an interest in the new area. On March 22, 1902, the *Grand Rapids Herald-Review* proclaimed: "There is no longer any doubt in the minds of those who have noted the explorations for iron ore on this end of the Mesaba range . . . that valuable mines in this vicinity will soon be among the shippers. The Great Northern Railway company commenced investigations some time ago [which were] . . . evidently satisfactory. . . . [and] immediately began to take options and buy up everything available on the range, and . . . has acquired title to thousands of acres of land."

But it was United States Steel (which leased Hill's holdings in 1906) that brought large-scale development to the area in the first decade of the 20th century. The move was largely the work of three men — Chester Congdon and Guilford G. Hartley of Duluth and Thomas F. Cole, who became president of United States Steel's Oliver Iron Mining Company in 1902. The Canadian-born Hartley, who lived for a time in Brainerd and represented that area in the 1883 Minnesota legislature, then moved to Duluth, where he engaged in a variety of businesses. With Congdon, who had been Henry Oliver's local attorney until 1901, Hartley saw the potential value of the lower-grade ore lands in Itasca County. In 1904 the two men formed the Canisteo Mining Company "with the intent of mining and marketing the sandy iron ore of that region." They reopened the Arcturus Mine and built a small experimental washing and concentrating plant near it. The process turned out to be inefficient, but the results were encouraging. This modest success, coupled with the persuasive talents of Hartley and Congdon, induced Cole to look more closely at the lower-grade ore lands in the Canisteo district, as the western Mesabi was frequently called. Until that time, the Oliver company, as well as most other firms, had been interested only in high-grade ores. Hartley and Congdon convinced Cole to "place his reputation behind the idea of the treatment of the lean ore as a business proposition." Cole, in turn, persuaded Gary and other United States Steel officials to invest some $10,000,000 in the sandy ores of the western Mesabi.[3]

The character of these ores was by then generally known. But diamond drilling had proven less effective than churn drilling in their exploration because large amounts of sand were present. Since diamond drilling had not been extensively carried out to secure continuous cores, specifics concerning the ore were still largely unavailable. Nevertheless continuing refinement of the washing process in the Arcturus plant encouraged Oliver officials to make the plunge in 1904. Congdon and Hartley sold their property to Oliver, and Cole brought the financial and technical resources of that firm to the western range. In 1910 United States Steel financed the erection of the first major beneficiating plant on the Mesabi on the east shore of Trout Lake. Nearby, on land purchased from lumberman William D. Washburn, the company also built Coleraine, a "model town" that owed its existence to Mesabi iron and to the persistence of Hartley, Congdon, and Thomas Cole, in whose honor it was named.[4]

Just north of Coleraine, Oliver laid out the Canisteo Mine, which had been discovered in the 1880s by David Adams. The first ore was shipped in 1907 — 5,483 tons of it. After 1910 the difficulties presented by the lean ores of the region were largely solved in the Trout Lake concentrating plant, which could treat over 70,000 tons of ore in a single working day. Chiefly as a result of this successful plant, the mine's production jumped to 339,057 tons in 1910.

The new town near the Trout Lake concentrator was "planned with deliberate care and commendable foresight" by John C. Greenway, general superintendent of Oliver's operations on the western Mesabi. Alabama born and Yale educated, Greenway had joined the Oliver enterprise after stints with Carnegie Steel and with Theodore Roosevelt's Rough Riders during the Spanish-American War. In a climate of opinion increasingly critical of big business in general and big steel in particular, Greenway had little difficulty in winning the approval and financial support of United States Steel for his plans. In fact, he "had a free hand in developing the model town as he saw fit. . . . it was his town," wrote a biographer. He created it and he ran it. The residents had no voice in Coleraine's government during the years it remained a company town.[5]

Land speculation in Coleraine was discouraged by Greenway's restrictions on the amount any one purchaser was allowed to buy; businesses were permitted to operate only in the downtown area; and company approval was required for each house site. Lots were donated to any church group wishing to build, and the company financed such municipal improvements as street grading, water and sewage systems, a fully equipped hospital, an athletic field, and a public park named for Edmund Longyear. According to Greenway's biographer, Coleraine "had an ethnic composition which set the town apart from others on the Mesabi Range" due in part to the superintendent's "typical turn of the century American middle class racial bias." A nationality chart prepared by the Oliver firm in 1908 indicated more "American-born . . . employees in the Canisteo District than anywhere else on the Range."[6]

Coleraine was only one of the communities in the new mining district. Older settlements, however, owed their beginnings to lumbering rather than mining, for unlike the eastern Mesabi, the western sector of the range had been settled before ore production got under way. As early as 1872 there were 17 lumber camps near Pokegama Falls and present Grand Rapids. Settled by fur traders and loggers, the latter Mississippi River village was incorporated in 1891, a few months before the Duluth and Winnipeg Railroad extended its line 72 miles from Cloquet to Grand Rapids, thus easing the transportation problem between those points. The population of Grand Rapids increased from a mere 277 in 1890 to 1,546 in 1895 and to 2,055 ten years later. The town's development followed the familiar pattern of other range villages, but in Grand Rapids the economy rested primarily on timber rather than ore.[7]

Some 20 miles to the north and east of Grand Rapids the village of Nashwauk also began as a lumbering center. Incorporated in 1902, its population of 684 in 1905 was composed of an ethnic mix of northern and southern European immigrants similar to that of older range towns. In 1903 the Eastern Railway branch of the Great Northern built ten miles of track to reach the nearby Hawkins Mine, which had opened the year before. In its first year of operation the Hawkins shipped only 5,892 tons of ore, but with the coming of the railroad its output increased to 108,048 tons the following year. The mine and the railroad gave Nashwauk a new lease on life, for the pine forests in the area were by then nearly exhausted.[8]

The establishment of additional towns between Nashwauk and Grand Rapids was more intimately tied to the development of mines, just as it had been on the eastern Mesabi. Bovey, situated on Trout Lake about seven miles northeast of Grand Rapids, was incorporated in 1904. It was named in honor of Charles A. Bovey, a Minneapolis lumberman who, in association with Hartley and Longyear, platted the townsite. So enticing were the advertisements, which promised "Beds of Iron Ore" near the new community, that 253 persons had moved to Bovey less than 12 months after its incorporation. The town acquired a newspaper in 1905, and a spur of the Duluth, Missabe and Northern Railroad relieved its isolation in 1906. Unlike tightly controlled nearby Coleraine, Bovey was known in its early years as a good saloon town (it had 17 of them by 1906). Moreover, it "was one of the few towns left in the state that still permitted open gambling." These attractions, coupled with the opening of the Walker Mine by Oliver, assured Bovey's success. The Walker shipped over 83,000 tons of washed ore to Lake Superior docks in 1907.[9]

The boom year for new villages in the Canisteo district was 1909 when Calumet, Coleraine, Marble, and Taconite were incorporated. The Oliver firm provided the impetus and much of the original financial backing for each of these company towns "in harmony with its general policy of trying to secure permanent, capable and reliable employes by ensuring them comfortable and attractive homes with the best educational advantages for themselves and families." Calumet and Marble, the easternmost of the new settlements, were situated near the Hill Mine, which began shipping in 1910 with 801,226 tons of ore. Taconite, five miles to the west, became home to employees of the nearby Holman Mine, which was opened by Hartley in 1904 and leased to Oliver the following year. One of the early shippers on the western Mesabi, the Holman produced 8,068 tons in 1907. By 1910 the Trout Lake beneficiating plant, the largest in the world at that time, was in operation, proving the feasibility of washing western Mesabi ores, and the Great Northern had completed a track between Grand Rapids and Nashwauk to serve the new towns. Thus within 20 years after the Merritt brothers discovered ore at Mountain Iron the mighty Mesabi had been fully explored. Dozens of thriving communities and operating mines straddled its now denuded hills. The pine forests were gone, and mining was fully established as the dominant industry throughout the length and breadth of the giant range.

The technological advances in treating lean ores that made mining feasible on the western Mesabi were soon to play an even more impor-

CUYLER ADAMS, for whom the
Cuyuna was named, was the
man most responsible for the
development of Minnesota's
third range in the early 1900s.
Photograph by Lee Brothers.

JOHN C. GREENWAY, general superintendent of
the Oliver Iron Mining Company's operations on
the western Mesabi, planned the new town of
Coleraine in the early 1900s. Arizona Historical
Society.

GUILFORD G. HARTLEY of Duluth was
one of the men who successfully interested United States Steel in the western Mesabi. St. Louis County Historical
Society Collection.

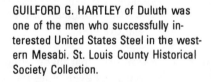

GEORGE CROSBY, who platted the town of Crosby in 1909, was a veteran mining man interested in both the Cuyuna and the western Mesabi. St. Louis County Historical Society Collection.

CROSBY in 1912 had become a well-established mining town on the Cuyuna Range. Photograph by J. J. Brindos.

AN EXPLORATORY SHAFT was sunk to 120 feet on the Cuyuna Range northwest of Brainerd in 1905 by Pickands Mather and Company. When the shaft flooded, the firm gave up the effort.

THE CUYUNA, unlike the other two Minnesota ranges, was settled before ore was discovered. This photograph of the Hans Anderson farm near Crosby was taken in 1908. Note the drilling tripod at right. Six years later the shaft of the Croft Mine was located between the house and the barn.

FIDDLERS entertained the miners at a location camp near Coleraine as operations got under way in 1906. Inga Nyman Collection.

THE BUSINESS DISTRICT of Bovey as it looked in 1906 (top below) and 1908. The blocks of Second Street pictured here had 23 store fronts in 1908, 13 of which were saloons. Many of them can be seen in both photographs. Upper below, Nute Collection.

THE MODEL TOWN of Coleraine was built by United States Steel at the north end of Trout Lake between 1905 and 1909. The brick building at right was the district headquarters of the Oliver Iron Mining Company, the operating division of United States Steel. Bovey may be seen in the distance at left and the water tower of the concentrating plant is barely visible beyond the trees above the first house at right. The photograph was taken from the roof of Greenway School about 1912.

ROOSEVELT AVENUE, the broad main street of Coleraine, was named for Theodore Roosevelt. It had substantial two-story buildings and electric lights by 1912.

A VIEW OF COLERAINE from Trout Lake in 1910 shows Cole Street in the foreground. The tall stack at left marked the Oliver Iron Mining shops.

HOUSES in such company towns as Coleraine and Marble were built by the Oliver Iron Mining firm in several designs. They were rented or sold to employees at reasonable prices. This photograph was taken in Coleraine about 1920.

241

THIS EXPERIMENTAL ore-washing plant was completed by Oliver Iron Mining on Trout Lake in 1907. After it successfully demonstrated the feasibility of removing sand from the lean ores of the western Mesabi, the company built the more substantial installation shown below.

THE TROUT LAKE BENEFICIATION PLANT near Coleraine is shown as it neared completion during the winter of 1909–10. The lean, sandy ores from both the Canisteo and Holman mines were treated here to raise the iron content to between 57 and 60 per cent, a level acceptable to blast furnacemen. The plant operated until 1973.

THE FIRST SHIPMENT of ore from the Cuyuna Range was hauled from the Kennedy Mine (top) over the Soo Line Railroad (bottom) in 1911. It took a month for the shipment to reach Superior, where it was again delayed waiting for the completion of the Soo Line dock. Both photographs from *Aitkin Independent Age*.

SUPERIOR HARBOR as it looked about 1960. Four Great Northern ore docks appear at the near left with two Northern Pacific Railroad docks just above them. The city of Superior is at left, and Minnesota Point in Duluth is visible in the upper right. Courtesy Douglas County Historical Society.

THE BOOM TOWN of Calumet on the western Mesabi Range was largely destroyed by fire in the summer of 1909 when this picture was taken. Note the logs in the street of what became the business district.

THE TOWN OF CUYUNA, like the range, was named for Cuyler Adams. Shown above as it looked about 1910, Cuyuna grew up near the Kennedy Mine. *Aitkin Independent Age*.

STEAM SHOVELS were used to load ore at the Canisteo Mine on the western Mesabi in 1907. First discovered by David Adams in 1886, the Canisteo did not begin really large-scale mining until 1909.

HYDRAULIC STRIPPING was sometimes cheaper than using steam shovels on the Cuyuna Range. Water under high pressure, shot from a rotating steel nozzle (seen below the pipe), washed the overburden to the center of the pit. An electric sand pump in the floating shack then forced it through the pipe to a disposal site. St. Louis County Historical Society Collection.

tant part in opening a third range some 60 miles to the south and west. Andrew Carnegie's 1892 prophecy that the Mesabi would not be "the last great deposit that Lake Superior is to reveal" proved to be true when Minnesota's smallest range — the Cuyuna — shipped its first ore in 1911.[11]

Extending some 68 miles northeast-southwest across Crow Wing and Aitkin counties, the Cuyuna was not opened until the second decade of the 20th century. Its presence, however, had been suspected as early as 1859. In February of that year a government surveyor laying down section lines north of Fort Ripley on the Mississippi noted with irritation: "It is utterly impossible to ascertain the true variation of any of the Random lines as the needle varies from 5° to 45° and dips a great deal — I am therefor compelled, to make my lines connect properly[,] to use back sights altogether." Surmising the cause of his problem, the surveyor wrote that there was "considerable local attraction . . . on the low marshes owing no doubt to the presence of Bog Iron, although, at this season, we could discover no indications of it." Other federal surveyors returned to the area in 1870 and 1871 when the Northern Pacific Railroad's construction crews laid tracks across Crow Wing County, but their reports made no mention of bog iron.[12]

Then in 1883 Roland D. Irving of the United States Geological Survey visited the region. Nine years later he published a report in which he theorized a possible connection between the geological formation of the Mesabi and what was to become known as the Cuyuna Range. With mining development escalating on the Vermilion and Mesabi during the 1890s, other geologists began to hint at the possible presence of iron ore to the southwest. Newton and Horace Winchell suggested in 1891 that the lower belt of the Vermilion formation "is possibly the same that appears on the Mississippi river at Pike Rapids in Morrison county." Thus in a situation exactly the reverse of that on the Mesabi, geologists predicted the existence of the Cuyuna before it was actually found, while on the Mesabi practical miners found ore where geologists thought it did not exist.[13]

The Cuyuna differed from the older ranges in other ways. For one thing its ore deposits did not reveal themselves in large outcroppings on the surface, a state of affairs that led a later geologist to describe the region as "geologically viewless." A layer of glacial drift from 14 to 300 feet thick hid the ore, and lakes, bogs, and pine forests covered the sandy, relatively flat surface. The ore lay in two formations: the North Range, an area five miles wide and twelve miles long near present-day Crosby and Ironton; the South Range, a deposit varying from one to seven miles wide running the length of the entire range. The only signs of the mineral wealth that lay beneath the lakes and forests were occasional pieces of float ore and the erratic performance of dip needles and magnetic compasses. The aberrations of the latter were noticed by early timber cruisers, one of whom is said to have sunk test pits as far north as Oxbow Rapids (now Ox Portage Rapids) in the Mississippi River in Aitkin County.[14]

A second difference between the Cuyuna and the older Minnesota ranges lay in the characteristics of the ore itself. Hard ores resembling those found on the Vermilion and soft ores resembling those on the Mesabi were present in greater "variety, texture, composition, and color" than in other Lake Superior districts. The richest ores were "medium soft reddish blue hematite . . . of Bessemer quality . . . analyzing as high as 67 to 68 per cent metallic iron and as low as .01 to .03 per cent phosphorus." Others were sandy and required washing to concentrate and increase their iron content. In addition, because many of the initial ore bodies were adjacent to lakes or lay near the Mississippi River, early developers worried about excessive water they might encounter in the mines. Some suggested draining the lakes before commencing serious operations. But Carl Zapffe, a contemporary mining expert, wrote in 1915 that actual experience had shown "nothing is now known that need cause any worrying over heavy pumping." Even more unusual were the manganese and manganiferous iron ores, found especially on the North Range. Later considered highly desirable, these deposits were at first "assiduously avoided by all who had any acquaintance with the desires of the blast furnace operators."[15]

A third notable difference to be found on the Cuyuna was in its settlement patterns. On the central and eastern Mesabi and the Vermilion frontiers test pits were sunk before roof beams were raised, but on the Cuyuna, settlement preceded the discovery of ore. Fort Ripley, on the western edge of the range, had been built in 1849–50 to maintain peace among the Ojibway, Dakota, and Winnebago Indians. Not far away at the junction of the Crow Wing and Mississippi rivers the fur-trading center of Crow Wing existed as early as the 1770s; augmented by settlers' as well as traders' families, it boasted a population of 600 in the 1860s. But its future was cut short in the next decade by the Northern Pacific Railroad's decision to cross the Mississippi farther to the northeast. After that Crow Wing dwindled and eventually became a ghost town.[16]

Brainerd, the largest community on the Cuyuna, sprang up at the point where the Northern Pacific elected to cross the Mississippi; it became the seat of Crow Wing County in 1871. Platted that year, the community evolved as the center of trade and railroad activities for a region in which logging was then the principal industry. Despite setbacks suffered as a result of Jay Cooke's banking failure and the panic of 1873, the town's population of 931 in 1875 increased to 7,110 ten years later. By the turn of the century half of the population of Crow Wing County resided in Brainerd. Smaller concentrations were to be found at Deerwood, a village on the county's eastern border, which became a mecca for hunters and fishermen, and at Aitkin about ten miles up the Mississippi which, like Brainerd, was settled as a result of the Northern Pacific's decision to build west through the area in 1871. By 1900 Aitkin had a population of 1,719 persons.[17]

A fourth difference discernible in the story of the Cuyuna lay in the sequence of its mineral development. During the settlement period, sporadic attempts were made to unearth the suspected ore. In 1875–76 Eli Griffin conducted a "systematic survey" on what came to be called the South Range. Lack of funds forced him to abandon the venture, but his interest in iron ore prospecting led to his later association with the Merritt enterprises on the Mesabi. In the summer of 1882 Henry Pajari, a Finnish immigrant who had previous mining experience in Michigan, carried out the first formal exploration. En route to take up a homestead in western Minnesota, he noticed specimens of iron ore near Deerwood. With financial backing from friends in Michigan, Pajari equipped himself, hired a young assistant, and began to plot lines of magnetic attraction. The two men made several attempts to sink test pits, but each time seepage from the surrounding marshes foiled their efforts. After three months Pajari, having exhausted his funds, gave up the search [18]

Repeated reports of magnetic attraction, crooked section lines, and compass deflections, coupled with news of ore discoveries on the Mesabi, may have been responsible for the step taken in 1893 by 40 residents of Brainerd. Each person contributed $10 to hire Peter G. Fogelstrom, a local contractor, to drill half a mile south of town where surface ore had been observed. Fifty feet down the drill struck a boulder, and the enterprise was abandoned. But according to Anna Himrod, the principal historian of the Cuyuna, "Fogelstrom himself believed that the iron ore would be found in the hills up the Mississippi River. And it was among these same hills that some real mines were developed."

In the early 1890s the man who was to be most responsible for the opening of those mines was wandering through the woods near Deerwood accompanied by his Saint Bernard dog. Cuyler Adams had several professions before he settled in the little town on Serpent Lake which was first named Withington and rechristened Deerwood in 1892. He was born in Illinois in 1852 and arrived in northern Minnesota 18 years later to spend a year trading and trapping with the Ojibway Indians near Vermilion Lake. After that he joined a survey crew of the Northern Pacific, which was building west from Duluth. Adams made his first big business deal three years after the panic of 1873 when he formed a syndicate with eastern capitalists Charles Francis and B. S. Russell to buy up Northern Pacific preferred stock at extremely depressed prices. Then, calling upon his experience as a surveyor, he selected 100,000 acres of rich farmland in eastern Dakota Territory (west of present Jamestown, North Dakota), trading the railroad certificates for the land at face value as provided in the stock agreement. There adjacent to land similarly acquired by Vermilion Range developer Charlemagne Tower, Adams, beginning in 1878, operated and managed a bonanza wheat farm known as Spiritwood. In the 1880s when bonanza farming began to decline, Spiritwood was subdivided, and Cuyler reportedly realized "nearly $100,000" from the venture.[19]

In 1882 the enterprising Adams returned to Minnesota, where he owned timberland in Crow Wing County and a summer home at Deerwood. He engaged in logging, sold railroad ties, and also operated a sawmill. Like other Deerwood residents, he noticed deflections of his compass while defining property lines. Curious, he began to make extensive surveys using both a terrestrial compass and a dip needle to map lines of maximum divergence. Gradually he developed an outline of what he believed to be an ore body, determined where the deposits seemed to be richest or nearest the surface, and located possible sites for future mining operations.

Adams then sought the financial backing of William C. White, an old North Dakota friend who also had a summer home at Deerwood, convincing him that drilling was warranted to locate marketable quantities of ore. White's nephew remembered the partnership as the particularly effective "combination of a man of Mr. Adam's inquiring and persistent type, with a man like my uncle, a lawyer whose enthusiasm was restrained by calm judgement and whose standing enabled him to obtain the money needed to test the ground by drilling." White attracted some Duluth associates to the venture, and on March 3, 1903, Adams, White, and William McGonagle, a Vermilion Range railroad pioneer, incorporated the Orelands Mining Company, capitalized at

$50,000. They began drilling south of Deerwood two months later.[20]

By that time others were also prospecting in Crow Wing County. The *Brainerd Dispatch* of February 27, 1903, reported, "There seems to be an undercurrent of excitement about town with regard to the probabilities that iron ore may be discovered in this section of the county. We understand there has been some mining experts who have been making a still hunt for indications of ore." By October 16 the paper was more optimistic, exclaiming that "prospecting for iron continues to enthuse a good many cruisers and speculators and new and more encouraging developments are reported daily."

In this favorable atmosphere the Orelands Mining Company appealed to local pride to sell its stock. Late in 1903, the company called a meeting of Brainerd businessmen, and to the delight of those assembled, Adams and White announced the discovery of ore. They described how it had been found by surveys and drilling, and they asked for the financial support of the local people to continue its development. Dr. Werner Hemstead, who attended the meeting, recalled that the two partners believed "New money ought to [be] raised at once and developement [sic] pushed to the limit, because as soon as it was known that an iron range had been discovered in Crow Wing County mining men and speculators would flock into the county and try to reap unearned benefits that follow the discovery of a new range. To forestall such a contingency Mr. Adams and Mr. White had organized a mining company to sell stock to raise new capital." The meeting had mixed success. Some businessmen took up the offer, but others may have been put off by a provision that assured control to the two organizers.[21]

The outside interests Adams and White had foreseen made their appearance in January, 1904. Foremost among them was the Oliver company, which announced plans to begin exploratory work on the Cuyuna Range in Aitkin County. Local speculators took this development as a stamp of approval from the principal Mesabi shipper. The exploratory announcement issued by Oliver officials stimulated sales of Orelands Mining stock and raised the funds needed to purchase additional equipment and to renew the search for ore. Unfortunately this optimistic atmosphere lasted only a few months. In April Oliver withdrew all of its employees and returned to its more profitable operations on the Mesabi. The company had put down a dozen drill holes near the village of Aitkin but declared that the results did not warrant any further investment of capital — a position it was to take off and on throughout the life of the range. Unlike the Vermilion and the Mesabi, the Cuyuna was to see little of the Oliver Iron Mining Company.

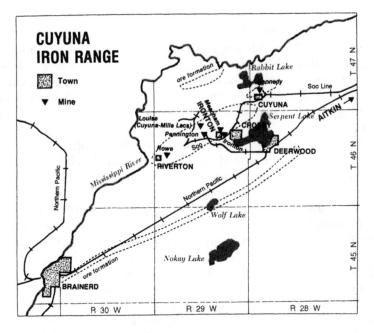

Instead it was to be developed and operated largely by independent entrepreneurs and by companies not associated with United States Steel — a sequence quite different from the patterns of the older ranges.[22]

The local entrepreneurs were undeterred by Oliver's withdrawal. A month after the company pulled out, William Rock and Roscoe C. Jamison, partners in a drilling outfit affiliated with Orelands Mining, located the Cuyuna's first marketable ore, but its quantity was too small for profitable development. Adams and his Orelands associates continued to work throughout 1904, drilling 22 test holes in the vicinity of Deerwood. Other newly formed mining companies such as the Gregory and the Brainerd became active and joined in the search.

By September, 1905, the new range had a name. Charles K. Leith, a geologist who had drawn a diagram of the hypothetical extension of the Mesabi ore formation in 1903, asked Adams' wife, Virginia B. Culver Adams, to name the range. She did so by combining the first syllable of her husband's given name with the name of their dog Una. Leith used "Cuyuna" to refer to the area before a meeting of the Lake Superior division of the American Institute of Mining Engineers in September, 1905.[23]

The first shaft on the Cuyuna was sunk that month in Nokay Lake Township, Crow Wing County, by the Hobart Iron Company, a division of the Cleveland firm of Pickands Mather and Company. The crews cut through 120 feet of glacial debris before striking iron ore. Unfortunately, they also encountered water and the venture was abandoned. The following year Oliver briefly returned to the new range, sinking a shaft near Rabbit Lake north of Deerwood. But this, too, flooded.[24]

The failure of the Pickands Mather and Oliver shafts frightened off other potentially interested firms and delayed development. Once again White and Adams assumed the initiative. Unable to raise enough money locally, White opened a new era in the story of the Cuyuna by seeking substantial outside investment capital. After visiting financiers in Pittsburgh and Cleveland, he reached S. A. Kennedy, president of the Rogers, Brown Ore Company in Buffalo, New York. This firm, which was already operating the Iroquois and Susquehanna mines on the Mesabi, leased property from the Orelands company in 1907 and began the construction of a shaft just south of Rabbit Lake. The successful effort became the Kennedy Mine in 1908. And there the first marketable iron ore produced on the Cuyuna Range was stock-piled.[25]

The company's efforts ushered in "lively days," as the *Duluth News Tribune* pointed out on October 25, 1908. With the exception of Oliver, "Nearly every big operator or mining concern known in Duluth is interested in the development of this range," the paper continued, listing 34 men and firms. Among them were Hartley and Congdon of the western Mesabi, John McAlpine and associates, North Star Iron Company, Northwestern Improvement Company (a subsidiary of the Northern Pacific), Pickands Mather, Robert Whiteside and Louis Rouchleau, and, of course, Rogers Brown. Another "big operator of the Cuyuna," according to the journal, was Franklin W. Merritt, the son of Napoleon, the second oldest of the Merritt brothers. Franklin, the paper said, had "showed up one of the largest bodies of ore in the state, running high in iron, with large quantities of manganese and very little phosphorous." Land values soared, and the report indicated that "fabulous prices" were being asked and received; the Northern Pacific reputedly paid "as high as $40,000 for a single 40 acre" tract.

Inadequate transportation facilities, like those on the Mesabi 16 years earlier, initially handicapped the Cuyuna's first mine. Machinery and supplies could reach Deerwood via the Northern Pacific, but from there they had to be moved by wagon or sled over a rough county road the final six miles to the Kennedy Mine. The *Brainerd Dispatch* of September 18, 1908, estimated that it cost as much to get coal from Deerwood to the Kennedy Mine as it did to bring the fuel from Pittsburgh to Deerwood.

To alleviate the transportation problem, Adams and two attorneys for the Minneapolis, St. Paul and Sault Ste. Marie — the Soo Line — incorporated the Cuyuna Iron Range Railway Company. They planned to build a line from the Mississippi River near Rabbit Lake to the docks at Duluth or Superior, and Adams busied himself with obtaining shipping contracts for the proposed road. Writing in October, 1909, to the Pittsburgh Coal Company, he said that he expected to secure "half the tonnage of Rogers Brown Ore Co. and Pickands Mather." But railroad men in Duluth apparently doubted that the Adams railroad would be built "all the way to the head of the lakes," considering it rather as a feeder line for the Soo. And they were right. On March 11, 1910, the *Brainerd Dispatch* announced that the Soo Line had purchased the completed Cuyuna road for $500,000. Before the end of the year Soo tracks were laid from Deerwood to Cuyuna, Crosby, and Ironton — new Cuyuna Range communities which, like so many older settlements on the Mesabi and Vermilion, owed their existence to iron ore.[26]

The oldest of the new mining towns was Cuyuna, platted in 1908. Situated about five miles northwest of Deerwood, the village's economy was bound up with the Kennedy Mine. Less than six miles to the southwest Ironton was platted in 1910 to serve such other early mines as the Meacham (later Armour No. 1), the Cuyuna-Duluth (later Ironton), and the Louise.[27]

The largest of the new iron communities was Crosby, "the metropolis of the Cuyuna Range country." located just east of Ironton on Serpent Lake. It was named for George H. Crosby, a veteran mining man described as "quiet and unassuming but at the same time shrewd and far sighted," whose interests on the western Mesabi included the Hawkins, LaRue, and Crosby mines near Nashwauk. In 1909 he platted the village of Crosby, which was incorporated the following year. Like Coleraine on the Mesabi, the community was planned with care. "Five miles of streets were cut out of the fields and forests," reported the *Crosby-Ironton Courier*, "[and] curbing and cement sidewalks . . . water mains and sanitary sewers installed. . . . 100 houses built to supply low-rent housing for the men coming to work in the new mines." According to one historian, George Crosby's "unwavering faith in the district, his contagious enthusiasm, his energy and resourcefulness" were as important to the development of the North Range as were Cuyler Adams' efforts in "recognition of the South Range and the Cuyuna District as a whole."

Running parallel to the South Range for most of its length was the Northern Pacific Railroad, which had the Cuyuna area largely to itself before the Soo Line arrived. The Northern Pacific owned alternate sections across parts of the range, and as early as 1910 (a few months after selling the Cuyuna Iron Range Railroad to the Soo Line) Adams was in touch with officials of the road. On September 20, 1911, he and two of his sons incorporated the Cuyuna Northern Railway Company capitalized at $1,000,000. The new line was to run between Deerwood and the North Range. A year later on October 20, according to Himrod, came the "gala day . . . [which] marked the opening of the Cuyuna Northern Railway; it was the date of the first shipment of ore from the Cuyuna-Mille Lacs mine . . . and the range entertained" a large excursion from Duluth. Less than three years later, the Cuyuna Northern was sold to the Northern Pacific, thus consolidating the latter road's foothold.[28]

The decision to build the Cuyuna Northern may have received added impetus from difficulties encountered by the new range's first ore shipment. In April, 1911, part of the 83,000 tons stock-piled at the Kennedy Mine was loaded on Soo Line cars bound for the holds of the "Alva" and the "Venezuela," ore carriers waiting for cargo at the docks in Superior. Unlike the triumphal first run of the Duluth and Iron Range Railroad from Tower to Two Harbors in 1884, the maiden Cuyuna shipment got off to a poor start. After a rousing send-off, "two monster engines . . . attached to 42 cars" headed east to Ironton and "then stopped for a month." Another abortive start was made, and the *Brainerd Dispatch* of May 4 said that such starts were "becoming as numerous as Sara[h] Bernhardt's farewell tours." More than a month after it left the stock pile, the ore reached Superior, where it again sat on a siding to await the completion of the Soo docks.[29]

Docking facilities at Superior had changed since the Duluth and Winnipeg became part of the Great Northern Railway system in 1896. The Soo Line dock was extended 800 feet in 1911 to accommodate the expected increase in ore traffic from the Cuyuna. Constructed of timber, this dock rose 78 feet above the waters of Allouez Bay; its width across the pockets was 58 feet. The Northern Pacific facility was even more impressive. Built of reinforced concrete, it initially stood 80 feet above the water, had 102 pockets, and stored 35,000 tons of ore. When completed in 1913, the Northern Pacific dock was the largest on the Great Lakes. Its over-all length reached 684 feet from shore, and two extensions, added in 1917 and 1926, tripled both the number of pockets and the storage capacity.

The increased size of the ore docks reflected the growth of Cuyuna shipments. From the 147,649 tons shipped from the Kennedy Mine in 1911, production jumped to 2,191,528 tons from 29 mines at the end of 1920, with a peak season in 1918 of 2,478,923 tons of ore from 27 mines. Part of the rapid growth can be attributed to the composition of the Cuyuna ores. Most of them were too high in phosphorous to be considered Bessemer quality, and they also contained a high percentage of ferruginous manganese ore. Containing 20 to 30 per cent manganese, they were used to some extent in the manufacture of special alloy steels as an important "deoxidizer" in the preparation of commercial steel. In September, 1912, the first manganiferous ore was sold by the Cuyuna-Mille Lacs Mine to the Lake Superior Iron and Chemical Company and treated at that firm's Ashland, Wisconsin, furnace. During World War I the United States was unable to obtain the quantities of manganese ore which it had previously imported from Russia and India, so the demand for such ore exerted a substantial influence on the Cuyuna district. Of all the Lake Superior ranges, it possessed the largest, richest, and most easily accessible reserves of ferruginous manganese ores, and its accelerated development was due to this fact. Furthermore, the product could be extracted by open-pit methods rather than the more costly underground mining techniques.[30]

For the first few years after their discovery, Cuyuna ores were mined by the methods used on the Mesabi. Churn drills and diamond drills were adapted to work at an angle — a necessary adjustment to suit the range's narrow, tilted deposits. Although much of the early mining was underground, because the deposits were too small to warrant stripping, in 1913 the Pennington Mine, developed near Ironton by the Tod Stambaugh interests, produced the first strip-mined ore. Hydraulic stripping was a Cuyuna innovation in which the overburden was washed away with pressure and picked up by a centrifugal sand pump. This technique was first employed in the Lake Superior district by the Rowe Mine near Riverton in 1913.[31]

As technology and transportation improved, the Cuyuna Range flourished. With land values soaring, Cuyler Adams complained in 1908 that he had "raked the country over with a fine tooth comb" and found it "exceedingly difficult to obtain any lands at all between here [Deerwood] and Rabbit Lake. Options for lease can be obtained but the farmers are pretty wild, and not inclined to talk purchase at all." The rise in population kept pace with land prices. Crow Wing County, for example, grew from 14,250 inhabitants to 24,566 in the first two decades of the century, while Aitkin County jumped from 6,743 people in 1900 to over 15,000 in 1920.[32]

Thus by the time the Cuyuna Range was being developed, its mine operators had difficulty obtaining iron ore property in a well-established area. Unlike the Mesabi and the Vermilion, the settlement frontier on the Cuyuna had been closed long before mining began. By the time entrepreneurs had devised ways to exploit the range's deposits, federal surveys had been completed, transportation links had been solidified, urban and rural dwellers were well entrenched, and land values had risen. Because the mining and settlement frontiers on the Vermilion and Mesabi ranges had moved hand in hand, pioneers there did not encounter quite the same problems. The economic patterns on the last of Minnesota's three ranges did not conform to those on the other two, and the Cuyuna never experienced the massive consolidations that characterized the Vermilion and the Mesabi.

The first of the ranges in the circle of iron ore ringing Lake Superior was discovered before the Civil War. As the Marquette, Gogebic, and Menominee gradually revealed their resources, the iron frontier shifted from Michigan to Wisconsin and then to Minnesota. In the last two decades of the 19th century the growing importance of the steel industry in the United States economy stimulated a wider search for new ore deposits. Several factors accounted for the rapid development and relative importance of northeastern Minnesota's iron reserves. The Vermilion, Mesabi, and some Cuyuna ores possessed qualities desirable for the Bessemer and open-hearth steelmaking technology of the era. Once the technology was adapted to accommodate the new ores, demand grew quickly. Improvements in rail and lake transportation, key factors in ending the isolation of the Vermilion and Mesabi mining frontiers, and more efficient mechanical devices for handling ore on the docks compensated for the handicap of distance between the new Lake Superior fields and the eastern sources of coke and limestone needed to process them. Important, too, was the sheer size of the Minnesota ranges which drew the attention of several large integrated corporations dominated by United States Steel after 1901. The creation of these corporate giants made available the necessarily large amounts of capital required to perfect more efficient mining techniques, withstand the impact of economic downturns, eliminate the waste of ruinous market speculation, and provide capable, professional management.

The panic of 1893, with its succeeding long years of depression, served as a turning point in the growth and organization of Minnesota's iron mining industry. Previously such individual explorers as George Stuntz on the Vermilion, John Mallmann and the Merritts on the Mesabi, and Cuyler Adams on the Cuyuna had broken the trail on successive mining frontiers. The boom psychology that accompanied the initial development of each new range attracted hundreds of other miners and speculators. But during the depression of the 1890s larger financial commitments were required. The concentration of holdings on the Mesabi increased as it had a few years earlier on the Vermilion, and local leadership fell by the wayside. Major corporations, most notably those led by Rockefeller and Carnegie, vertically integrated back to their raw materials, acquiring mining property, expanding the essential transportation network, and introducing skilled management and seemingly unlimited capital to the ranges. These corporate giants operated on a scale previously unattainable. The small, independent mining companies, led by enterprising individual pioneer entrepreneurs, could not function in the deteriorating economic situation nor compete in the ruthless world of big business. Unable to match the financial capacities of the large integrated firms, many were slowly absorbed or squeezed out.

At the end of the initial decades of the 20th century, Minnesota's iron range frontiers were all but closed. The golden age of the individual entrepreneur waned with changing business mores, increasing government controls, the rise of labor unions, and the depletion of mineral lands, setting the stage for a future more corporate and less personal in character.[33] The formative era of individual enterprise that ended on the Vermilion about 1888, on the eastern Mesabi about 1901, and on the western Mesabi a decade later, reached its climax on the Cuyuna by 1911. This frontier period had seen the founding of new villages and cities in a northern wilderness, the opening of major mines, the establishment of railroad systems to carry the ore to Lake Superior, the construction of the first large ore docks at Two Harbors, Superior, and Duluth, and the evolution of specially designed steamships to transport the ore down the Great Lakes. Although they evolved in succeeding years, the basic patterns on the Minnesota ranges were cast in the general outlines they were to retain until the high-grade natural ores were exhausted in the mid-20th century.

By that time Minnesota's iron ranges had supplied the steel sinews during two world wars and had continued to support an expanding industrial component in the American economy. From 1884 through 1950, 84,487,601 tons of high-grade ore were mined on the Vermilion, in 1890–1950, 1,718,766,320 tons on the Mesabi, and in 1911–50, 69,897,814 tons on the Cuyuna.[34] Not until the 1950s did significant changes take place and the patterns of the late 19th and early 20th centuries give way in Minnesota to a new kind of frontier — a technological frontier devised by scientists that made possible the utilization of the low-grade ores and taconite bypassed in earlier years.

The Vermilion Range maintained its level of high-grade ore shipments until 1958 and then declined swiftly. The last ore was taken from its stock pile in 1963, and the Soudan Mine became a unique state park two years later. From an all-time record of 3,714,000 tons in 1953 Cuyuna output dropped below 2,000,000 tons for the first time in 18 years during 1958; it has fallen steadily ever since. According to 1977 figures, the latest readily available, Cuyuna production had ceased and only 159,250 tons were shipped from stock piles. Attempts to concentrate the low-grade iron and manganese-bearing ores remaining there have thus far been unsuccessful.

In contrast, post-World War II demand for the Mesabi's ores continued at substantial levels. Although the 1953 season established the all-time record of 75,953,000 tons, output generally remained above 45,000,000 tons annually. The district experienced a slump from 1958 to 1964, but even then production did not drop below 33,900,000 tons. Labor felt the downturn, however, and employment dropped 23 per cent from 18,000 to 14,000 workers. Nonetheless Mesabi mines continued to provide more than half of the nation's total output, a position the giant range had held since early in the 20th century. The dominance of the Mesabi is evident when cumulative figures through 1977 are analyzed. Total national output topped 5,400,000,000 tons with the Mesabi providing 58 per cent, or more than 3,000,000,000 tons. It also furnished 70 per cent of the ores of the entire Lake Superior district and 93 per cent of those mined in Minnesota.

That the Mesabi Range continues to provide raw material for American steelmakers in the 1970s can be attributed to the impact of the new technology of taconite in reviving Minnesota's iron-mining industry. Heavy demand for steel during two world wars resulted in enormous ore consumption that largely depleted the state's high-grade deposits. Millions of tons of low-grade ore, passed over by generations of miners, remained. How could they be made usable? Beginning as early as the second decade of the 20th century, technical advances in mining and metallurgy focused, at first unsuccessfully, on these vast deposits.[35]

Taconite as it comes from the ground possesses little commercial value. This extremely hard rock, containing only 20 to 25 per cent iron, must be crushed, separated, sifted, and rolled in an elaborate and expensive process of beneficiation. Edward W. Davis, professor of mines, and his assistants at the Mines Experiment Station of the University of Minnesota, pioneered the development of a feasible way of extracting the fine particles of iron embedded in the hard rock. Beginning in 1913 Davis worked for decades to perfect a method and convince the industry of its practicality. In 1946 Reserve Mining Company announced plans to construct a beneficiating plant at Silver Bay and to build a new town on the shores of Lake Superior 50 miles north of Duluth.

By the time Reserve shipped the first taconite pellets in April, 1957, other firms either intended to build or had begun construction of additional facilities. Ten years later shipments of taconite pellets surpassed those of natural ores, and their percentage of the total Minnesota output has increased steadily in the intervening years. In 1979 eight taconite plants were in operation or under construction in northeastern Minnesota. Although the hurdle of satisfactorily disposing of taconite wastes in an environmentally sound and economically feasible manner remains to be overcome, the iron-ore industry is booming once again in northeastern Minnesota. Already mining operations have been re-established there in a more stable economic atmosphere than anyone would have thought possible in the early 1950s.

The vision of the early developers — Tower, the Merritts, Cuyler Adams, and others — provided the initial impetus for the discovery and development of the iron ranges of northeastern Minnesota. Thousands of people now call that region home and continue to derive employment from the resources in its earth. Laurence A. Rossman, former editor of the *Grand Rapids Herald-Review*, summarized the historical sequence when he wrote: "The history of minerals . . . has been marked by a general pattern. First comes discovery. Then comes the depletion of those minerals which are cheapest to mine and highest in quality. Then follows exhaustion and abandonment. Communities. founded in high hope often end in deep despair. In the meantime, the search for that which is rich and cheap proceeds. When search proves fruitless and the best has been exhausted, men turn to that which has been poorer and with science and invention attack remaining metals, build plants, recreate communities, expand employment and rebuild an industry whose resources may have almost unlimited life." That is exactly what happened on the iron mining frontiers of the Vermilion, Mesabi, and Cuyuna ranges of northern Minnesota.[36]

REFERENCE NOTES

CHAPTER 1 — OPENING THE LAKE SUPERIOR HINTERLAND
— *pages 1 to 15.*

[1] Elden Johnson, *The Prehistoric Peoples of Minnesota*, 9–11 (*Minnesota Prehistoric Archaeology Series*, no. 3 — 2nd ed., St. Paul, 1978); Roy W. Drier and Octave J. Du Temple, eds., *Prehistoric Copper Mining in the Lake Superior Region: A Collection of Reference Articles* (n.p., 1961); James B. Griffin, ed., *Lake Superior Copper and the Indians: Miscellaneous Studies of Great Lakes Prehistory* (University of Michigan Museum of Anthropology, *Anthropological Papers*, no. 17 — Ann Arbor, 1961); William P. F. Ferguson, "Michigan's Most Ancient Industry: The Pre-Historic Mines and Miners of Isle Royale," in *Michigan History*, 7:155–162 (July–October, 1923); Lawrence J. Burpee, ed., *Journals and Letters of Pierre Gaultier de Varennes de la Vérendrye and His Sons*, 153 (Toronto, 1927); Selwyn Dewdney and Kenneth E. Kidd, *Indian Rock Paintings of the Great Lakes*, 17, 21 (2nd ed., Toronto, 1967).

[2] Griffin, ed., *Lake Superior Copper*, 32–39; Reuben G. Thwaites, ed., *The Jesuit Relations and Allied Documents: Travels and Explorations of the Jesuit Missionaries in New France, 1610–1791*, 1:33, 50:265 (Cleveland, 1896, 1899). The most famous of the large copper rocks is the 6,000-pound Ontonagon Boulder, later exhibited at the Smithsonian Institution; F. Clever Bald, *Michigan in Four Centuries*, 232–234 (New York, 1954).

[3] Louise P. Kellogg, *The French Régime in Wisconsin and the Northwest*, 350 (Madison, 1925); Joseph E. and Estelle L. Bayliss and Milo M. Quaife, *River of Destiny: The Saint Marys*, 311 (Detroit, 1955).

[4] Kellogg, *French Régime*, 351–357.

[5] Jonathan Carver, *Travels through the Interior Parts of North America, in the Years 1766, 1767, and 1768*, 138–140 (Reprint ed., Minneapolis, 1956); Louise P. Kellogg, *The British Régime in Wisconsin and the Northwest*, 107–113 (Madison, 1935).

[6] For a thorough discussion of the material here and below, see Chapters 1 and 7, note 64, of a forthcoming study by William E. Lass tentatively entitled "Minnesota's Boundary with Canada," to be published in 1980 by the Minnesota Historical Society.

[7] Howard Jones, *To the Webster-Ashburton Treaty: A Study in Anglo-American Relations, 1783–1843*, 134–136 (Chapel Hill, N.C., 1977); Hunter Miller, ed., *Treaties and Other International Acts of the United States*, 4:366 (Washington, D.C., 1931). Thomas Le Duc, "The Webster-Ashburton Treaty and the Minnesota Iron Ranges," in *Journal of American History*, 51:476 (December, 1964), concluded that neither Webster nor Ashburton knew of the iron deposits. See also Lass, "Minnesota's Boundary with Canada," Chapter 7.

[8] William W. Folwell, *A History of Minnesota*, 1:90–97 (Reprint ed., St. Paul, 1956); Elliott Coues, ed., *The Expeditions of Zebulon Montgomery Pike*, 1:146 (Reprint ed., Minneapolis, 1965).

⁹ Folwell, *Minnesota*, 1:102; Ralph H. Brown, ed., "With Cass in the Northwest in 1820," and Bertha L. Heilbron, "Lewis Cass, Exploring Governor," in *Minnesota History*, 23:126–134 (June, 1942), 31:93–97 (June, 1950).

¹⁰ Lewis Cass to John C. Calhoun, October 21, 1820, in Clarence E. Carter, ed., *The Territorial Papers of the United States (Michigan)*, 11:65–69 (Washington, D.C., 1943); James Allen, "Report . . . of H. B. Schoolcraft's Exploration of the Country at and beyond the Sources of the Mississippi, on a Visit to the Northwestern Indians in 1832," in *American State Papers: Military Affairs*, 5:330 (Washington, D.C., 1860); Henry R. Schoolcraft, *Summary Narrative of an Exploratory Expedition to the Sources of the Mississippi River*, 127, 134, 226 (Philadelphia, 1855).

¹¹ J[oseph] N. Nicollet, *Report Intended to Illustrate a Map of the Hydrographical Basin of the Upper Mississippi River*, 63 (26 Congress, 2 session, *Senate Documents*, no. 237 — serial 380); Martha C. Bray, ed., *The Journals of Joseph N. Nicollet: A Scientist on the Mississippi Headwaters With Notes on Indian Life, 1836–37*, 5–10, 15, 99–103 (St. Paul, 1970); Warren Upham, *Minnesota Geographic Names: Their Origin and Historic Significance*, 147, 503 (Reprint ed., St. Paul, 1969); "The Spelling of Mesabi," in *Skillings' Mining Review*, vol. 20, August 15, 1931, p. 6. In 1892 the U.S. Board of Geographic Names adopted "Mesabi" as the official spelling of the name of the range.

¹² For this and the paragraph below, see Herman J. Viola, *Thomas L. McKenney, Architect of America's Early Indian Policy: 1816–1830*, 135–153 (Chicago, 1974); Charles J. Kappler, ed., *Indian Affairs: Laws and Treaties*, 2:268–273 (Washington, D.C., 1904).

¹³ Edmund J. Danziger, Jr., "They Would Not Be Moved: The Chippewa Treaty of 1854," in *Minnesota History*, 43:175–185 (Spring, 1973); Kappler, ed., *Laws and Treaties*, 2:648–652, 685–690; Folwell, *Minnesota*, 1:470–478; Samuel T. Dana, John H. Allison, and Russell N. Cunningham, *Minnesota Lands*, 81, 108 (Washington, D.C., 1960). See also Chapter 3, below.

¹⁴ Bald, *Michigan*, 200–202; Kappler, ed., *Laws and Treaties*, 2:542–545; Lew A. Chase, "Fort Wilkins, Copper Harbor, Mich.," and James Fisher, "Fort Wilkins," in *Michigan History*, 4:608–611 (April–July, 1920), 29:155–165 (April–June, 1945). Established on May 28, 1844, Fort Wilkins remained in use until August 30, 1870. It has since been restored as a historic site. Thomas Friggens, "Fort Wilkins: Army Life on the Frontier," in *Michigan History*, 71:221–250 (Fall, 1977).

¹⁵ Angus Murdoch, *Boom Copper: The Story of the First U.S. Mining Boom* (New York, 1943); William B. Gates, Jr., *Michigan Copper and Boston Dollars: An Economic History of the Michigan Copper Mining Industry* (Cambridge, Mass., 1951); Robert J. Hybels, "The Lake Superior Copper Fever, 1841–47," and Clark F. Norton, "Early Movement for the St. Mary's Falls Ship Canal," in *Michigan History*, 34:97–119, 224–244, 309–326 (June, September, and December, 1950), 39:257–280 (September, 1955).

¹⁶ Edsel K. Rintala, *Douglass Houghton: Michigan's Pioneer Geologist*, 1, 18, 33, 50–81 (Detroit, 1954).

¹⁷ Bald, *Michigan*, 237; Horace E. Burt, "William Austin Burt: Inventor," in *Michigan History*, 6:175–185 (January, 1922).

¹⁸ Peter White, "The Iron Region of Lake Superior," and Philo M. Everett, "Recollections of the Early Explorations and Discovery of Iron Ore on Lake Superior," in *Michigan Pioneer and Historical Collections*, 8:149, 11:161–174 (Lansing, 1885, 1887); Ernest H. Rankin, "The Founding of the Port of Marquette," in *Inland Seas*, 32:13–16 (Spring, 1976).

¹⁹ White, in *Michigan Pioneer and Historical Collections*, 8:158; James E. Jopling, "The Marquette Range — Its Discovery, Development and Resources," in American Institute of Mining Engineers, *Transactions*, 27:541–555 (New York, 1898); Ernest H. Rankin, "Marquette's Iron Ore Docks," in *Inland Seas*, 23:233 (Fall, 1967); D. H. Merritt, "History of the Marquette Ore Docks," in *Michigan History*, 3:424–430 (July, 1919). Walter Havighurst, *Vein of Iron: The Pickands Mather Story* (Cleveland, 1958), the only major account of this influential company, does not cover this early period. For a

comprehensive geological review, see Charles R. Van Hise and William S. Bayley, *The Marquette Iron-Bearing District of Michigan* (U.S. Geological Survey, *Monographs*, vol. 28 — Washington, D.C., 1897). All figures in this book unless otherwise specified refer to long tons of 2,240 pounds each.

[20] Charles Moore, ed., *The Saint Marys Falls Canal: Exercises at the Semi-Centennial Celebration*, 108–113, 118–129 (Detroit, 1907); Bruce Catton, *Michigan: A Bicentennial History*, 119–122 (New York, 1976).

[21] Samuel P. Ely, "Historical Address Delivered July Fourth, 1876," in *Michigan Pioneer and Historical Collections*, 7:166–169 (Lansing, 1886); Edmund A. Calkins, "Railroads of Michigan since 1850," in *Michigan History*, 13:7, 10 (Winter, 1929; Harlan Hatcher, *A Century of Iron and Men*, 62–67, 71–73 (Indianapolis, 1950).

[22] Louise P. Kellogg, "The Rise and Fall of Old Superior," in *Wisconsin Magazine of History*, 24:3–19 (September, 1940); William R. Marshall, "Henry Mower Rice," in *Minnesota Historical Collections*, 9:654–658 (1901); Superior city charter in *Superior Chronicle*, May 13, 1858; Philip R. Cloutier, "John C. Breckinridge, Superior City Land Speculator," in *Kentucky Historical Society Register*, 57:12–19 (January, 1959); Henry Cohen, *Business and Politics in America from the Age of Jackson to the Civil War: The Career Biography of W. W. Corcoran*, 159–201 (Westport, Conn., 1971); Grover Singley, *Tracing Minnesota's Old Government Roads*, 23, 24 (*Minnesota Historic Sites Pamphlet Series* no. 10 — St. Paul, 1974).

[23] John A. Bardon, "A Short History of Fond du Lac, Superior, and Duluth," typed copy, State Historical Society of Wisconsin, Madison; Walter Van Brunt, *Duluth and St. Louis County, Minnesota: Their Story and People*, 1:66, 68, 73, 95, 114, 117 (Chicago and New York, 1921). On north shore settlements below, see Jessie C. Davis, *Beaver Bay: Original North Shore Village*, 16–20 (Duluth, 1968); Julius F. Wolff, Jr., "Some Vanished Settlements of the Arrowhead Country," in *Minnesota History*, 34:177–179 (Spring, 1955); R[obert] B. McLean, *Reminiscences of Early Days of the Head of the Lakes*, 4, 8 (Duluth, [1913?]).

[24] Van Brunt, *Duluth and St. Louis County*, 1:92, 137, 139, 162; *U.S. Census, 1860, Population*, 259; *Minnesota State Census*, 1865, p. 105.

[25] Peter Temin, *Iron and Steel in Nineteenth-Century America: An Economic Inquiry*, 125–132 (Cambridge, Mass., 1964); Alan Birch, *The Economic History of the British Iron and Steel Industry, 1784–1879*, 319–330 (New York, 1968).

[26] Louis M. Hacker, *The World of Andrew Carnegie: 1865–1901*, 337–341 (Philadelphia, 1968); John N. Boucher, *William Kelly: A True History of the So-Called Bessemer Process*, 1–76, 102–113 (Greensburg, Pa., 1924).

[27] J[ames] C. Carr and W[alter] Taplin, *History of the British Steel Industry*, 98–103 (Cambridge, Mass., 1962); Lilian G. Thompson, *Sidney Gilchrist Thomas: An Invention and Its Consequences*, 126–143 (London, 1940); Birch, *Economic History*, 378–386.

[28] Lake Superior Iron Ore Association, *Lake Superior Iron Ores*, 123, 125 (Cleveland, 1938). Unless otherwise indicated the 1938 edition of this work has been cited throughout these notes. See also Lewis Beeson and Victor F. Lemmer, *The Effects of the Civil War on Mining in Michigan* (Lansing, Mich., 1966).

[29] *Lake Superior Iron Ores*, 125, 127, 308.

[30] *National Cyclopaedia of American Biography*, 13:125 (New York, 1906); Victor S. Clark, *History of Manufactures in the United States, 1860-1914*, 2:74 (Reprint ed., New York, 1949); Bernhard C. Korn, "Eber Brock Ward: Pathfinder of American Industry," 124, 158–160, 185–189, Ph.D. thesis, Marquette University, 1942; Fritz Redlich, *History of American Business Leaders*, 1:95 (Ann Arbor, Mich., 1940).

[31] Charles T. Jackson, *Report on the Geological and Mineralogical Survey of the Mineral Lands of the United States in the State of Michigan*, 477–479 (31 Congress, 1 session, *Senate Executive Documents*, no. 1 — serial 551); J[ohn] W. Foster and J[osiah] D. Whitney, *Report on the Geology of the Lake Superior Land District: The Iron Region*, part 2, p. 30, 52, 57 (32 Congress, special session, *Senate Executive Documents*, no. 4 —

serial 609); Hermann Credner, *Die Vorsilurischen Gebilde der Oberen Halbinsel von Michigan in Nord-Amerika*, 516–554 (Berlin, 1869). On the geology, see William S. Bayley, *The Menominee Iron-Bearing District of Michigan* (U.S. Geological Survey, *Monographs*, vol. 46 — Washington, D.C., 1904); John L. Buell, "Menominee Range," in Lake Superior Mining Institute, *Proceedings*, 11:38–49 (Ishpeming, Mich., 1905).

[32] Alvah L. Sawyer, *A History of the Northern Peninsula of Michigan and Its People*, 1:284–287 (Chicago, 1911). Construction was actually carried out by the Menominee River Railway Company, a subsidiary of the Chicago and Northwestern.

[33] *Lake Superior Iron Ores*, 168, 169; American Iron Ore Association, *Iron Ore 1977*, 10 (Cleveland, 1977).

[34] *Lake Superior Iron Ores*, 322.

[35] David Dale Owen, *Report of a Geological Survey of Wisconsin, Iowa, and Michigan*, 149–155, 425–447 (Philadelphia, 1852); Eleanor I. Shapiro, "Colonel Charles Whittlesey," in *Michigan History*, 28:384–389 (July–September, 1944); I[ncrease] A. Lapham, "The Penokee Iron Range," in Wisconsin State Agricultural Society, *Transactions*, 5:391–400 (Madison, 1860).

[36] "The Gogebic Range," in Lake Superior Mining Institute, *Proceedings*, 10:158–162 (Ishpeming, Mich., 1904); "History of the Gogebic Range Shipments," in *Skillings' Mining Review*, vol. 55, April 16, 1966, p. 6; "Captain Joseph Sellwood," in Lewis Publishing Company, *Minnesota*, 192 (Chicago, 1915); Railway & Locomotive Historical Society, *The Railroads of Wisconsin, 1827–1937*, 9, 17 (Boston, 1937).

[37] *Lake Superior Iron Ores*, 190, 308.

[38] Birch, *Economic History*, 371–378; Carr and Taplin, *British Steel Industry*, 31–35; William T. Hogan, *Economic History of the Iron and Steel Industry in the United States*, 2:408–413 (Lexington, Mass., 1971); Temin, *Iron and Steel*, 138–152.

CHAPTER 2 — THE VERMILION RANGE DISCOVERED — *pages 16 to 38*

[1] Burpee, ed., *La Vérendrye Journals*, 146, 150–153, 357.

[2] John McLoughlin, "Description of the Indian Country from Fort William to Lake of the Woods," [1821?], copy in St. Louis County Historical Society Collections, Minnesota Historical Society (MHS) Northeast Regional Center, University of Minnesota, Duluth (hereafter abbreviated as SLC), original in Louis R. Masson Collection, McGill University Library, Montreal.

[3] Owen, *Report*, 213; Walter B. Hendrickson, "David Dale Owen's Geological Survey of Minnesota," in *Minnesota History*, 26:222–233 (September, 1945).

[4] Owen, *Report*, 250, 312, 314, 334, 417. The Gunflint district, a narrow belt of unproductive iron formation, extends for over 100 miles from the U.S.-Canadian boundary northeast to the Animikie Range near Thunder Bay, Ont. Between 1850 and 1880 a number of geologists recorded iron-bearing rocks near Gunflint Lake. In 1886 Henry Mayhew of Grand Marais and several Minneapolis businessmen organized the Gunflint Lake Iron Company. Three years later this firm succeeded in transporting mining equipment to the Paulson Mine on Gunflint Lake by hauling it over what later became known as the Gunflint Trail; *Vermilion Iron Journal* (Tower), September 12, 1889. In 1892 the company completed construction of the Port Arthur, Duluth and Western Railroad (often referred to as "Poverty, Distress and Want" by its employees) to serve the mine. The line ran 85.5 miles from Port Arthur (now Thunder Bay), Ont., to the Paulson property. The company never located iron ore in marketable quantities. It abandoned all operations and sold the railroad under foreclosure in August, 1899.

See Newton H. Winchell, Geological Notes, vol. 52, July 5–6, 8, 21, 24, 1893, in

Winchell Papers, MHS; Henry V. Poor, *Manual of the Railroads of the United States, 1893*, 1127, *1899*, 892 (New York); Minnesota Geological Survey, *Annual Report*, 1887, pp. 65–87 (St. Paul, 1888); N[ewton] H. Winchell and Warren Upham, *The Geology of Minnesota*, 4:481–490 (Minnesota Geological and Natural History Survey, *Final Report* — St. Paul, 1899), hereafter cited as *Final Report*; Frank F. Grout, Robert P. Sharp, George M. Schwartz, *The Geology of Cook County, Minnesota*, 78 (Minnesota Geological Survey, *Bulletins*, no. 39 — Minneapolis, 1959); J. E. Gill, "Gunflint Iron-Bearing Formation, Ontario," in Canada Department of Mines, Geological Survey, *Summary Report, 1924*, Part C, 28–88 (Ottawa, 1926); Thomas M. Broderick, "Economic Geology and Stratigraphy of the Gunflint Iron District, Minnesota," in *Economic Geology*, 15:422–452 (July-August, 1920).

⁵ Winchell and Upham, *Final Report*, 1:82 (Minneapolis, 1884); J. Wesley Bond, *Minnesota and Its Resources*, 79 (Redfield, N.Y., 1853).

⁶ Dwight E. Woodbridge, "The Mesabi Iron Ore Range," in *Engineering and Mining Journal*, 79:122 (January 14, 1905).

⁷ Alexander Ramsey, "Message," in Minnesota Territory, *House Journal*, 1849, p. 15. Ramsey's role in early Minnesota mining ventures and townsites is worthy of further study, for he and his brother Justus were quietly involved in a number of such efforts. See, for example, Chapter 1, note 23, above.

⁸ Here and below, see Alice E. Smith, "Caleb Cushing's Investments in the St. Croix Valley," in *Wisconsin Magazine of History*, 28:7–19 (September, 1944); James T. Dunn, *The St. Croix: Midwest Border River*, 78–98 (Reprint ed., St. Paul, 1979); Lucile M. Kane, *The Waterfall That Built A City: The Falls of St. Anthony in Minneapolis*, 16–20 (St. Paul, 1966); Superintendent of Mineral Lands, *Report*, in 28 Congress, 1 session, *Senate Executive Documents*, vol. 11, no. 175, pp. 1–4 (serial 461).

⁹ Ramsey, "Message," in Minnesota, *House Journal*, 1860, p. 180.

¹⁰ Charles L. Anderson and Thomas Clark, *Report on Geology and Plan for a Geological Survey of the State of Minnesota*, 9, 11, 14, 24–26 (St. Paul, 1861). Clark had purchased 5/144ths of Beaver Bay Township from Earle S. Goodrich; quit claim deed, [March 26, 1857], Thomas Clark Papers, MHS.

¹¹ Henry A. Swift, "Message," in *Minnesota Executive Documents*, 1863, p. 23 (the message was delivered on January 11, 1864); Minnesota, *General Laws*, 1864, p. 111. The most ambitious of the copper ventures was the North Shore Mining Company at French River; see Henry M. Rice, "Mineral Region of Lake Superior," in *Minnesota Historical Collections*, 2:181 (1867); Van Brunt, *Duluth*, 1:162.

¹² *Superior Gazette*, November 5, 1864; Hanchett, *Report of the State Geologist*, 6, 7 (St. Paul, 1865).

¹³ Minnesota, *General Laws*, 1865, p. 84. Eames held the position for two years before the legislature refused to make further appropriations. Apparently Thomas Clark had shown an interest in the appointment; see Miller to Clark, December 23, 1864; Marshall to Clark, March 14, 1866, both in Clark Papers.

¹⁴ Henry H. Eames, *Report of the State Geologist on the Metalliferous Region Bordering on Lake Superior*, 11, 16, 19 (St. Paul, 1866), and *Geological Reconnoissance of the Northern, Middle and Other Counties of Minnesota*, 31 (St. Paul, 1866). Eames reached the Vermilion over a trail from Beaver Bay; for the route, see Winchell and Upham, *Final Report*, 6:plates 68 and 77 (St. Paul, 1901).

¹⁵ For the information here and in the three paragraphs below, see Eames, *Report of the State Geologist*, 6, 10, 19–23; George H. Primmer, "Pioneer Roads Centering at Duluth," and David A. Walker, "Lake Vermilion Gold Rush," in *Minnesota History*, 16:295–297, 44:49–54 (September, 1935, Summer, 1974). Stuntz's original map and field notes of the Vermilion Trail are in SLC. On the Bois Fort Treaty, see Kappler, ed., *Laws and Treaties*, 2:650, 916; *Superior Gazette*, November 4, 1865, February 10, 1866.

¹⁶ William E. Culkin, "George Riley Stuntz and His Times," 1935, and Burleigh K. Rapp, "The Life of George R. Stuntz," 1958, unpublished manuscripts, copies in MHS and SLC; Van Brunt, *Duluth*, 1:66.

[17] On Posey (sometimes spelled Posi), see Stuntz, "The Great Iron Range," in William F. Leggett and Frederick J. Chipman, *Duluth and Environs*, 67 (Duluth, 1895); a typed version by Stuntz entitled "Discovery of Iron Ore on the Vermilion Range" is in SLC. See also Dwight E. Woodbridge and John S. Pardee, eds., *History of Duluth and St. Louis County Past and Present*, 1:229–235 (Chicago, 1910); *Superior Gazette*, February 20, 1864. *Register of Officers and Agents, Civil, Military, and Naval in the Service of the United States*, 1867, p. 154, 1869, p. 177 (Washington, D.C., 1867, 1869) confusingly listed Posey's places of service as Wisconsin and Canada. Van Brunt, *Duluth*, 1:140, places him among the early settlers on Minnesota Point.

[18] Stuntz recalled the work described here and below in the *Vermilion Iron Journal*, November 5, 1891, and in Van Brunt, *Duluth*, 1:345. The trip is described in *Superior Gazette*, December 16, 1865. On the specimens sent to the Smithsonian, mentioned below, see Albert H. Chester, "The Iron Region of Northern Minnesota," in Minnesota Geological Survey, *Annual Report*, 1882, p. 156 (St. Paul, 1883); Helen M. White, "Report on Hematite Ore Sample 5021 in the Smithsonian Institution," typed manuscript, 1963, MHS.

[19] On Stone, see Lake Superior Mining Institute, *Proceedings*, 16:259 (Ishpeming, Mich., 1911); Van Brunt, *Duluth*, 1:184. In 1852 Stuntz was surveying the boundary between the state of Wisconsin and Minnesota Territory at the time Sargent was surveyor-general of the district of Iowa, Wisconsin, and Minnesota Territory. According to Stuntz, it was through Sargent that he reached Cooke in 1869 and convinced him of the value of Minnesota lands. U.S. Surveyor General, "Field Notes of the Survey of Township Lines," Book 148, in Minnesota Secretary of State's office, St. Paul; *Register of Officers and Agents*, 1851, p. 139 (Washington, D.C., 1851); Stuntz, in Woodbridge and Pardee, eds., *St. Louis County*, 1:232.

[20] Henrietta M. Larson, *Jay Cooke: Private Banker*, 249 (Reprint ed., New York, 1968); Poor, *Manual of Railroads, 1874–75*, 645; John L. Harnsberger, "Land, Lobbies, Railroads, and the Origins of Duluth," in *Minnesota History*, 37:89–93 (September, 1960). Van Brunt, *Duluth*, 1:165–167, and Woodbridge and Pardee, eds., *St. Louis County*, 1:242, offer interesting firsthand reminiscences of Cooke's trip.

[21] Here and below, see Harnsberger and Cecil H. Meyers, "Financing a Frontier City: The Pioneer Banks of Duluth," both in *Minnesota History*, 37:93–95, 119–121; Van Brunt, *Duluth*, 1:184, 200; Lester B. Shippee, "The First Railroad Between the Mississippi and Lake Superior," in *Mississippi Valley Historical Review*, 5:121–142 (September, 1918); Larson, *Jay Cooke*, 273–279; *United States Census*, 1870, *Population*, p. 181 (Washington, D.C., 1872). On the canal, see June D. Holmquist and Jean A. Brookins, *Minnesota's Major Historic Sites: A Guide*, 167–170 (Revised ed., St. Paul, 1972). On the Zenith City phrase, see Woodbridge and Pardee, eds., *St. Louis County*, 1:216.

[22] Larson, *Jay Cooke*, 383–411; Van Brunt, *Duluth*, 1:243.

[23] Hal Bridges, *Iron Millionaire: Life of Charlemagne Tower*, 135, 299 (Philadelphia, 1952), a well-written biography to which the present author is indebted.

[24] Meyers, in *Minnesota History*, 37:121; Bridges, *Iron Millionaire*, 23, 26, 38, 44, 48, 51–55, 93–97. See also Grace Lee Nute, "Charlemagne Tower, Developer of the Vermilion Iron Range," in *Gopher Historian*, vol. 6, April, 1952, p. 16.

[25] Bridges, *Iron Millionaire*, 99–103, 108, 296. For the Northern Pacific reorganization plans, see Larson, *Jay Cooke*, 416.

[26] Bridges, *Iron Millionaire*, 141, 299; *The Northern Pacific Railroad's Land Grant, and the Future Business of the Road*, 3 (New York, 1870), Northern Pacific pamphlet collection, no. 33, in MHS. Wilkeson served as secretary of the Northern Pacific in 1871–86 and again in 1887–89. His later letters to Tower concerning railroad matters are in Secretary's Letter Books, 1868–1921, vol. 128, Northern Pacific Railway Company Records, MHS.

[27] *Lake Superior Iron Ores*, 322; Bridges, *Iron Millionaire*, 142, 143, 146, 300. Anson Northup was hired to furnish the provisions.

[28] Bridges, *Iron Millionaire*, 144, 300.

[29] Quoted in Bridges, *Iron Millionaire*, 148, 300. See also Albert H. Chester, "Exploration of the Iron Regions of Northern Minnesota During the Years 1875 and 1880," pp. 2, 3, typed copies in MHS and SLC. Northup had already taken 20 men into the area just north of Colby Lake, where Chester's group joined them. The Indian agency was located in Township 62 North, Range 16 West, section 22. Such locations are hereafter abbreviated as T62N, R16W.

[30] Bridges, *Iron Millionaire*, 150. Wieland surveyed T60N, R13W in January and March, 1872, and T60N, R12W in March and October, 1872. This area is near present Babbitt. Humason worked in T59N, R14W (near present Hoyt Lakes) during September and October, 1873. See U.S. Surveyor General, "Field Notes of the Survey of the Subdivision and Meander Lines," Books 59, 65, 74, in Minnesota Secretary of State's office.

[31] On the development of geological knowledge of the area, see J. Morgan Clements, *The Vermilion Iron-Bearing District of Minnesota*, 64–128, 130 (United States Geological Survey, *Monographs*, no. 45 — Washington, D.C., 1903), also published in 57 Congress, 2 session, *House Executive Documents*, no. 433 — serial 4515; Newton H. Winchell, "Progress of Opinion as to the Origin of the Lake Superior Iron Ores," Geological Society of America, *Bulletin*, 23:317–328 (July 15, 1912); Frederick L. Klinger, "Geology and Ore Deposits of the Soudan Mine, St. Louis County, Minnesota," Ph.D. thesis, University of Wisconsin, 1960; P. K. Sims and G. B. Morey, eds., *Geology of Minnesota: A Centennial Volume*, 79–81, 172–176 (St. Paul, 1972); Jerome F. Machamer, "Geology and Origin of the Iron Ore Deposits of the Zenith Mine, Vermilion District, Minnesota," in Minnesota Geological Survey, *Special Publication Series*, no. 2, pp. 5–15 (Minneapolis, 1968).

[32] Geologically, "taconite," a word which was coined by Newton H. Winchell, refers to the general iron formation of northeastern Minnesota and specifically to the magnetic ores of the Mesabi Range. These very old rocks contain only 20 to 35 per cent iron and were thus of little interest during the initial exploitation of higher grade ores. See Edward W. Davis, "Taconite: The Derivation of the Name," in *Minnesota History*, 33:282 (Autumn, 1953); P. K. Sims and G. B. Morey, *Geologic Sketch of the Tower-Soudan State Park*, 7–12, 16–20 (Minnesota Geological Survey, *Education Series*, no. 3 — Minneapolis, 1966); Michael Eliseuson, *Tower Soudan: The State Park Down Under*, 3–5 (*Minnesota State Park Heritage Series*, no. 1 — [St. Paul], 1976).

[33] Edmund C. Bray, *Billions of Years in Minnesota: The Geological Story of the State*, 18–50 (St. Paul, 1977); and "How Did Nature Make Minnesota's Iron?" in *Gopher Historian*, vol. 6, April, 1952, p. 3, provide nontechnical discussions.

[34] Here and below, see Bridges, *Iron Millionaire*, 150, 151, 300. Chester's reports to Tower may have been delivered verbally; they have not been located in the Charlemagne Tower Papers, Columbia University. The MHS has a microfilm copy of portions of these papers. Chester's findings were not published until 1882. See Minnesota Geological Survey, *Annual Report*, 1882, pp. 156–160. See also *Duluth News Tribune*, February 22, 1914, magazine sec., p. [8]. John Mallmann recalled the blast in a letter to Newton H. Winchell, September 27, 1890, printed in N. H. and H. V. Winchell, *Iron Ores of Minnesota*, 175 (Minnesota Geological Survey, *Bulletins*, no. 6 — Minneapolis, 1891). See also Stuntz, in *Vermilion Iron Journal*, November 5, 1891.

[35] Bridges, *Iron Millionaire*, 151, 300; Chester, "Explorations of the Iron Regions," 4.

[36] Chester, in Minnesota Geological Survey, *Annual Report*, 1882, p. 161; Stone to Tower, February 2, 1876, Chester to Tower, May 26, 1876, with accompanying map, and Tower to Stone, December 30, 1875, March 14, 1876, letter book, 404, 580 — all in Tower Papers. The latter was quoted in part in Bridges, *Iron Millionaire*, 152, 300. Chester defended his judgment, claiming his examination of the small portion of the Mesabi Range he saw was completely accurate. Twenty-five years later Newton Winchell supported the professor, writing that "Chester's report on that part of the Mesabi Range

was unfavorable, and nothing has transpired since to invalidate his conclusions"; Winchell, "Sketch of the Iron Ores of Minnesota," in *American Geologist*, 29:155 (March, 1902). Since the mid-20th century the area has been a major source of raw material for taconite plants.

[37] Stone to Tower, March 20, June 10, 1876, and Tower to Stone, June 16, 1876, letter book, 773, Tower Papers.

[38] For ore prices, see *Lake Superior Iron Ores*, 322. On the Michigan rumors, see Bridges, *Iron Millionaire*, 155; Tower to Munson, January 5, 1880, and Tower to Stone, August 30, October 11, 1880, Tower Papers. See also Minnesota Geological Survey, *Annual Report*, 1878, p. 22 (Minneapolis, 1879).

[39] Minnesota, *General Laws*, 1872, pp. 86–88; F. Garvin Davenport, "Newton H. Winchell, Pioneer of Science," in *Minnesota History*, 32:214–225 (December, 1951).

[40] Minnesota Geological Survey, *Annual Report*, 1878, pp. 9, 23.

[41] United States, *Statutes at Large*, 17:91–96, 465.

[42] Stone's letters, here and below, are partially quoted in Bridges, *Iron Millionaire*, 157–160, 301, 302; Stone to Tower, December 17, 1879, Tower Papers. See also Frémont P. Wirth, *The Discovery and Exploitation of the Minnesota Iron Lands*, 38–41 (Cedar Rapids, Ia., 1937); Paul W. Gates, *History of Public Land Law Development*, 219, 238–247, 394–434 (Washington, D.C., 1968). The pre-emption laws were repealed by Congress in 1891.

[43] On the abuses, especially in the Duluth land district, see Gates, *Public Land Law*, 441; Wirth, *Minnesota Iron Lands*, 41–55. Beginning in 1886 the government canceled the Tower company's title to over 5,000 acres; see Bridges, *Iron Millionaire*, 248–253.

[44] Stone contract, January 26, 1880; Marshall to Tower, March 1, 1883, Tower Papers — partially quoted in Bridges, *Iron Millionaire*, 160, 161, 165, 251, 302, 310. Marshall, a salaried officer of Tower's railroad by 1884, was paid $2,400; see Duluth and Iron Range Railroad, Statement of Salaried Officers, September 20, 1884, Tower Papers.

[45] U.S. Surveyor General, "Field Notes of Subdivision Survey," T62N, R15W, Book 84.

[46] United States, *Statutes at Large*, 9:352, 519, 12:3; Minnesota, *General Laws*, 1862, pp. 121–133; Gates, *Public Land Law*, 321–335.

[47] Gates, *Public Land Law*, 334. An interesting explanation of the term "swamp lands" may be found appended to a certified copy (April 12, 1881) of the Minnesota legislature's 1875 act pertaining to the Duluth and Iron Range Railroad land grant filed in the Tower Papers. As used in this legislation, the note reads, the term "most nearly takes the place of the term 'meadow lands' . . . and is applied by surveyors who, entering a new section of land, and finding water standing upon any part of it, give the name 'swamp land' to the whole section." This frequently comprises some of "the very best agricultural and timber lands" in the state.

[48] Bridges, *Iron Millionaire*, 165, 302; Stone to Tower, April 4, 1881, Tower Papers. Scrip, a certificate issued by the government to certain groups of individuals such as soldiers and half-breeds, entitled the holder to claim a certain allotment of land. The claimants frequently sold it. See Wirth, *Minnesota Iron Lands*, 58. Stone later insisted that he "never used a piece" of *Sioux* scrip; see Stone to Tower, February 23, 1889, Tower Papers.

[49] Bridges, *Iron Millionaire*, 161, 302; Chester, "Exploration of the Iron Regions," 8; Woodbridge and Pardee, eds., *St. Louis County*, 1:235; Stuntz, in Leggett and Chipman, *Duluth and Environs*, 70.

[50] Chester, "Exploration of the Iron Regions," 6, 7; Bridges, *Iron Millionaire*, 161–164.

[51] Chester, "Exploration of the Iron Regions," 8, 9; Stuntz to Stone, September 11, 1880, partially quoted in Bridges, *Iron Millionaire*, 163, 302. Total expenses came to nearly $6,000; see Chester to Tower, January 29, March 15, 19, 1881, Tower Papers. Chester also collected specimens for the U.S. Geological Survey.

[52] Chester, in Minnesota Geological Survey, *Annual Report*, 1882, pp. 161–167; Newton Winchell, "The Discovery and Development of the Iron Ores of Minnesota," in *Minnesota Historical Collections*, 8:30 (1898).

[53] Partially quoted in Bridges, *Iron Millionaire*, 165, 302. Stone's receipt for $5,000, dated January 24, 1880, is in Tower Papers.

[54] Bridges, *Iron Millionaire*, 166–168, 302; *Biographical Directory of the American Congress, 1774–1971*, 632 (92 Congress, 1 session, *Senate Document*, no. 92–8 — Washington, D.C., 1971); Sawyer, *History of the Northern Peninsula*, 2:990–993.

[55] Minnesota, *House Journal*, extra session, 1881, p. 283; *Senate Journal*, extra session, 1881, p. 260; *General Laws*, 1881, extra session, 55. This statute remained in effect until declared unconstitutional by the attorney general in 1896 and repealed by the legislature the following year; *General Laws*, 1897, p. 41. Stone's effectiveness as a lobbyist was probably enhanced by his earlier service in the House, where he represented the 29th district; Minnesota, *Legislative Manual*, 1877, p. 100.

CHAPTER 3 — MINES AND A RAILROAD ON THE VERMILION
— *pages 49 to 72*

[1] Rice, in *Minnesota Historical Collections*, 2:182.

[2] Tower to Stone, January 10, 20, 1881, letter book, 635, 657, Tower Papers; Articles of incorporation, Book E, 581. Unless otherwise indicated all articles of incorporation cited may be found in Minnesota Secretary of State's office, St. Paul. See also Minnesota Railroad Commissioner, *Annual Report*, 1881, p. 92 (St. Paul, 1882); Minnesota, *General Laws*, 1881, p. 23, 1883, p. 3.

[3] On the Ontonagon and Duluth incorporators, see Edward W. Davis, *Pioneering with Taconite*, 7–12 (St. Paul, 1964); Articles of incorporation, Book C, 207; and Chapter 4 below. See also Minnesota, *Special Laws*, 1875, p. 287. On Graves, see Warren Upham and Rose B. Dunlap, *Minnesota Biographies, 1655–1912*, 273 (*Minnesota Historical Collections*, vol. 14 — 1912).

[4] Minnesota, *Special Laws*, 1876, p. 315, 1878, p. 526. On Spalding, who moved from Michigan's copper country to Duluth in 1869, see his "An Autobiographical Reminiscence," 30–35, typed manuscript, MHS.

[5] Smith to Stone, January 17, 1882, Tower Papers, partially quoted in Bridges, *Iron Millionaire*, 175, 303.

[6] Duluth and Iron Range Railroad Company (hereafter abbreviated as D&IR), Board of Directors' Minutes, March 1, 1882, Tower Papers. The episode is described in greater detail in Bridges, *Iron Millionaire*, 175–177.

[7] Bridges, *Iron Millionaire*, 170; Articles of incorporation, Book H, 46. A copy may also be found in the Tower Papers.

[8] On the various proposed routes, see R. H. Lee, Report, November 17, 1880, December 1, 1882, J. B. Fish to Stone, October 9, 1880, and Stuntz to Chester, August 16, 1880, Tower Papers. See also Bridges, *Iron Millionaire*, 177. According to Minnesota Railroad Commissioner, *Annual Report*, 1884, p. 199 (St. Paul, 1884), 67.5 miles of track were built from Two Harbors to Tower. On the Duluth extension, see Poor, *Manual of Railroads, 1887*, 810, and p. 60, below.

[9] Minnesota, *House Journal*, 1883, pp. 23, 231; Spalding, "Reminiscence," 32; Minnesota, *Special Laws*, 1883, p. 203. Newspapers in Duluth, St. Paul, and Minneapolis quickly took sides on the issue; for a summary of opposing viewpoints, see *St. Paul Daily Globe*, November 3, 1882.

[10] Here and below, see Minnesota, *Senate Journal*, 1883, pp. 371, 459; Tower to Stone, March 8, 1883, Tower Papers. A graphic description of the Senate fight, as well as

Tower's letter of thanks to Gilfillan, appears in Bridges, *Iron Millionaire*, 179–184, 304, and *St. Paul Daily Globe*, March 2, 1883. For biographical data on Gilfillan and Billson, respectively, see Marion D. Shutter and J. S. McLain, eds., *Progressive Men of Minnesota*, 136 (Minneapolis, 1897), and Albert N. Marquis, ed., *The Book of Minnesotans*, 45 (Chicago, 1907). On the St. Paul offices of the Duluth and Iron Range, see Minnesota Railroad Commissioner, *Annual Report*, 1883, p. 73 (Minneapolis, 1884). The Duluth and Winnipeg later became part of the Great Northern system; see Chapter 5, note 8, below.

[11] Bridges, *Iron Millionaire*, 170, 184, 185, 190, 304; D&IR Board of Directors' Minutes, April 24, May 30, 1883, Tower Papers. On Tower, Jr., see Dumas Malone, ed., *Dictionary of American Biography*, 18:607 (New York, 1936).

[12] Bridges, *Iron Millionaire*, 185, 304; Tower, Jr., to Tower, May 16, 1883, Tower Papers.

[13] Copies of Lee's degrees from Lawrence Scientific School, Cambridge, Mass., and a series of telegrams, Lee to Tower, Jr., are in the Tower Papers. For his work with Chester, see Chapter 2, above. On relations between Tower, Jr., and Lee and between Tower and his son, see Bridges, *Iron Millionaire*, 187–194. A printed prospectus of the company's plans entitled *The Vermilion Iron District* (St. Paul, 1883) is in the MHS library.

[14] These events were recalled by two participants, Thomas Owens, in the *Duluth News-Tribune*, February 6, 1938, p. 4, and William McGonagle, in "Early Recollections of the Duluth & Iron Range Rail Road," typed manuscript, in MHS, printed in *Lake County* (Two Harbors) *Chronicle*, July 26, 1923, pp. 1, 11. Both Owens and McGonagle subsequently served the D&IR and its successor, the Duluth, Missabe and Iron Range, well into the 20th century. See Sarah S. McGonagle, *William Albert McGonagle: A Biographical Memoir*, 9–13 (New York, 1935); William E. Culkin, "William Albert McGonagle," in *Minnesota History*, 11:413–420 (December, 1930); Owens obituary, *Duluth News-Tribune*, October 1, 1944, p. 1. The "Three Spot," and other equipment are preserved near the Lake County Historical Society's museum in the depot at Two Harbors.

[15] Lee to Tower, Jr., August 3, 1883, Tower Papers; Bridges, *Iron Millionaire*, 194, 199, 214, 305, 306.

[16] On Morcom, see Lake Superior Mining Institute, *Proceedings*, 14:197 (Ishpeming, Mich., 1909). On the colonists here and below, see Elisha J. Morcom, Jr., "The Discovery and Development of the Iron Ore Industry," 2, 3, typed paper, 1926, MHS; Morcom, Jr., in *Ely Miner*, August 10, 1934, p. 4; "Tower, Minn. & Soudan," in "Sketches of Range Towns," undated manuscript, in Will J. Massingham Papers, MHS; "Reminiscences of William Blamey," 2, typed paper, 1928, unpublished manuscripts file, SLC. Blamey was a member of the second group of colonists.

[17] Bridges, *Iron Millionaire*, 204, 205, 210, 305, 306; Stone to Tower, April 5, 1884, and maps of Tower in 1884 and 1893, in Tower Papers. Reports on the number of houses vary; see Morcom to Tower, Jr., December 31, 1884, Tower Papers, and "Tower, Minn. & Soudan," 11, Massingham Papers. On Owens, see [Charles E. Ellis], *Iron Ranges of Minnesota: Historical Souvenir of the Virginia Enterprise*, [17] (Virginia, Minn., 1909).

[18] Bridges, *Iron Millionaire*, 204–206, 305. See also Tower, Jr., to Tower, April 5, 1884, Tower Papers; Morcom, Jr., "Discovery," 3, 6, and notes 53, 54, below.

[19] Morcom to Tower, Jr., December 31, 1884, Tower, Jr., to Tower, March 29, 1884, Tower to Stone, April 3, 1884, Tower Papers; Bridges, *Iron Millionaire*, 209, 306; Franklin Prince to Tower, Jr., July 10, 1884, vertical file, Iron Ore Mining — Minnesota Iron Co., SLC; *Duluth Tribune*, March 14, 1884.

[20] On construction problems, see Bridges, *Iron Millionaire*, 214–219, 306; Lee to Tower, Jr., May 2, 1884, Ely to Tower, Jr., May 15, 1884, Tower Papers; D&IR, *Transportation of Iron Ore*, 3 (Duluth, 1927).

[21] For participants' descriptions of the day, see Tower, Jr., to Tower, August 2, 1884,

Tower Papers; Morcom, Jr., "Discovery," 4; Owens, in *Duluth News-Tribune*, February 6, 1938, p. 4. See also Bridges, *Iron Millionaire*, 220–223. Just as a fir tree signals the beginning, a broom is placed in the last car when a pit or shaft is closed down permanently, signifying a clean sweep.

[22] Tower to George and Samuel Ely, August 16, 1884; Carnegie Brothers & Company to George and Samuel Ely, August 24, 1884, Morcom to Tower, Jr., December 31, 1884, Tower Papers; *Lake Superior Iron Ores*, 199. On the ore dock, see D&IR, *Transportation of Ore*, 3; Thomas Owens, "Early Iron Ore Operations in Minnesota," 8–10, unpublished manuscripts file, 1932, SLC. See also Chapter 9, note 36, below.

[23] Ely to Tower, May 15, 1884, Tower Papers; Bridges, *Iron Millionaire*, 231, 307.

[24] Bridges, *Iron Millionaire*, 225, 229, 232, 307. The railroad spent $1,996,633.04 on construction and $458,305.20 for equipment. Tower contributed $1,750,000 in return for mortgage bonds at par according to D&IR, Annual Report, 1884, Tower Papers.

[25] Ely to Stone, February 7, 1883, partially quoted in Bridges, *Iron Millionaire*, 226, 307; Stone to Tower, February 19, 1883, Tower Papers; biographical sketch (typed) in Samuel P. Ely Papers, Baker Library, Harvard University. For the terms of Ely's 1883 purchase, see Bridges, 227.

[26] Ely, enclosed in Stone to Tower, July 28, 1884, Tower Papers; George and Samuel Ely to Tower, August 15, 1884, quoted in Bridges, *Iron Millionaire*, 230–232; *Lake Superior Iron Ores*, 199. A copy of the Ely contract, April 19, 1884, is in Tower Papers.

[27] Bridges, *Iron Millionaire*, 233–237, 308.

[28] Bridges, *Iron Millionaire*, 237, 308. On the loan arrangements with Silverman, see also memorandum of "A Meeting of the Board of Directors of the D&IRR," December [n.d.], 1884, Tower Papers.

[29] Bridges, *Iron Millionaire*, 238, 308; Lee to Tower, Jr., November 9, 1884, Tower Papers.

[30] Bridges, *Iron Millionaire*, 242, 243, 308, 309; Clark, *History of Manufactures*, 2:194.

[31] *Lake Superior Iron Ores*, 322; Bridges, *Iron Millionaire*, 246, 309.

[32] *Minneapolis Tribune*, August 17, 1886; Minnesota Railroad and Warehouse Commission, *Annual Report*, 1887, part 2, p. 133 (St. Paul, 1887); Lee to Tower, Jr., November 27, 1885, May 4, 1886, Tower Papers; Bridges, *Iron Millionaire*, 265, 311. The Duluth and Northern was incorporated by nine men from Duluth and the Twin Cities, including R. S. Munger, Luther Mendenhall, and George H. Christian, in August, 1885; Articles of incorporation, Book M, 251.

Early in 1885 Stone announced that the D&IR would request from the legislature treatment equal to that granted other lines, saying it was a "simple matter of justice to a company that has in good faith expended in the state during the past eighteen months over $2,500,000." The legislature responded with an amendment to the 1875 act aiding the railroad and reiterating the proviso that the track to Duluth be built by March 3, 1888. See *St. Paul Daily Globe*, February 11, 1885; Minnesota, *General Laws*, 1885, p. 258.

On the selection of land grant sections by the D&IR and the lawsuits concerning them, see *Duluth Daily News*, January 15, December 18, 1890. By June, 1888, about 85,000 acres had been awarded. The rest were received by 1894. Minnesota Iron Company, *Annual Report*, June 11, 1888, p. 5 (Chicago, 1888), copy in MHS; Bridges, *Iron Millionaire*, 266, 311.

[33] Bridges, *Iron Millionaire*, 194–197, 305. See also John Birkinbine, "The Iron Mines of Minnesota," in *Iron Age*, vol. 39, April 14, 1887, p. 29. Minnesota Manuscript Census Schedules, Lake County, 1885, show a total population of 238 men and 112 women; all manuscript censuses cited in this volume are in the Minnesota State Archives, MHS. A brief analysis reveals that 23 per cent of the people were from Canada, 14 per cent were born in Sweden, 13 per cent in Michigan, 9 per cent in Norway, and 8 per cent in Minnesota. Two Chinese men are also listed. Two Harbors became a city on February 26, 1907; see Upham, *Minnesota Geographic Names*, 294.

³⁴ Tower to Stone, July 10, 1883, and Tower to Tower, Jr., August 22, 1883, quoted in Bridges, *Iron Millionaire*, 195, 305.

³⁵ Birkinbine, in *Iron Age*, vol. 39, April 14, 1887, p. 29; *Vermilion Iron Journal*, January 1, 1891; *U.S. Census*, 1890, *Population*, p. 191. According to D&IR, *Transportation of Iron Ore*, 10, the dock was only 36 feet high and had 46 pockets. For a description of the five parallel double docks in 1900, see John L. Morrison, "Two Harbors in 1900," in *Iron Trade Journal*, December 27, 1900, pp. [3–5]. See also Chapter 9, page 216, below.

³⁶ Minnesota Manuscript Census Schedules, St. Louis County, 1885; Upham, *Minnesota Geographic Names*, 490; Sisters Bernard Coleman and Verona LaBud, *Masinaigans: The Little Book, A Biography of Monsignor Joseph F. Buh*, 168 (St. Paul, 1972). See also Timothy L. Smith, "Factors Affecting the Social Development of Iron Range Communities," 9, unpublished manuscript, 1963, MHS.

³⁷ "Tower, Minn. & Soudan," in Massingham Papers. Timber resources were responsible for the founding of nearby Winton, for example; see Ely Roaring Stoney Days Committee, *A Souvenir Booklet*, 56–62 (Ely, 1958).

³⁸ *Vermilion Iron Journal*, July 19, 26, 1888, January 1, 1891; Van Brunt, *Duluth*, 1:381. See also O. A. Wiseman, "The Lumbering Industry at Tower," unpublished manuscripts file, 1926, SLC.

³⁹ *Duluth Daily News*, April 13, 1890; *Vermilion Iron Journal*, January 7, 1892; Anthony C. Schulzetenberge, "Life on the Vermilion Range Before 1900," 21, typed manuscript, 1963, copy in MHS library.

⁴⁰ On the incorporation of Tower, see Minnesota, *Special Laws*, 1889, p. 105.

⁴¹ Birkinbine, in *Iron Age*, vol. 39, April 14, 1887, p. 29.

⁴² Henry R. Mussey, *Combination in the Mining Industry: A Study of Concentration in Lake Superior Ore Production*, 95–102 (Columbia University, *Studies in History, Economics and Public Law*, vol. 23, no. 3 — New York, 1905).

⁴³ Henry H. Porter, *H. H. Porter: A Short Autobiography*, 10, 11, 13, 15, 19, 21–23, 31 (Chicago, 1915); Bridges, *Iron Millionaire*, 267, 268, 311. On Union Steel as a customer, see Ely letter enclosed in Stone to Tower, July 28, 1884, Tower Papers.

⁴⁴ Bridges, *Iron Millionaire*, 268, 311; Porter, *Autobiography*, 35; *New York Times*, July 12, 1887; Tower, Jr., to Tower, November 19, 1886, Tower Papers.

⁴⁵ Ely to Tower, Jr., January 29, February 14, 1887, quoted in Bridges, *Iron Millionaire*, 270, 311.

⁴⁶ Silverman to Tower, August 25, 1886, Tower, Jr., to Silverman, August 30, 1886, Tower Papers. The latter stated the belief that Porter and Jay Morse were behind the proposal. Ely to Tower, August 25, 1886, indicated that Samuel Ely had seen Silverman's letter of that date and quoted Morse as saying "the proposition to buy out the Minnesota Iron Company came from Silverman alone."

⁴⁷ On the meeting, see Tower, Jr., to Tower, February 9, 1887, and Stone to Tower, Jr., April 20, 1887, Tower Papers, both partially quoted in Bridges, *Iron Millionaire*, 269, 273, 311. The Bridges quotation on Tower's attitude is on p. 271.

⁴⁸ Bacon to Morse, April 21, 1887, in Bridges, *Iron Millionaire*, 272, 311. Bacon's election to the presidency is noted in *Duluth Daily News*, October 18, 1891.

⁴⁹ Ely to Morse, April 28, 1887, quoted in Bridges, *Iron Millionaire*, 272; Railroad and Warehouse Commission, *Annual Report*, 1887, part 2, p. 133.

⁵⁰ Sales agreement, May 3, 1887, in Minnesota Mining & Railroad Syndicate Papers, Baker Library, Harvard University; Tower to Albert Tower, May 7, 1887, in Albert Tower Papers, MHS; *Minneapolis Daily Tribune*, May 11, 1887; *New York Times*, May 17, 1887; Bridges, *Iron Millionaire*, 273, 311.

⁵¹ Lee to Tower, May 27, June 12, 1889, Tower Papers; Bridges, *Iron Millionaire*, 277–279, 312. Stone was not included in the new arrangements. He had come to regard Tower, Jr., as his bitter enemy as early as 1886. See Bridges, *Iron Millionaire*, 254–264, 274, 280. Stone remained active in mining circles after his association with Tower ended; he met a tragic death on October 25, 1900, when he was asphyxiated by gas in a room at

Duluth's Kitchi Gammi Club. *Duluth Daily News*, December 1, 1890, January 19, 1891, and *Duluth News Tribune*, April 25, 1890, October 26, 1900, p. 8.

On Tower, Jr.'s, diplomatic service, see *Dictionary of American Biography*, 18:607. The "powerful, grasping, combative syndicate" was charged with forcing Tower to sell his property by Frank Wilkeson, a journalist who was the son of Samuel Wilkeson, Tower's long-time associate in the Northern Pacific. In articles which appeared in the *New York Times* of July 10, 12, 1887, the journalist dwelled on fraudulent land claims and criticized the Rockefellers as "monopolists" and "robber barons of old" who were "covertly menacing" to the "plucky individual enterprise" of Tower.

[52] Minnesota Iron Company, *Annual Report*, 1888, p. 3, 5, 6; Bridges, *Iron Millionaire*, 279–281.

[53] Winchell and Winchell, *Iron Ores of Minnesota*, 177; *Vermilion Iron Journal*, January 26, 1888; *Lake Superior Iron Ores*, 199. According to Morcom, Jr., "Discovery," 6, the Stone and Breitung pits were the first to be developed.

[54] For descriptions of mining methods on the Vermilion here and below, before and after 1900, see Winchell and Winchell, *Iron Ores of Minnesota*, 176–186; Charles E. Van Barneveld, *Iron Mining in Minnesota*, 186–204 (University of Minnesota, School of Mines Experiment Station, *Bulletins*, no. 1 — Minneapolis, 1913); John H. Hearding, "Mules," undated typed manuscript, MHS. On the Soudan, see also D. H. Bacon, "The System of Filling at the Mines of the Minnesota Iron Company, Soudan, Minn.," in American Institute of Mining Engineers, *Transactions*, 21:299–304 (New York, 1892); Kenneth Duncan, "The Soudan Mine and the Minnesota Iron Company," in *Skillings' Mining Review*, vol. 56, November 4, 1967, pp. 22–24. In 1963 the Oliver Iron Mining Division of U.S. Steel deeded the mine to the state of Minnesota, which in July, 1965, opened it as the unique Tower-Soudan State Park; see Holmquist and Brookins, *Minnesota's Major Historic Sites*, 171–174.

[55] William B. Phillips, "Iron and Steel," in *Mineral Industry*, 1:272–274 (1892), an annual statistical publication; Minnesota Bureau of Labor, *Biennial Report*, 1901–02, in Minnesota, *Executive Documents*, 1903, vol. 2, p. 879 (St. Paul, 1903).

[56] U.S. Surveyor General, "Field Notes of the Subdivision Survey," T61–63N, Book 60; *Vermilion Iron Journal*, February 9, March 15, 1888; Porter, *Autobiography*, 35. The state of Wisconsin named Pattison State Park, some 20 miles south of Superior, in their honor. On Martin Pattison, see Woodbridge and Pardee, eds., *St. Louis County*, 2:865.

[57] Winchell and Winchell, *Iron Ores of Minnesota*, 196–198; Mussey, *Combination in the Mining Industry*, 95; Van Brunt, *Duluth*, 1:365–370; *Vermilion Iron Journal*, February 9, 1888; *Duluth Daily News*, August 3, 1891; *Duluth Daily Tribune*, November 10, 1890; *Lake Superior Iron Ores*, 199 (1938), 227 (1952); "Chandler Iron Mine Concludes Long and Useful Career," in *Skillings' Mining Review*, vol. 31, November 28, 1942, p. 8; *Ely Miner*, September 9, 1937, p. 1.

[58] Construction can be followed in *Vermilion Iron Journal*, especially October 13, December 15, 1887; see also Minnesota Iron Company, *Annual Report*, 1888, p. 4.

[59] F. W. McKinney & Company, *Spalding on the Shagawa*, 12, 16 (Duluth, [1888?]). For a glowing description of early Ely, see *Vermilion Iron Journal*, March 15, 1888. Some sources, including Upham, *Minnesota Geographic Names*, 482, state that the town was named for Arthur rather than Samuel Ely, but others indicate that not only the village but also the first child born there were named for Samuel; see, for example, Van Brunt, *Duluth*, 1:382.

[60] [Ellis], *Iron Ranges of Minnesota*, [70–74]; Van Brunt, *Duluth*, 1:383–389; Coleman and LaBud, *Masinaigans*, 175.

[61] Van Brunt, *Duluth*, 1:383, 385; *U.S. Census*, 1890, *Population*, 1:204. By 1900 Ely's population had risen to 3,717; by 1910 to 3,972; see Van Brunt, 1:390.

[62] On the D&IR extension to the Zenith Mine, for example, see *Duluth Daily News*, August 7, 1892. The other exceptions occurred in 1924, 1932–35, and 1938. See *Lake Superior Iron Ores*, 199, 308 (1938), 227 (1952); on the suspension, see *Vermilion Iron*

Journal, August 3, 10, 1893; Minnesota Bureau of Labor, *Biennial Report*, 1899–1900, in Minnesota, *Executive Documents*, 1900, vol. 3, p. 996 (St. Paul, 1900); *Minnesota State Census*, 1895, pp. 258, 259.

⁶³ *Vermilion Iron Journal*, June 23, July 7, 1892. In this period "Austrian" was a general term applied to any immigrant from the lands encompassed by the Austria-Hungary empire; at this time those on the Vermilion Range were largely Slovenes. Earlier the Tower enterprise had suffered a strike by dock hands and sawmill workers at Two Harbors in 1886; Bridges, *Iron Millionaire*, 247.

⁶⁴ Tower obituary, in *New York Daily Tribune*, July 26, 1899, and others collected in Tower Papers.

CHAPTER 4 — THE MESABI — A NEW EL DORADO — *pages 73 to 98*

¹ The Mesabi surpassed the Marquette shipments as early as 1895; see *Lake Superior Iron Ores*, 308; Orlin M. Sanford, quoted in *Duluth Daily News*, June 25, 1890. For an early reliable description, see Charles K. Leith, *The Mesabi Iron-Bearing District of Minnesota*, 20–24, 191–194, 206–208 (U.S. Geological Survey, *Monographs*, vol. 43 — Washington, D.C., 1903).

² Leith, *Mesabi Iron-Bearing District*, 22; Grace Lee Nute, ed., *Mesabi Pioneer: Reminiscences of Edmund J. Longyear*, 30 (St. Paul, 1951).

³ Eames, *Geological Reconnoissance*, 18, 56.

⁴ Otto E. Wieland, "Early Beaver Bay and Its Part in the Discovery of Iron," 3, typed paper, 1938, MHS; Helen Wieland Skillings, *We're Standing on Iron! The Story of the Five Wieland Brothers, 1856–1883*, 49–51 (Duluth, 1972).

⁵ On the activities of the Ontonagon Pool here and three paragraphs below, see Davis, *Pioneering with Taconite*, 8–13. Economic and character references for these men are in Dun & Bradstreet, Inc. Papers, Michigan no. 54, Ontonagon County, in the Baker Library, Harvard University.

⁶ On the D&IR, see Chapter 3, above.

⁷ In December, 1905, Mesaba stockholders sold all their remaining property, 8,840 acres, to George A. St. Clair of Duluth, receiving nearly $10.00 per acre for it. See Davis, *Pioneering with Taconite*, 13, 16.

⁸ Chester, in Minnesota Geological Survey, *Annual Report*, 1882, pp. 156–160. On Geggie, see *Duluth Daily Tribune*, April 25, 1890; Geggie to Dwight E. Woodbridge, October 22, 1904, SLC, copy in MHS. Geggie pointed out that "this was before the discovery of soft ore at Mountain Iron." On Mallmann here and below, see Lee to Tower, Jr., August 27, 1888, Tower Papers; Minnesota Geological Survey, *Annual Report*, 1888, p. 89, 1889, p. 7 (St. Paul, 1889, 1890); Winchell, in *Minnesota Historical Collections*, 8:33; see also notes 32, 33, below.

⁹ Biographical data on the Merritt family are scattered and often contradictory. The most reliable sources are: Andrus R. and Jessie L. Merritt, "The Story of the Mesabi," 2–13, a typed manuscript reminiscence, 1934, in MHS, hereafter cited as Andrus, "Story"; Alfred Merritt, "Reminiscences of Early Days of the Head of the Lakes," a typed manuscript, 1915, in SLC, published with minor changes under the title, "Autobiography," in Van Brunt, *Duluth*, 3: 1086–1095 (Chicago and New York, 1921). It will be cited hereafter as Alfred, "Reminiscences." Also useful is *A Biographical History with Portraits of Prominent Men of the Great West*, 220 (Lewis J.), 492 (Cassius), 588 (Andrus), 610 (Leonidas), 632 (Napoleon), 656 (Alfred) (Chicago, 1894).

The only full-length, published volume is Paul De Kruif, *Seven Iron Men* (New York, 1929). It is highly romanticized and factually unreliable. De Kruif's seven Merritts were:

Leonidas, Alfred, Cassius, Lewis J., Napoleon, John E. (eldest son of Lucien), and Wilbur J. (eldest son of Jerome), but Glen J. Merritt, "Life of Alfred Merritt," 8, undated typed paper in the author's possession (copy in MHS), maintained that the three brothers — Leonidas, Alfred, and Cassius — "were the only Merritt sons who ever had anything to do with these iron discoveries."

A total of ten children were born to Lewis and Hephzibah or Hephzibeth, whose name is variously spelled in her own signatures as well as by family members. Two of them, Annis and Eugene, died in childhood of scarlet fever, according to Andrus, "Story," 2, 69. See also Arthur P. Burris, comp., *Burris Ancestors*, 80–95 (Minneapolis, 1974), including the Merritt line, and a genealogical table of four generations in *Duluth Daily Commonwealth*, May 24, 1893.

For some time the Merritts operated a hotel in Oneota; Merritt Ledger Book, 1858–70, "Cash from transients," under accounts for Merritt House, 175, 213, in Merritt Family Papers, SLC.

[10] Alfred, "Reminiscences," 1, 7; Andrus, "Story," 10, 15, 17, 40, 71, 81–83, 105–114, 127, 169; *Biographical History with Portraits*, 495, 588, 632–635. Both Andrus and Alfred stated that Jerome and Lucien did not make the trip to Minnesota with the others in 1856. Napoleon had preceded the group to Oneota and was on hand to greet his mother when she arrived. Lucien served briefly at the Methodist Episcopal Church in Two Harbors from October, 1889, to September, 1890, and at the Central Avenue Methodist Church in Oneota from 1890 to 1895; see Morrison, in *Iron Trade Journal*, December 27, 1900, p. [35] and *90th Anniversary Central Avenue Methodist Church 1856–1946*, [10] ([Duluth, 1946?]), copies in MHS.

[11] Alfred, "Reminiscences," 12; Andrus, "Story," 63–69, 71, 95. See also Franklin A. King, "The Remarkable Merritt Family," in *Missabe Iron Ranger*, house organ of the Duluth, Missabe and Iron Range Railroad, May, 1967, p. 2.

[12] Alfred, "Reminiscences," 12; Andrus, "Story," 95–97, 115. The "Chaska," 67 feet long and 17 feet wide, sank in February, 1874, according to information at the Marine Museum, Duluth. Government piers were built at Superior between 1867 and 1875; see 41 Congress, 1 session, *House Executive Documents*, no. 1, part 2, vol. 2, p. 40. On the extent of Lon's business enterprise, see Lon to Anak A. Harris, October 25, 1894, and *List of Lands in The Duluth Land District of Minnesota owned by Eaton & Merritt* (Duluth, 1884), Merritt Papers.

[13] Andrus, "Story," 83–95, 97, 98; Alfred, "Reminiscences," 13; *Biographical History with Portraits*, 656; Glen, "Life of Alfred," 7. On Isle Royale, see Andrus to "Dear Brothers and Sisters," and to "Dear friends at home," June 18, 21, 1874, Merritt Papers.

[14] Andrus, "Story," 99, 117, 124, 125; Alfred, "Reminiscences," 14; Charles A. Norcross, "Missabe: The Story of a Modern Industrial Spoliation," 30, [1912], Merritt Papers, copy in MHS; *Duluth Daily Commonwealth*, May 24, 1893. On the Merritt timber holdings, see, for example, bills of sale, November 26, 1879, January 14, 1880, Merritt Papers, and a volume of the records of Eaton & Merritt, SLC. On the public offices of Alfred and Cassius, see Glen, "Life of Alfred," 7; *Biographical History with Portraits*, 492. Lon refused to run for the legislature in 1890 but served in the Minnesota House in 1893 and as Duluth commissioner of public utilities, 1914–17, and commissioner of finance, 1921–25; see *Duluth Daily News*, October 17, 1892; *Dictionary of American Biography*, 12:571 (New York, 1933); W. F. Toensing, *Minnesota Congressmen, Legislators, and other Elected State Officials: An Alphabetical Check List, 1849–1971*, 81 (St. Paul, 1971).

[15] Andrus, "Story," 125.

[16] [Leonidas Merritt], "Outline History of the Discovery and Development of the Missabe Iron Range and the Merritts' Connection with the Same," 1–3, typed manuscript, Merritt Papers; Alfred, "Reminiscences," 11; Andrus, "Story," 49. The Merritts' interest in the Vermilion is suggested by the articles of incorporation for a Vermillion Range Iron Company, dated June 10, 1889, Merritt Papers.

[17] Norcross, "Spoliation," 34–36, Merritt Papers; Edmund J. Longyear, "Explorations

on the Mesabi Range," in American Institute of Mining Engineers, *Transactions*, 27:537, 541, reprinted in Nute, ed., *Mesabi Pioneer*, 95–100.

[18] Andrus, "Story," 133. A pace equaled 2.64 feet and 2,000 paces constituted a mile, according to Nute, ed., *Mesabi Pioneer*, 37.

[19] Andrus, "Story," 128–132. Minnesota Constitution, article 8, section 2, provided that "the principal of all funds arising from sales . . . of lands, or other property, granted or entrusted to this state in each township for educational purposes, shall forever be preserved inviolate . . . and the income arising from the lease or sale of said school lands shall be distributed to the different townships throughout the state." See William Anderson and Albert J. Lobb, *A History of the Constitution of Minnesota With the First Verified Text*, 233 (University of Minnesota, *Studies in the Social Sciences*, no. 15 — Minneapolis, 1921); Samuel G. Iverson, "The Public Lands and School Fund of Minnesota," in *Minnesota Historical Collections*, 15:290, 293, 297–299 (1915). Lon told a congressional committee in 1911 that he had "induced Auditor Brayden [*sic*] . . . to introduce a lease law . . . and I stayed there all winter and got that law through in the last days . . ."; see 62 Congress, 2 session, House of Representatives Committee on Investigation of U.S. Steel Corporation, *Hearings*, 3:1890 (Washington, D.C., 1912), hereafter cited as Stanley Committee, *Hearings*. The committee was chaired by Augustus O. Stanley.

[20] Minnesota, *General Laws*, 1889, p. 68; Wirth, *Minnesota Iron Lands*, 49–51, 172; Roy O. Hoover, "Leonidas Merritt and the Braden Bill," tape of paper presented October 20, 1978, MHS annual meeting, in MHS. Minnesota acquired a substantial sum to aid public education during the initial years of iron mining operations. Between fiscal years 1890–1912, for example, the state received $443,665.77 from mineral leases and contracts and $2,059,991.85 from ore royalties. The Mountain Iron Mine lands that should have contributed to the school fund did not do so. On the omission of this mine and for revenue figures, see Matthias N. Orfield, *Federal Land Grants to the States With Special Reference to Minnesota*, 219, 228–234 (University of Minnesota, *Studies in the Social Sciences*, no. 2 — Minneapolis, 1915).

[21] Stanley Committee, *Hearings*, 3:1887, 1890; Winchell and Winchell, *Iron Ores of Minnesota*, 350–355; Norcross, "Spoliation," 37–41, Merritt Papers. One volume of township maps of the Mesabi Range marked and annotated with respect to ore deposits in Itasca and St. Louis counties by Alfred and Cassius Merritt is in the MHS; a number of other such maps are in the Merritt Papers. The Merritt explorations are described by Lon in a letter to Anak Harris, October 25, 1894, Merritt Papers. For biographical data on John E., who served in the Minnesota legislature from 1923 to 1931, see his obituary in *Aitkin County* (Aitkin) *Republican*, July 7, 1932, p. 1. On acquisition of mineral land, below, see Wirth, *Minnesota Iron Lands*, 172. See also Lewis Merritt to "Dear Brothers and Nephew," October 1, 1897, Merritt Papers.

[22] On May 16, 1892, the Merritts established the Iron Exchange Bank in Duluth. Within a year it held nearly $500,000 in deposits and played a major financial role in the family's Mesabi operations before it was forced to close in 1895; see Meyers, in *Minnesota History*, 37:124. On logging in the area, see R. Newell Searle, *Saving Quetico Superior: A Land Set Apart*, 12 (St. Paul, 1977); Folwell, *Minnesota*, 4:15 (Reprint ed., St. Paul, 1969); Agnes M. Larson, *History of the White Pine Industry*, 277 (Minneapolis, 1949). On the upswing of Duluth, see *Minnesota State Census*, 1885, Appendix A, 65 (St. Paul, 1885); *U.S. Census*, 1890, *Population*, 1:203, 360; *R. L. Polk's Duluth Directory*, 1885–86, 1889–90.

[23] Horace V. Winchell, "The Mesabi Iron Range," in Minnesota Geological Survey, *Annual Report*, 1891, p. 114 (Minneapolis, 1893). The Winchells continued to be impressed with the extent and value of Mesabi deposits. In 1897 Newton testified that he had advised State Auditor Braden "to hold on to" state-owned lands as potentially profitable; see Minnesota, *House Journal*, 1897, appendix, 94. See also Andrus, "Story," 149–151; Stanley Committee, *Hearings*, 3:1891.

[24] Alfred, "Reminiscences," 15; Andrus, "Story," 149. Neil McInnis, a storekeeper at

Tower, sold the Merritts "a lot of supplies, principally groceries," and shipped them to Alfred's crew via dog teams; see *Eveleth News*, December 19, 1906, p. 2.

25 Articles of incorporation, Book Z, 401. The other partners were James T. Hale and Roswell H. Palmer of Duluth. For a map of the Merritts' holdings in this area, see Hansen Evesmith, "Mountain Iron Mine," 4, in "Sketches of the History of Mines of the Mesaba Iron Range of Minnesota," 1932, typed copies in MHS and SLC. Evesmith was assistant to Cassius C. Merritt, treasurer of the Duluth, Missabe and Northern.

26 Norcross, "Spoliation," 43; Merritt Papers. For brief sketches of Chase, see Marquis, *Book of Minnesotans*, 85; Upham and Dunlap, *Minnesota Biographies*, 118.

27 Stanley Committee, *Hearings*, 3:1889; Andrus, "Story," 152.

28 For a recent geological assessment, see Sims and Morey, *Geology of Minnesota*, 204–217. See also Newton H. Winchell, "The Iron Ore Ranges of Minnesota, and Their Differences," in Minnesota Academy of Science, *Bulletins*, 5:61–66 (September, 1911); William H. Emmons, "The Iron Ores of Minnesota," in *Journal of Geography*, 14:179–181 (February, 1916); George M. Schwartz and George A. Thiel, *Minnesota's Rocks and Waters: A Geological Study*, 242, 244–248 (Revised ed., Minneapolis, 1963).

29 Leith, *Mesabi Iron-Bearing District*, 195, 198, 206, 208, 218, 223. As a general rule miners found the top ore of poorer quality because it contained a higher percentage of phosphorous and therefore did not meet Bessemer standards. The highest grade ore and the most easily marketable product lay in the heart of the deposit.

30 Nute, ed., *Mesabi Pioneer*, 14.

31 Stanley Committee, *Hearings*, 3:1889; John E. Merritt, "Speech Made at the 40th Anniversary of the Discovery of Mountain Iron Mine, June 30, 1931," 3, typed copy, MHS. For a brief history of the Mountain Iron Mine, abandoned in 1956 and declared a National Historic Landmark, see Holmquist and Brookins, *Minnesota's Major Historic Sites*, 175–178. On the early pits at Mountain Iron, see *Duluth Tribune*, November 5, 1890.

32 Nute, ed., *Mesabi Pioneer*, 5, 10, 12–35; Evesmith, "Mesaba Station," in "Sketches." Mallmann had assisted Stuntz on the Chester expedition of 1875 and managed an exploration in 1888 for a group of investors which included Charlemagne Tower, Jr., and Richard H. Lee. See Van Brunt, *Duluth*, 1:353, 374, 395, 398; Lee to Tower, Jr., August 27, 1888, Tower Papers. On Mallmann's later work, see, for example, Evesmith, "Siphon Mine," in "Sketches"; *Duluth Daily News*, February 23, October 23, November 3, 1890; *Duluth News Tribune*, February 10, 1901.

33 Nute, ed., *Mesabi Pioneer*, 13; E. J. Longyear, "Explorations on the Mesabi Range," in American Institute of Mining Engineers, *Transactions*, 27:537–539 (July, 1897); Kirby Thomas, "Test Drilling on the Mesabi Iron Range," and A. L. Gerry, "Iron Mining on the Mesabi Range," in *Engineering and Mining Journal*, 75:896, 79:319, 94:693 (June 13, 1903, February 16, 1905, October 12, 1912).

34 Nute, ed., *Mesabi Pioneer*, 5–14, 102–109; Edgar K. Soper, "The Iron Ranges of Minnesota," in *Engineering and Mining Journal*, 91:769 (April 15, 1911); Van Barneveld, *Iron Mining in Minnesota*, 29, 33.

35 Nute, ed., *Mesabi Pioneer*, 7; John G. Cohoe, "Recollections of Early Mesaba," 3, unpublished manuscripts file, 1939, SLC.

36 Van Brunt, *Duluth*, 1:422; *Minnesota State Census*, 1895, pp. 50, 126, 192, 258 (St. Paul, 1895); *Duluth Evening Herald*, February 12, 1898. By 1900 only 62 residents remained; *U.S. Census*, 1900, *Population*, p. 226.

37 *Duluth News Tribune*, March 6, 1892, June 4, 1893; Nute, ed., *Mesabi Pioneer*, 46–48; [Ellis], *Iron Ranges of Minnesota*, [79].

38 Nute, ed., *Mesabi Pioneer*, 35. The condition of the Mesabi Trail was a frequent source of concern and complaint; see *Duluth Tribune*, March 11, April 13, 1892.

39 "Mount Iron, Minn.," in "Sketches of Range Towns," undated, Massingham Papers; *Duluth News Tribune*, June 4, 1893. See also [Ellis], *Iron Ranges of Minnesota*, [79]; "Mountain Iron," in *Skillings' Mining Review*, vol. 47, October 18, 1958, pp. 4, 26.

40 *Duluth News Tribune*, June 19, 20, 1893; Minnesota Manuscript Census

Schedules, St. Louis County, 1895; *Minnesota State Census*, 1895, pp. 50, 192, 258; *Skillings' Mining Review*, vol. 47, October 18, 1958, p. 27. For later data, see Smith, "Social Development of Iron Range Communities," 10–12.

41 Andrus, "Story," 157; Articles of incorporation, Book A-2, p. 70; Cohoe, "Early Mesaba"; "Biwabik," in *Skillings' Mining Review*, vol. 47, June 14, 1958, p. 4; Evesmith, "Biwabik," 3, in "Sketches"; on Wilbur J., see obituary in *Minneapolis Star-Journal*, September 17, 1931, p. 1; Leonidas Merritt to Members and Shareholders, September 26, 1890, in Alexander Chambers Papers, MHS.

42 Nute, ed., *Mesabi Pioneer*, v, 102–109.

43 Here and below, see *Duluth News Tribune*, December 6, 1892, June 4, 19, 20, 1893; Van Brunt, *Duluth*, 1:431–433, 438; "Biwabic (to 1915)," 2, 8, in "Sketches of Range Towns," Massingham Papers; Coleman and LaBud, *Masinaigans*, 178; Minnesota Manuscript Census Schedules, 1895; *Minnesota State Census*, 1895, pp. 49, 192, 258. On the naming of Merritt, see also "Biwabik," in *Skillings' Mining Review*, vol. 47, June 14, 1958, p. 5. Biwabic is the Ojibway word for iron; F[rederic] Baraga, *Dictionary of the Otchipwe Language*, 146 (Reprint ed., Minneapolis, 1966). On the extension of the Duluth and Iron Range Railroad, see Chapter 5, note 26.

44 McCaskill was a Canadian timber cruiser and ore explorer who assisted in many early test-pit operations on the Mesabi; see Van Brunt, *Duluth*, 1:397, 444, and *Duluth Daily News*, February 21, 1892. On the McKinleys, the McKinley Townsite Company, and the development of the village, see Articles of incorporation, Book J-2, p. 175; Van Brunt, *Duluth*, 1:433–435; *Duluth Tribune*, January 19, March 28, June 22, 1892; [Ellis], *Iron Ranges of Minnesota*, [80]; *Polk's Duluth Directory*, 1888–89, p. 377, 1892–93, p. 460, 1895–96, p. 434. For the Merritts' role, below, see Andrus, "Story," 157. On May 25, 1892, John, William, and Duncan McKinley with others incorporated the McKinley Iron Company. Although the Oliver Mining Company developed the McKinley Mine in 1907, the brothers by then had lost control of the property to John D. Rockefeller; Articles of incorporation, Book A-2, p. 91 and Book F-2, p. 418; Van Brunt, 1:452, 453. On the mines below, see Leith, *Mesabi Iron-Bearing District*, 28.

45 See, for example, Marion S. Cann, *Is This Hand Worth Playing? A Problem for Investors*, 74 (Duluth, 1892) for a chart of major firms, their officers, and selected financial data. The booklet was published by the Duluth Stock Exchange. For biographical glimpses of Hibbing, see his obituary in *Duluth News Tribune*, July 31, 1897; John H. Hearding, "Ore Capital Named After Frank Hibbing," undated, unpublished manuscripts file, SLC; "Frank Hibbing," in Lewis Publishing Company, *Minnesota*, 115–119; Theodore Christianson, *Minnesota: The Land of Sky-Tinted Waters*, 3:43 (Chicago, 1935); Winfield S. Downs, ed., *Encyclopedia of American Biography*, 10:141–145 (New series, New York, 1934).

46 *Duluth Tribune*, October 15, 1890; Articles of incorporation, Book E-2, p. 170. Capitalized at $5,000,000, the Lake Superior Iron Company eventually controlled leases on portions of more than 100 quarter sections. See Chapter 6, below; Evesmith, "Lake Superior Iron Company," 1, in "Sketches"; *Duluth News Tribune*, March 17, 20, June 4, 1893. On the Hull-Rust-Mahoning pit, now a National Historic Landmark, see Holmquist and Brookins, *Minnesota's Major Historic Sites*, 178. Other mines close to the Hibbing group listed here included the Day, Penobscot, and Stevenson; *Lake Superior Iron Ores*, 207, 237, 245.

47 Hearding, "Ore Capital," 2; "Hibbing," [1905], in "Sketches of Range Towns," Massingham Papers; Woodbridge and Pardee, eds., *St. Louis County*, 2:698; *Mesaba Ore and Hibbing News*, August 22, 1903, pp. 1, 2; Minnesota Manuscript Census Schedules, 1895; *Minnesota State Census*, 1895, pp. 50, 192, 258; Paul H. Landis, *Three Iron Mining Towns: A Study in Cultural Change*, 125 (Ann Arbor, Mich., 1938). On the railroad to Hibbing, see Chapter 5, note 21. On Chisholm, below, see [Ellis], *Iron Ranges of Minnesota*, [58]; Van Brunt, *Duluth*, 496–500; Woodbridge and Pardee, eds., *St. Louis County*, 2:722, 790; *Minnesota State Census*, 1905, p. 42 (St. Paul, 1905).

48 *Duluth News Tribune*, March 28, May 21, 1892, June 19, 20, 1893; "Virginia,

Minnesota (to 1913)," 2, 9, 15, in "Sketches of Range Towns," Massingham Papers; Woodbridge and Pardee, eds., *St. Louis County*, 2:687; Van Brunt, *Duluth*, 2:586; [Ellis], *Iron Ranges of Minnesota*, [6]. Virginia burned for the second time on June 7, 1900. On the extension of the railroad to Virginia, see Chapter 5, p. 108, below. On A. E. Humphreys, see *Who Was Who, 1897–1942*, 606 (Chicago, 1942). Humphreys was born in the part of Virginia that became West Virginia in 1863.

[49] Minnesota Manuscript Census Schedules, 1895; *Minnesota State Census*, 1895, pp. 50, 193, 259; [Ellis], *Iron Ranges of Minnesota*, [5]. On Virginia's lumber industry, see Larson, *White Pine Industry*, 235, 362, 400; Centennial Booklet Committee, *The Virginia Story: Historical Souvenir Booklet*, 19, 20 (n.p., [1949]). On the Missabe Mountain Company, see Articles of incorporation, Book D-2, p. 309; Evesmith, "Missabe Mountain Mine," 1, in "Sketches"; and Chapter 5, below.

[50] Landis, *Three Iron Mining Towns*, 20, 23, 64, 125. On violent deaths in range towns, see, for example, *Virginia Enterprise*, April 29, 1904, p. 8.

[51] Evesmith, "Biwabik Mine," 4, "Cincinnati Mine," 1, in "Sketches." On the Cincinnati, see also *Duluth Tribune*, January 19, March 6, 27, April 6, June 21, 1893.

[52] On Adams, see Century Publishing and Engraving Company, *Encyclopedia of Biography of Minnesota*, 243 (Chicago, 1900); [Ellis], *Iron Ranges of Minnesota*, [41]; Articles of incorporation, Book H-2, p. 102; Leslie J. Tobin, "Eveleth Firsts," 2, typed paper, 1937, MHS; Van Brunt, *Duluth*, 2:522. Mines shipping in the Virginia area by 1901 included the Alpena, Auburn, Columbia, Minnewas, and Sauntry; *Lake Superior Iron Ores*, 252–254. For more on the development of the western Mesabi, see Chapter 10, below.

[53] Here and below, see Minnesota Manuscript Census Schedules, 1895; *Minnesota State Census*, 1895, pp. 50, 192, 258; *St. Paul Pioneer Press*, November 9, 1922, p. 1; Woodbridge and Pardee, eds., *St. Louis County*, 2:702–704; Van Brunt, *Duluth*, 2:516, 522, 523; Landis, *Three Iron Mining Towns*, 28, 31; Tobin, "Eveleth Firsts," 1; McInnis, in *Eveleth News*, December 19, 1906, p. 2.

[54] John H. Hearding, "Pioneer Mining Man Describes Early Days in Eveleth District," in *Skillings' Mining Review*, vol. 12, August 18, 1923, p. 1; *Lake Superior Iron Ores*, 252; W. R. Van Slyke, "The Discovery of Iron Ore: The Mines of Eveleth and the Development of Eveleth," 1–4, 9, 10, typed paper, 1937, MHS. The three mines named here, plus the Commodore, Franklin, Norman, and Union, were all located near present Franklin, a mining community that developed into a village in 1915; Van Brunt, *Duluth*, 2:709. On the Adams Mining Company, see *Duluth News Tribune*, June 4, July 20, November 5, 13, 1893. For the extension of railroads to Eveleth, see Chapter 5, notes 21, 26.

CHAPTER 5 — THE MESABI GIANT GOES TO MARKET. — *pages 99 to 118*

[1] Alfred, "Reminiscences," 15; Norcross, "Spoliation," 50, 51, Merritt Papers; Alfred testimony, Stanley Committee, *Hearings*, 3:1845. Frank A. King, *The Missabe Road*, 45 (San Marino, Calif., 1972) maintained that the Merritts did approach the D&IR but were turned down. Norcross was commissioner of the Nevada Bureau of Agriculture and Irrigation; apparently at Andrus Merritt's instigation, he wrote an account of the Merritts. See, for example, Lon to Norcross, February 15, 1912; Norcross to Andrus, October 3, 1912, and to Lon, March 31, 1912, Merritt Papers.

[2] Andrus, "Story," 160, 161, attributed the eastern trip to Cassius rather than Leonidas and claimed that a satisfactory report from the expert would have meant "a fifty-fifty division" by which Carnegie furnished $3,000,000 for "constructing a railway

from the different mines to Duluth Harbor, and to build an iron-ore dock. In return Carnegie was to own one-half" of the railroad, dock, and mining property. "In addition he was to furnish money to secure and develope other mines, our experience and labor to off-set his capital." Lon recalled the eastern journey in Stanley Committee, *Hearings*, 3:1892–1894. See also Norcross, "Spoliation," 51, Merritt Papers; Joseph F. Wall, *Andrew Carnegie*, 593 (New York, 1970).

³ *Duluth Daily News*, February 21, 1892, quoting *American Economist*. See also *Duluth Daily News*, May 27, 1891, for an earlier example of Duluth's interest in a Mesabi railroad.

⁴ Articles of incorporation, Book B-2, 330, Book H, 546; Duluth, Missabe and Northern Railway Company, "Synopsis of Secretary's Records, 1891–97," p. 1, in the Duluth, Missabe and Iron Range Railway headquarters, Duluth. The incorporators included Leonidas, Alfred, Andrus, Cassius, and Napoleon Merritt, Kelsey Chase, James T. Hale, S. R. Payne, and R. H. Palmer. Duluth, Missabe and Northern is abbreviated hereafter as DM&N.

⁵ King, *Missabe Road*, 46; *Duluth Daily News*, October 16, 1891; Directors' Minutes, August 19, October 27, 1891, in DM&N, "Synopsis," 1, 2, 4. The British-American Trust Company was interested in acquiring DM&N bonds, but did not do so according to Directors' Minutes, August 30, 1892, February 7, 1893, in "Synopsis," 10, 22.

⁶ Born in Canada in 1837, Grant moved from Ohio to Faribault in 1863 and soon went into railroad contracting. He quickly accumulated a fortune laying track in Minnesota, Wisconsin, Iowa, the Dakotas, Montana, and the Canadian West. In 1892–93 he was mayor of Faribault. From 1895 to 1910 he was the leading investor in the Orinoco Company, formed to develop the iron ore and other resources of Venezuela. See Century Publishing and Engraving Company, *Encyclopedia of Biography of Minnesota*, 1:447; Hugh J. McGrath and William Stoddard, *History of the Great Northwest and Its Men of Progress*, 408 (Minneapolis, 1901); Gretchen Kreuter, "Empire on the Orinoco: Minnesota Concession in Venezuela," in *Minnesota History*, 43:198–212 (Summer, 1973). For a copy of the Grant contract, January 28, 1892, see DM&N and American Steel Barge Company Papers — both in Charles W. Wetmore Papers, Baker Library, Harvard University. The MHS has a microfilm copy of some of these papers cataloged under American Steel Barge Company. The Grant contract appears on roll 3. See also DM&N, "Plans and Specifications for Constructing the Duluth, Missabe and Northern Railway," [1891], a manuscript in MHS; Minutes of Stockholders' Annual Meeting, February 2, 1892, in DM&N, "Synopsis," 5.

⁷ Traffic agreement, April 14, 1892, in Wetmore Papers (roll 3), in Leonidas Merritt *v.* Duluth, Missabe and Northern Railway Company, *Paper Book*, Exhibit A, 490–508 (Duluth, 1895), in offices of the clerk of St. Louis County District Court, Case no. 9431, Duluth, and in Duluth, Missabe and Iron Range Railway headquarters, Duluth; Xerox and microfilm copies in MHS. It was signed by Kelsey Chase and S. R. Payne for the DM&N and H. J. Boardman and Charles G. Heim for the Duluth and Winnipeg. On the route of the Duluth and Winnipeg, see Benjamin Whiteley, "Brief Historical Sketch of the Duluth and Winnipeg Railroad Enterprise, 1878–1896," [7, 8], in Great Northern Railway Company Records, MHS. See also Directors' Minutes, April 22, 1892, in DM&N, "Synopsis," 9; Andrus, "Story," 163, and Chapter 3, above.

⁸ Traffic agreement, April 14, 1892, Wetmore Papers (roll 3); Articles of incorporation, Book D, 240, Book W, 81. The loss of DM&N traffic and the panic of 1893 hurt the Duluth and Winnipeg. In January, 1893, through the efforts of Donald Grant, the road was acquired by William Van Horne of the Canadian Pacific, which operated it until 1897. It went into receivership in 1894, and in 1897 it was sold to James J. Hill to become part of the Great Northern system. Its acquisition marked Hill's entrance upon the Mesabi iron ore shipping scene. See Walter Vaughan, *Sir William Van Horne*, 244–252, 254, 257, 274–276 (*Makers of Canada Series*, vol. 10 — London and Toronto, 1926); Joseph W. Thompson, "The Genesis of the Great Northern's Mesabi Ore Traffic," in

Journal of Economic History, 16:551–557 (December, 1956); Minnesota Railroad and Warehouse Commission, *Annual Report*, 1890, pp. 317, 322, 1892, p. 413 (Minneapolis, 1891, 1893). See also Chapter 9, page 218, below.

⁹ For the rivalry between the port cities, see, for example, *Superior Daily Leader*, June 30, October 13, 1891. On the land company, see Articles of incorporation, Book S, 328, Wisconsin Secretary of State, Corporation Division, Madison; Superior City Statistician, *Annual Report*, 1892, pp. 13–19 (Superior, Wis., 1893); *Duluth Tribune*, April 17, 1892. On the July 22, 1891, contract with the Superior land company and the agreement of October 7, 1891, see histories of the Duluth and Winnipeg Railroad Company in Great Northern Records. See also Minnesota Railroad and Warehouse Commission, *Annual Report*, 1893, p. 355 (Minneapolis, 1894).

¹⁰ Contract between Duluth and Winnipeg and Duluth and Winnipeg Terminal Company, July 25, 1892; Whiteley, "Sketch of the Duluth and Winnipeg," 10, both in Great Northern Records. On the construction of the dock, see *Superior Daily Leader*, November 2, 1892. In 1893 the terminal company extended the dock an additional 600 feet and added 126 ore pockets, according to the *Superior Evening Telegram*, February 22, 1893. Three years later a final addition increased the number of pockets to 250 and gave the dock a total capacity of 37,500 tons. The *Superior Inland Ocean* of December 12, 1896, reported that the company spent $413,616 for all dock construction.

¹¹ Kimberly lease, April 23, 1892, copy in Wetmore Papers. See also Andrus, "Story," 161. For additional local comment on the Kimberly leases of this and other mines, see *Duluth Tribune*, April 24, May 3, 1892; *Duluth Daily News*, May 2, 3, 1892; *Duluth News Tribune*, December 5, 1892. Stockholders of the Kanawha Mine turned down a proposal to lease it to Kimberly, according to *Duluth Daily News*, May 14, 1892. For the subsequent history of the Biwabik Mine, see Norman J. Setnicker, "The Development of the Biwabik Mine from 1892–1920," master's thesis, University of Minnesota Duluth, 1968. The DM&N track to the Biwabik Mine was authorized by the board of directors on October 28, 1891, and constructed in 1892, according to DM&N, "Synopsis," 5, 8.

¹² *Dictionary of American Biography*, 14:19 (New York, 1934).

¹³ June D. Holmquist, "Convention City: The Republicans in Minneapolis, 1892," in *Minnesota History*, 35:64–76 (June, 1956); Henry O. Evans, *Iron Pioneer: Henry W. Oliver, 1840–1904*, 198–203 (New York, 1942).

¹⁴ On the Cincinnati Mine, see *Duluth Tribune*, June 21, 22, 1892; *Duluth Daily News*, June 21, 1892; Evesmith, "Cincinnati Mine," 1, 2, in "Sketches." On the Missabe Mountain, see Oliver-Merritt lease, August 1, 1892, copies in Merritt Papers; Andrus, "Story," 170; *Duluth Daily News*, August 4, 7, 1892; Evans, *Iron Pioneer*, 203. For the adverse reactions of Carnegie and Frick, see Wall, *Andrew Carnegie*, 596, and Chapter 9, below.

¹⁵ *Duluth Tribune*, July 7, 1892; King, *Missabe Road*, 48. The payment of bonds to Grant for completion of the main line was approved by the directors on January 4, 1893, according to DM&N, "Synopsis," 12.

¹⁶ St. Louis County Historical Society, *Retrace by Rail the 'Iron Trail:' 70th Anniversary of the First Ore Shipment*, [18] (Duluth, 1962), a pamphlet in MHS which reproduced a handwritten invoice of November 11, 1892, listing expenses of the October 18 run totaling $41.05.

¹⁷ The DM&N had access to Duluth's Union Depot because of an arrangement between the Duluth and Winnipeg and the St. Paul & Duluth; Minnesota Railroad and Warehouse Commission, *Annual Report*, 1892, p. 413. See also *Duluth News Tribune*, November 1, 1892; *Superior Daily Leader*, November 2, 1892; *Superior Times*, November 19, 1892; *Lake Superior Iron Ores*, 253. A somewhat different recent account may be found in King, *Missabe Road*, 49. Many historical reports of this first shipment are both brief and confused, failing to distinguish between the October and November events. For example, Van Brunt, *Duluth*, 1:421, stated "The shipment of October 17, 1892, was what might be termed a demonstration shipment, no attempt being made to

load another train until the following spring" — a statement disproved by contemporary newspapers; see also Joseph W. Thompson, "An Economic History of the Mesabi Division of the Great Northern Railway Company to 1915," 280, Ph.D. thesis, University of Illinois, 1956. Twenty-nine analyses of the Mountain Iron shipments made by C. F. Joyce between November 1 and 18, 1892, averaged 63.8 per cent iron and 0.034 per cent phosphorous, more than meeting Bessemer standards. Copies of the Joyce records are in Merritt Papers and Wetmore Papers (roll 3). See also note 28, below.

[18] For the views of other visitors, see *Duluth Tribune*, January 11, March 4, 27, May 16, 1892; *Duluth Daily News*, February 21, March 15, 1892. On Carnegie's visit, see also *Duluth News Tribune*, October 1, 1893.

[19] E. D. Reis to Henry W. Oliver, January 2, 1893, copy in Wetmore Papers (roll 1); Leith, *Mesabi Iron-Bearing District*, 294; Temin, *Iron and Steel*, 197; *Superior Sunday Forum*, January 20, 1895; *New York Daily Tribune*, December 2, 1895. According to Norcross, "Spoliation," 93, Merritt Papers, a Kimberly furnace blew up in the spring of 1893. As late as April 16, 1892, the *Engineering and Mining Journal*, 53:419, reported that the Mesabi ores were "low grade" and not of Bessemer quality. According to John E. Merritt, "Speech," 5, the first Mesabi ore was smelted in Kimberly's laboratory and "made into a ferrule for the tip of a cane, the cane made from a pine root taken from the tree that gave us the location of the ore on the Biwabik. Mr. Kimberly brought it to Duluth and presented it to Leonidas Merritt." For a good account of technical furnace data, see Hogan, *Iron and Steel Industry*, 1:27–31, 211–217, 2:391–402.

[20] *Duluth Tribune*, January 19, April 13, 1892.

[21] Minnesota Railroad and Warehouse Commission, *Annual Report*, 1893, pp. 288, 295. The line was extended to Hibbing in 1894 and to Eveleth in 1895; Minnesota Railroad and Warehouse Commission, *Annual Report*, 1894, p. 267, 1895, p. 321 (St. Paul, 1895); Evesmith, "Missabe Mountain Mine," 3, and "Cincinnati Mine," 2, in "Sketches"; Richard S. Prosser, *Rails to the North Star*, 130 (Minneapolis, 1966). On the Merritts' loyalty to Duluth, below, see Alfred, "Reminiscences," 17; Lon to Hephzibah Merritt, December 16, 1892, Merritt Papers. For their plan to locate the ore docks, below, see Alexander McDougall to Wetmore, January 17, 1893, Wetmore Papers (roll 3).

[22] On the committees' proposals here and below, see *Duluth Tribune*, February 1, 21, 24, March 19, 20, April 9, 1892; *Duluth Daily News*, July 16, 22, 1891, February 21, 24, April 22, 1892.

[23] *Duluth Daily News*, February 2, April 22, 1892; Minutes of Stockholders Meeting, January 20, 1893, in DM&N, "Synopsis," 13.

[24] Alfred, "Reminiscences," 16; *Duluth News Tribune*, December 22, 1892, May 26, 1893. The vote, reported in the issue of May 27, 1893, was:

	Yes	No
City of Duluth	1,409	242
West Duluth	471	92
Mesabi Range townships (Biwabik, 81–3;		
Mountain Iron, 268–12; Virginia, 186–7)	535	22
Vermilion Range townships (Tower, 10–65; Breitung, 44–109)	54	174
Remainder of St. Louis County	12	3
TOTAL	2,481	533

The DM&N pledged the St. Louis County bonds to Rockefeller, according to Directors' Minutes, December 1, 1893, in DM&N, "Synopsis," 35.

[25] Andrus Merritt testimony, in Leonidas Merritt *v.* Duluth, Missabe and Northern Railway Company, *Paper Book*, 305; Norcross, "Spoliation," 71, Merritt Papers; Andrus, "Story," 170–172. See also note 8, above.

[26] On the activities of the D&IR and the Minnesota Iron Company, see Mussey, *Combination in the Mining Industry*, 117; *Duluth Daily News*, February 21, April 15,

August 7, October 3, 1892; *Duluth News Tribune*, March 24, 1893. On extension of the branch lines, see D&IR, Board of Directors' Minutes, April 13, June 6, September 29, 1892, August 28, September 15, 1893, in Duluth, Missabe and Iron Range Railway headquarters, Duluth; Minnesota Railroad and Warehouse Commission, *Annual Report*, 1893, p. 277, 1894, p. 245, 1895, p. 298; Setnicker, "Development of the Biwabik Mine," 47. On November 16, 1893, the Minnesota Iron Company leased the Commodore Mine, a transaction "specially significant in that it is the first decided step the company has taken on the Mesaba." Evesmith, "Commodore Mine," 2, in "Sketches."

[27] Andrus, "Story," 172. In 1854 at the age of nine McDougall moved with his family from Scotland to the small Ontario settlement of Collingwood on Georgian Bay in Lake Huron. After limited schooling and a short stint as a blacksmith apprentice, he shipped out on Great Lakes schooners. Janet C. Sanborn, ed., *The Autobiography of Captain Alexander McDougall*, 3, 5, 7, 8, 16 (Cleveland, 1968). On the D&IR ore docks, see *Duluth News Tribune*, December 20, 26, 1892.

[28] A handwritten bill of lading for whaleback 102, dated November 16, 1892, was reproduced in *Skillings' Mining Review*, vol. 13, May 31, 1924, p. 1. The original is in the Western Reserve Historical Society, Cleveland. Forty-four whalebacks were launched between 1889 and 1897, all but six from shipyards in West Superior, Wis. At present the last survivor, the "Meteor," is preserved as a museum on the lake shore at Superior. See Sanborn, ed., *Autobiography of McDougall*, 77; Ryck Lydecker, *Pigboat, the Story of the Whaleback*, 27 (Duluth, 1973); John H. Wilterding, Jr., *McDougall's Dream: The American Whaleback*, 13 ([Green Bay, Wis.], 1969). According to Sanborn, ed., page 32, whaleback 101 first carried 1,200 tons of Vermilion ore in 1888 from Two Harbors to Cleveland.

[29] Sanborn, ed., *Autobiography of McDougall*, 33, 39–41. A copy of the articles of incorporation, December 22, 1888, is in Wetmore Papers (roll 2). In the planning to form American Steel Barge, Rockefeller held 15 of the 100 shares, according to Joseph L. Colby to Wetmore, December 5, 1888, Wetmore Papers (roll 2), which also discussed McDougall's patents.

[30] Scattered and unsatisfactory biographical material on Wetmore was gleaned from correspondence, telegrams, and business records in Wetmore Papers and in the Rockefeller Family and Associates Archives, North Tarrytown, N.Y. Wetmore later served as a director of the Bethlehem Steel Company and as the first president of the Montana Power Company. Poor health forced his retirement in 1913, and he lived the remaining six years of his life in England and Washington, D.C. A short sketch appeared in *National Cyclopaedia of American Biography*, 18:21 (New York, 1922). See also Chapter 7, note 1, below.

[31] Wetmore to Hoyt, November 21, 1892, Wetmore Papers (roll 1).

[32] Hoyt to Wetmore, December 15, 1892, Wetmore Papers (roll 1).

[33] Andrus, "Story," 173–176; *Duluth News Tribune*, January 30, 1893; Stanley Committee, *Hearings*, 3:1894. McDougall also became suspicious because of the continuous presence of Joseph L. Greatsinger, D&IR president, in the offices of Chase. Upon learning of the offer to sell the entire operation, McDougall worked hard to convince the Merritts of the value of their holdings and encouraged them "to hold on, and fight it out . . . and operate all themselves." See McDougall to Wetmore, January 12, 1893, Wetmore Papers (roll 3). Names of stockholders and the number of shares held by each on December 24, 1892, are listed in Leonidas Merritt v. Duluth, Missabe & Northern, *Paper Book*, Exhibit 14, p. 529; Chase's testimony is on p. 326. The vote tally, as recorded in the Minutes of DM&N Stockholders Meeting, January 20, 1893, may be found on p. 296.

[34] Handwritten memorandum, undated, discussing this portion of the Chase-Merritt conflict in Wetmore Papers (roll 3).

[35] These legal maneuvers and financial transactions can be followed in *Duluth News Tribune*, January 25, 26, 27, 30, 31, February 2, 3, 4, 8, 1893; *New York Times*, February

1, 1893. The five promissory notes, dated January 30 and February 2, 1893, appeared as Plaintiffs' Exhibits A–E in Alfred and Leonidas Merritt v. American Steel Barge Company, 11th Judicial District, Case no. 9262, in the office of the clerk of St. Louis County District Court, Duluth, and in the Federal Records Center, Kansas City, Mo., microfilm in MHS. On the DM&N stock issue, its new officers, and dismissal of the Grant injunction, see Directors' Minutes, January 30, February 3, 7, 1893, in DM&N, "Synopsis," 15–23. Wetmore was appointed second vice-president, according to the minutes of May 5, 1893, in "Synopsis," 26. See also Chapter 6, page 139, below.

[36] Merritt-Wetmore and DM&N-American Steel Barge Company contracts, both dated December 24, 1892, in Merritt Papers. The former is also printed in Leonidas Merritt v. Duluth, Missabe and Northern, *Paper Book*, Exhibit C, 512–521. The latter is also in Wetmore Papers (roll 1). On the Hibbing branch, see Norcross, "Spoliation," 76, Merritt Papers, and note 21, above.

[37] Alfred and Andrus Merritt testimony, in Leonidas Merritt v. Duluth, Missabe and Northern, *Paper Book*, 229, 242, 263, 300. On Cassius, see Norcross, "Spoliation," 81, Merritt Papers.

[38] For the progress of construction, see *Duluth News Tribune*, March 9, 28, June 16, 29, 1893. A cost breakdown totaling $397,614 prepared by C. H. Martz, chief engineer, December 3, 1892, is in Wetmore Papers (roll 3). He estimated the cost of grading at $203,000; 31 miles of tracklaying at $124,000; sidings at $34,246; 81,840 ties at $.20 each or $16,368; stations at $10,000; clearing 20 miles of right of way at $6,000; and one 140-foot truss-span bridge at $4,000.

[39] On the ore dock, see *Duluth News Tribune*, January 7, 23, 27, February 25, March 2, 30, April 17, 1893; *New York Times*, February 6, 1893; *Mississippi Valley Lumberman*, vol. 23, June 9, 1893, p. 4.

[40] The letter is in Merritt Papers. It is quoted in part in King, *Missabe Road*, 52.

[41] *Duluth News Tribune*, January 1, 1894.

[42] Directors' Minutes, October 24, 1895, in DM&N, "Synopsis," 63; *Duluth News Tribune*, March 14, 1896, October 30, 1897; *Duluth Evening Herald*, January 6, April 9, 1896, December 13, 1897; *Iron Trade Review*, vol. 39, September 27, 1906, p. 14. The Merritts' original dock was dismantled in 1907; construction of the first steel and concrete ore dock began in 1912. See *Marine Review*, vol. 35, May, 1907, p. 30, vol. 42, February, 1912, p. 60. The present dock was still the largest in the area, although part of it was used for merchandise rather than ore, according to Frank King of the Duluth, Missabe and Iron Range headquarters staff, December 18, 1978.

[43] Figures compiled by the author from *Polk's Duluth Directory*, 1886–87, 1889–90, 1893–94; Secretary of State, *Annual Report*, in Minnesota, *Executive Documents*, 1891, vol. 1, pp. 153–161 and 1892, vol. 1, pp. 570–574 (St. Paul, 1891, 1892).

[44] Leonidas Merritt, "Answer to the Gates Pamphlet," 16, a 1912 manuscript in Merritt Papers. In December, 1892, the Merritt family owned 54.3 per cent of the Missabe Mountain Mine stock, 42.5 per cent of the Biwabik, and 42 per cent of the Mountain Iron, according to calculations based upon lists of stockholders in Leonidas Merritt v. Duluth, Missabe and Northern, *Paper Book*, Exhibits 16, 18, 20, pp. 533–539, 544–550, 555–557. They also owned majority stock in the Great Northern, Great Western, and Shaw mining companies, according to Norcross, "Spoliation," 64–66, Merritt Papers. On Murray's participation, see *Paper Book*, Exhibit C, 521.

CHAPTER 6 — THE PANIC OF 1893 — *pages 127 to 143*

[1] Here and below, see Charles Hoffmann, *The Depression of the Nineties: An Economic History*, 10, 54, 68–70 (*Contributions in Economics and Economic History*,

no. 2 — Westport, Conn., 1970). Rendigs Fels, *American Business Cycles, 1865–1897*, 98–112 (Chapel Hill, N.C., 1959) discusses the Jay Cooke panic. On 1857, see Chapter 1, note 24, above.

² The quotations appear in Hoffmann, *Depression of the Nineties*, 54–63. See also Samuel Rezneck, *Business Depressions and Financial Panics: Essays in American Business and Economic History*, 177–198 (New York, [1968]). Beginning early in 1893 the nation's economy fluctuated in wavelike fashion, falling to a low point in June, 1894, rising briefly in the fall of 1895, and falling again in early 1897, before beginning an upward climb that continued into the first decade of the 20th century. See Fels, *American Business Cycles*, 184–219.

³ Robert F. Martin, *National Income in the United States, 1799–1938*, 60 (New York, 1939); U.S. Bureau of the Census, *Historical Statistics of the United States, Colonial Times to 1957*, 365 (Washington, D.C., 1960). According to the latter, p. 416, Bessemer steel production fell 22.8 per cent from 4,168,435 tons in 1892 to 3,215,686 in 1893. The year 1893 also witnessed a 45.8 per cent decline in the number of operating blast furnaces from 253 to 137; H. S. Fleming, "Iron and Steel," in *Mineral Industry*, 2:349 (1893).

⁴ Percentages were calculated from figures given in Crowell & Murray, *The Iron Ores of Lake Superior*, 48 (4th ed., Cleveland, 1920); Bureau of the Census, *Historical Statistics*, 365.

⁵ *Lake Superior Iron Ores*, 329. Lake Erie ports holding reserves included Ashtabula, Cleveland, Fairport, Huron, Lorain, Sandusky, and Toledo, Ohio; Erie, Pa., and Buffalo, N.Y. Ashtabula and Cleveland between them stock-piled 63.8 per cent of the total. Leith, *Mesabi Iron-Bearing District*, 293, contains a breakdown of Lake Superior iron ore prices, 1891–1901, divided into various Bessemer and non-Bessemer classifications for each range.

⁶ Here and below, see Phillips, in *Mineral Industry*, 1:272–275 (1892); *Duluth News Tribune*, June 26, July 1, 24, August 8, 1893, January 2, 1894; *Lake Superior Iron Ores*, 199; *New York Times*, December 25, 1893. On the Soudan, see Chapter 3, above.

⁷ *Lake Superior Iron Ores*, 254.

⁸ Crowell & Murray, *Iron Ores of Lake Superior*, 48, 49; Bureau of the Census, *Historical Statistics*, 365.

⁹ Lewis E. Young, *Mine Taxation in the United States*, 30, 54 (University of Illinois, *Studies in the Social Sciences*, vol. 5, no. 4 — Urbana, 1917); Wirth, *Minnesota Iron Lands*, 184. A summary of taxes collected may be found in Minnesota State Auditor, *Biennial Report*, 1933–34, p. 255 ([St. Paul, 1934]). For the 20th century, see Roy G. Blakey, *Taxation in Minnesota*, 234–276 (University of Minnesota, *Studies in Economics and Business*, no. 4 — Minneapolis, 1932). A federal income tax of 2 per cent on incomes above $4,000 a year was levied by Congress in 1894 but it was invalidated by the U.S. Supreme Court the following year; Ray Ginger, *Age of Excess: The United States From 1877 to 1914*, 170, 173–175.(New York, 1965).

¹⁰ Mussey, *Combination in the Mining Industry*, 125; G. O. Virtue, *The Minnesota Iron Ranges*, 386 (U.S. Bureau of Labor, *Bulletin*, no. 84 — Washington, D.C., 1909); Minnesota Bureau of Labor, *Biennial Report*, 1902, vol. 2, p. 859 (St. Paul, 1902).

¹¹ Leith, *Mesabi Iron-Bearing District*, 283.

¹² "Cost of Equipping a Mesabi Underground Mine," in *Engineering and Mining Journal*, 46:549 (September 20, 1913). Julius Moersch and J. W. Allen, "Mine Inspections," in Minnesota Bureau of Labor, *Biennial Report*, 1902, vol. 2, p. 883, listed the following Mesabi mines exclusively utilizing underground techniques in 1902: Adams, Agnew, Albany-Utica, Burt, Chisholm, Clark, Corsica, Elba, Franklin, Genoa, Hawkins. Hull, La Belle, Lincoln, Malta, Penobscot, Pettit, Pillsbury, Roberts, Rust, Sellers, Sharon, Spruce, and Union. For underground methods on the Vermilion, see Chapter 3, above. See also Leith, *Mesabi Iron-Bearing District*, 262.

¹³ Woodbridge, in *Engineering and Mining Journal*, 79:266. For open pits on the Vermilion, see Chapter 3, above.

14 Setnicker, "Development of the Biwabik," 42, 56; *Duluth Tribune*, June 29, 1892.

15 *Duluth News Tribune*, August 7, 1893; *Superior Sunday Forum*, June 7, 1896; C. E. Bailey, "Mining Methods on the Mesabi Range," in American Institute of Mining Engineers, *Transactions*, 27:530–532 (July, 1897); Kirby Thomas, "Mining Ore in the Mesabi District," in *Scientific American Supplement*, 57:23700 (May 4, 1904); Dwight E. Woodbridge, "Mining with the Steam-Shovel," and E. K. Soper, "Iron Mining in Minnesota," in *Mining and Scientific Press*, 92:417, 101:768 (June 23, 1906, December 10, 1910).

16 Charles R. Van Hise and Charles K. Leith, *The Geology of the Lake Superior Region*, 498 (U.S. Geological Survey, *Monographs*, vol. 52 — Washington, D.C., 1911).

17 Kirby Thomas, "Mining Methods in the Vermilion and Mesabi Districts," in *Mining World*, 21:202 (August 27, 1904). A technically accurate description of steam-shovel mining on the Mesabi during the initial decades may be found in Van Barneveld, *Iron Mining in Minnesota*, 131–152. See also Albert H. Fay, "Steam Shovel Work on Mesabi Range," and C. M. Haight, "Steam Shovel Operation," in *Engineering and Mining Journal*, 91:420–423, 92:359 (February 25, 1911, February 14, 1914); Woodbridge, in *Mining and Scientific Press*, 92:417.

18 Woodbridge, in *Engineering and Mining Journal*, 79:466.

19 "The Wonderful Iron Mines of Lake Superior," in *Scientific American*, 101:449 (December 11, 1909); Thomas Owens, "Early Iron Ore Operations in Minnesota," 9, typed paper, 1932, copy in MHS; Crowell & Murray, *Iron Ores of Lake Superior*, 31, 32. On the dump cars, below, see Evesmith, "Biwabik Mine," 6; A[nton] Tancig, "The Evolution of Equipment in Minnesota Iron Mining Industry, 1883–1953," in *Skillings' Mining Review*, vol. 42, August 15, 1953, p. 2.

20 On the use of crushers by the Minnesota Iron Company, see F. W. Denton, "Methods of Iron Mining in Northern Minnesota," in American Institute of Mining Engineers, *Transactions*, 27:349–352 (February, 1897). See also Tancig, in *Skillings' Mining Review*, vol. 42, August 8, 15, 1953, pp. 1–4.

21 F. W. Denton, "Open-Pit Mining, With Special Reference to the Mesabi Range," in Lake Superior Mining Institute, *Proceedings*, 3:85, 89 (Ishpeming, Mich., 1895); Van Hise and Leith, *Geology of the Lake Superior Region*, 498.

22 Nelson P. Hulst, "Methods of Mining Iron Ore in the Lake Superior Region," in Engineers' Society of Western Pennsylvania, *Proceedings*, 15:99–102 (1899); "Milling System of Mining Mesabi Iron," in *Engineering and Mining Journal*, 96:545 (September 20, 1913).

23 Bailey, in American Institute of Mining Engineers, *Transactions*, 27:531; Van Barneveld, *Iron Mining in Minnesota*, 152–155.

24 Horace V. Winchell, "The Mines of the Minnesota Iron Company," in *Engineering Magazine*, 13:891 (September, 1897); Woodbridge, in *Engineering and Mining Journal*, 79:267, 365.

25 Woodbridge, in *Engineering and Mining Journal*, 79:267; Leith, *Mesabi Iron-Bearing District*, 285.

26 Leith, *Mesabi Iron-Bearing District*, 280, 282, 284; Thomas, in *Scientific American Supplement*, 57:23700; Woodbridge, in *Engineering and Mining Journal*, 79:267.

27 According to Wetmore, he suggested that the Merritts "purchase 3000 shares of treasury stock . . . at par for cash, this being the only way in which additional stock could be issued while the injunction was in force"; undated memorandum, Wetmore Papers (roll 3). On Wetmore's funds, see Directors' Minutes, February 1, 1893, in DM&N, "Synposis," 17; Leonidas Merritt *v.* DM&N, "Receipts and Disbursements," in *Paper Book*, 310–312, listing his payments to the road in 1893. Only $150,000 was received from Wetmore from February to June, 1893. On the five promissory notes, see Chapter 5, note 35. See also Norcross, "Spoliation," 111, Merritt Papers; Andrus, "Story," 182; Frederick T. Gates to Rogers, December 24, 1892, and memoranda, December 19, 24, 1892, February 13, 1893, as well as Wetmore to Rockefeller, December 29, 1892 — all in Rockefeller Archives. In the latter, Wetmore offered to obtain

more bonds if Rockefeller chose to exercise the option 30 days from December 24, 1892, or "within six months after the completion of the [rail]road."

On Rockefeller's unwillingness to become further involved, see Gates, memorandum, February, 1893, stressing the financier's belief "that it was unwise for him to make any further investments in Iron Ore Mines"; a series of telegrams, January 2 through August 21, 1893, transmitted over commercial and Standard Oil private telegraph lines; memoranda, Gates to Wetmore, January 18, March 3, 1893; Gates to Charles and Joseph Colby, March 8, 1893 — all in Rockefeller Archives. See also Frederick T. Gates, *The Truth About Mr. Rockefeller and the Merritts*, 7 [New York, 1911]. The final quotation may be found in Leonidas Merritt, "Outline History of the Discovery and Development of the Missabe Iron Range and the Merritts Connection with the Same," June 7, 1894, p. 8, Merritt Papers. On the other hand, Rockefeller was not totally disinterested in the Mesabi. He asked for and received information on the furnace tests of the first shipments of Mesabi ore in January, 1893. Franklin Rockefeller to John D. Rockefeller, January 4, 1893, and Reis to Oliver, January 13, 1893 (copy) — both in Rockefeller Archives.

[28] Alfred testimony and Rockefeller deposition in Rockefeller *v.* Alfred Merritt, U.S. Circuit Court of Appeals, 8th Circuit, Case no. 195, *Transcript of Record*, 352, 414 (1895). The original documents in this case are in Federal Records Center, Kansas City, Mo.; the MHS has a copy of the *Transcript*. The initial suit, Alfred Merritt *v.* Rockefeller, was brought in St. Louis County District Court, Duluth, from which it was appealed; microfilm in MHS. See also Gates, *Truth*, 8; Norcross, "Spoliation," 95, 97, Merritt Papers. Details on the Hibbing purchase are in *Duluth News Tribune*, March 11, 17, 20, June 4, 1893; Evesmith, "Lake Superior Iron Company," 1–3, in "Sketches." Many of the mines on this property were later operated by the Oliver Iron Mining Company. The McKinley Mine negotiations are outlined in a series of telegrams (copies) from Leonidas Merritt to Wetmore, May 9, 11, 12, 16, 18, 1893, and Lon to Lizzie, May 24, 1893 — all in Merritt Papers.

[29] Gates, *Truth*, 9; Articles of incorporation, Book G-2, 107; *Duluth News Tribune*, July 5, 16, November 5, 1893. On the lease of the Lone Jack Mine, see Wetmore to Lon, July 3, 1893; Leonidas, "Answer to Gates," 28, Merritt Papers. On the Rathbun, see Wirth, *Minnesota Iron Lands*, 192.

[30] Here and below, see Norcross, "Spoliation," 88, Merritt Papers; Alfred and Leonidas Merritt *v.* American Steel Barge Company, Complaint, 3, 4, 72, and Hoyt testimony, 88–90; list of collaterals pledged on Rockefeller loans, 1893, Merritt Papers.

[31] On Lon's departure for New York, see his telegram to Wetmore (copy), May 17, 1893, Merritt Papers. An undated list in these papers shows Alfred in New York August 17–27, November 16–27; Andrus May 19–29, June 20–early July; and John E., E. T., T. A., N. B., and Hulett Merritt present at various times. The payroll records of the DM&N are in the possession of the Duluth, Missabe and Iron Range Railroad, Duluth. See also Alfred testimony, in Rockefeller *v.* Alfred Merritt, *Transcript*, 351; Lon and Alfred testimony, in Leonidas Merritt *v.* DM&N, *Paper Book*, 120, 263; Stanley Committee, *Hearings*, 3:1853; Gates, *Truth*, 7.

[32] Norcross, "Spoliation," 89, 94, 95, Merritt Papers; memoranda between Wetmore, Colby, and Hoyt, and penciled copies of proposals for financing and organizational structure, in Rockefeller Archives; Articles of incorporation, in State Historical Society of Wisconsin, Archives Division, Madison; Gates, *Truth*, 7, 8, states that Joseph Colby told him about the plan on March 16, 1893, and that it collapsed on April 11, 1893. On the Aurora and Tilden mines, see *Lake Superior Iron Ores*, 185, 187.

[33] Leonidas testimony, Leonidas Merritt *v.* DM&N, *Paper Book*, 116, 173; Norcross, "Spoliation," 89, 91–94, Merritt Papers; Gates, *Truth*, 8.

[34] On the bond issue, see Chapter 5, note 24. On Gogebic production, see *Lake Superior Iron Ores*, 190, 308. See also Norcross, "Spoliation," 96, 97, Merritt Papers; Joseph L. Colby to Leonidas, June 10, 1893, and copies of the preliminary Wetmore-Colby agreement, June, 1893, in Wetmore Papers, and as Defendant's Exhibits nos. 1, 2, in Rockefeller *v.* Alfred Merritt, *Transcript*, 576–580.

35 Alfred and Leonidas Merritt v. American Steel Barge Company, Complaint, 4–8, Answer, 17; Allan Nevins, *John D. Rockefeller: The Heroic Age of American Enterprise*, 2:367 (New York, 1940); Alfred to Lon, July 30, 1893, Defendant's Exhibit no. 39, in Rockefeller v. Alfred Merritt, *Transcript*, 613–615, in which Alfred referred to "reports Started by the Porter crowd"; Gates, *Truth*, 8. A succinct description of the brothers' financial problems is in Hulett to Lon, June 30, 1893, Merritt Papers. The letter concludes "we need [money] *awful bad* now." Leonidas claimed that the "ructions so graphically described by Gates, were incited by the Minnesota Iron Company or Rockefeller agents," and that after he complained to Gates about them, they stopped immediately. Leonidas, "Answer to Gates," 22, Merritt Papers.

CHAPTER 7 — A MERGER AND A NEW PARTNER — *pages 144 to 162*

1 For discussions of the role of the promoter as "the director and coordinator" of "industrial metamorphosis," see Gabriel Kolko, *The Triumph of Conservatism: A Reinterpretation of American History, 1900–1916*, 20 (Chicago, 1963); Mark Sullivan, *Our Times: The United States 1900–1925*, 2:314–320 (New York and London, 1927). According to the obituary in Harvard University, *Bulletin*, September 25, 1919, Wetmore's career included the presidency of the North American Company in 1893 and later the Detroit Edison Company — both public utilities firms.

2 The quotations are from Ralph W. and Muriel E. Hidy, *History of Standard Oil Company (New Jersey): Pioneering in Big Business 1882–1911*, 1:207–230 (New York, 1955). It seems probable that the Merritts were aware of Rockefeller's reputation, for the Ohio court's decision, for example, drew editorial comment in the *Duluth Tribune*, March 4, 1892.

3 On the May journey, see Hulett to Messrs. Robinson and Flinn, May 6, 1893, in Lewis J. Merritt and Son letter book, Merritt Papers. On Gates, see *Dictionary of American Biography*, 7:182; *Philadelphia Public Ledger*, January 21, 1912, p. 3; *A Service in Memory of Frederick Taylor Gates* (Montclair, N.J., February 19, 1929), pamphlet in MHS.

4 Allan Nevins, "Frederick T. Gates and John D. Rockefeller," and Frederick T. Gates, "The Memoirs of Frederick T. Gates," both in *American Heritage*, April, 1955, pp. 66–68, John D. Rockefeller, *Random Reminiscences of Men and Events*, 116, 117 (New York, 1909).

5 Gates, *Truth*, 9; Alfred, Napoleon, Cassius, Andrus, James, and Lewis Merritt to Leonidas, telegram (copy), July 11, 1893, and Andrus to Wetmore, July 18, 1893, both in Rockefeller Archives. Napoleon, Lewis, and Hulett later revoked their powers of attorney, saying "it had outlived its usefulness and done all that it was intended for it to do." See Napoleon and Lewis to Gates, November 13, 1893, Rockefeller Archives. Lewis and Hulett to Leonidas, November 13, 1893, Merritt Papers, explained that they thought the revocation advisable because "it will make it easier for you in the future to drive better trades." More formal authorizations of family members appear as Defendant's Exhibits nos. 45, 46, in Rockefeller v. Alfred Merritt, *Transcript*, 619–621.

6 Copies of various versions of preliminary agreements, July 1, 14, 15, 27, August 12, 1893, are in Rockefeller Archives, in Wetmore Papers (roll 3), and in Rockefeller v. Alfred Merritt, Plaintiff's Exhibits nos. 1, 4, and Defendant's Exhibits nos. 12, 30–32, in *Transcript*, 541–549, 570–572, 585–587, 606–610. The amounts transmitted to DM&N are listed in Leonidas Merritt v. DM&N, *Paper Book*, 310–312; Alfred's testimony appears on pp. 256–258. Additional pertinent testimony by Alfred may be found in Rockefeller v. Alfred Merritt, *Transcript*, 353.

7 Hulett to Lon, July 1, 1893, Merritt Papers. The telegrams are printed in Rockefeller v. Alfred Merritt, *Transcript*, 581, 605. Copies of articles of incorporation, July 21,

1893, and major financial transactions creating the Consolidated may be found in Wetmore Papers (roll 3); a series of telegrams between Gates and Rockefeller, July 31, August 1, 3, 9, 10, 1893; Wetmore note, July 24, 1893; Wetmore to Rockefeller, November 23, 1893 — all in Rockefeller Archives. On the Lake Superior Iron Company, see Chapter 6, page 139, above. See also Gates, *Truth*, 9.

[8] Lon and Wetmore were also involved in a loan of $73,000 from the Southern National Bank for which Wetmore used DM&N stocks and bonds and Merritt securities and notes to cover his own losses. On this and Wetmore's insolvency, see Lon's testimony, in Leonidas Merritt *v.* DM&N, *Paper Book*, 178–185, 213, 214, 527. Colby and Hoyt protested "that Wetmore's reckless methods have destroyed largely the credit of the Barge Company," according to Gates to Rockefeller, September 23, 1893, Rockefeller Archives. As a result of Wetmore's activities, Alfred Merritt found himself the endorser for $400,000 worth of collateral and Rockefeller already held as collateral a majority interest in the DM&N as well as over 15,000 shares of stock in various Merritt mining companies by August 28, 1893. See Chapter 8, below; Alfred testimony and Rockefeller deposition, in Rockefeller *v.* Alfred Merritt, *Transcript*, 360, 421, 422; Gates testimony, in Leonidas Merritt *v.* DM&N, *Paper Book*, 365–370; and Leonidas, "Answer to Gates," 29–31, 36, Merritt Papers. In the latter, Lon described Wetmore's "embezzlement" of $100,000 belonging to the Mountain Iron Company "to pay personal loans made to him by Rockefeller" and other Wetmore maneuvers.

[9] Wetmore to Rockefeller (draft), probably written between July 30 and August 5, 1893, Wetmore Papers (roll 3). Some paragraphs have been inserted for clarity. According to the *Duluth News Tribune*, March 3, 1894, "Mr. Wetmore, it is alleged, took certain other stock which the Merritts put into his hands and traded it to Mr. Rockefeller for more negotiable security." The North American Company, incorporated in New Jersey in 1890, was a public utilities holding company; *Poor's Manual of Industrials*, 406 (New York, 1910).

[10] Colby testimony, in Rockefeller *v.* Alfred Merritt, *Transcript*, 321, 335.

[11] Leonidas to Gates, August 9, 1893, and final contract, August 28, 1893 — both in Rockefeller Archives. Other versions of the contract (with different figures) appear as Plaintiff's Exhibit no. 3 in Rockefeller *v.* Alfred Merritt, *Transcript*, 104, 559–569, and in Wetmore Papers (roll 3).

[12] Rockefeller's contributions are described in Wetmore Papers, West Superior Iron & Steel Company Papers, and Samuel P. Ely Papers — all in Baker Library, Harvard University. The Ely Papers contain annual reports of the Spanish-American Iron Company; the West Superior Iron & Steel Papers have lists of shareholders, financial statements, and mortgages. Memoranda, loan agreements, and supporting documents on the $500,000 and $150,000 amounts below are in Rockefeller Archives. See also Gates, *Truth*, 8, 9, 18; Stanley Committee, *Hearings*, 3:1856.

[13] Gates, *Truth*, 11; Nevins, *Rockefeller*, 2:359–427. For the conspiracy theory, see Lon and Alfred testimony, in Stanley Committee, *Hearings*, 3:1845–1932; Andrus, "Story," 265–269; Leonidas, "Answer to Gates," Merritt Papers.

[14] The Rockefeller version is stated in Gates, *Truth*, 11, and repeated without qualification in Nevins, *Rockefeller*, 2:369.

[15] Leonidas testimony, in Rockefeller *v.* Alfred Merritt, *Transcript*, 223–229. However, in Leonidas Merritt *v.* DM&N, Lon testified that the meeting occurred "early in July"; see *Paper Book*, 210.

[16] Stanley Committee, *Hearings*, 3:1900; Andrus, "Story," 184. See also Leonidas, "Answer to Gates," 4, 26–28, and Leonidas, "Memoranda of an Interview with J. D. Rockefeller," undated — both in Merritt Papers. These versions of the interview also differ slightly in their details.

[17] Rockefeller *v.* Alfred Merritt, Plaintiff's Exhibit no. 6B, *Transcript*, 573; Lon to Lizzie, June 6, 1893, and Leonidas, "Answer to Gates," 26, Merritt Papers.

[18] Rockefeller *v.* Alfred Merritt, Rockefeller deposition, *Transcript*, 407–409.

[19] Rockefeller, *Random Reminiscences*, vi, 115, 116, 120–122. On Rockefeller's reti-

cence, see Gates, *Truth*, 11; Nevins, *Rockefeller*, 1:73, 2:695. For further evidence concerning the July date of the interview, see Chapter 8, note 35, below.

20 Gates, *Truth*, 10; memoranda and notes for dictation, undated and May 1–August 31, 1893, and copy of Consolidated contract, August 28, 1893, Rockefeller Archives. Charles and William Scheide were Standard Oil officials.

21 Gates, *Truth*, 10, 11. These figures included the five major mining properties: Biwabik, Missabe Mountain, Mountain Iron, Rathbun, and Shaw. The balance of the $23,354,000 can be accounted for by adding the values of lessor properties: Great Northern, Great Western, and McKinley. Sources for both sides agreed that the Adams and the Lone Jack, previously acquired by the Merritt-Wetmore syndicate, had been turned over to Rockefeller earlier; see memorandum of agreement, July 14, 1893, Rockefeller Archives; Gates, *Truth*, 9; Leonidas, "Answer to Gates," 28, Merritt Papers.

22 Rockefeller *v.* Alfred Merritt, undated letter, Merritts to Gates, Plaintiff's Exhibit no. 2, *Transcript*, 549–559, offers somewhat different figures. The figures here are in Norcross, "Spoliation," 141, and varied ones appear on 142, 148, Merritt Papers.

23 Gates, *Truth*, 10; Norcross, "Spoliation," 140–150, Merritt Papers; Consolidated contract, August 28, 1893, Rockefeller Archives.

24 On the Minnesota Iron Company, see the firm's annual report in *Duluth News Tribune*, July 24, 1893.

25 These figures were assembled from American Steel Barge documents in Wetmore Papers (roll 3). See also U.S. Commissioner of Corporations, *Report on the Steel Industry*, 1:147–149 (Washington, D.C., 1911), and Consolidated contract, August 28, 1893, Rockefeller Archives.

26 Lewis J. to "Andy," August 20, 1893, Merritt Papers; Andrus, "Story," 187, 188. The Merritts' attorney at the time was Joseph B. Cotton.

27 Lon to Andrus, August 7, 1893, in Rockefeller *v.* Alfred Merritt, Defendant's Exhibit no. 6, *Transcript*, 582; Leonidas, undated memorandum no. 5, p. 2, typed copy, and Leonidas, "Outline History," June 7, 1894, p. 10 — both in Merritt Papers.

28 Leonidas to E. G. Chapman, August 30, 1893, Rockefeller *v.* Alfred Merritt, Defendant's Exhibit no. 18, *Transcript*, 596; other similar examples of Leonidas' attitude at this time may be found in Defendant's Exhibits nos. 13–16, *Transcript*, 587–595. See also *Duluth News Tribune*, August 3, September 7, 1893; Leonidas, Alfred, and Andrus to Rockefeller, November 22, 1893, Rockefeller Archives.

29 Lewis J. to Andrus, November 21, 1893, in Lewis J. Merritt & Son letter book, Merritt Papers.

CHAPTER 8 — THE MERRITTS *VERSUS* ROCKEFELLER — *pages 170 to 201*

1 Norcross, "Spoliation," 150–152, Merritt Papers; Leonidas to Hudson Wilson, September 22, 1893, Defendant's Exhibit no. 16, in Rockefeller *v.* Alfred Merritt, *Transcript*, 597–601.

2 Gates to H. L. Satterlee, September 11, 1893, and to Leonidas Merritt, September 19, 1893, Rockefeller Archives; Gates, *Truth*, 13. Norcross, "Spoliation," 151, Merritt Papers, wrote: "The question naturally arises whether the motive behind this benevolent intrigue was not a close calculation on the number of adverse shares required to unsettle the Merritts in control of the consolidation." For another explanation by Lewis Merritt, see p. 198, below.

3 Norcross, "Spoliation," 174–178, Merritt Papers; Directors' Minutes, October 18, 1893, January 3, 1894, in DM&N, "Synopsis," 33, 37. A copy of the board's resolution, January 3, 1894, and Wetmore's reply are in Wetmore Papers. For local comment on Wetmore's removal, see *Duluth News Tribune*, November 27, 1893.

[4] Leonidas to Frank Jenkins, October 20, 1893, Defendant's Exhibit no. 20, in Rockefeller v. Alfred Merritt, *Transcript*, 597–601. On DM&N earnings in 1894, see Chapter 9, below.

[5] Leonidas to Gates, November 3, 1893, Defendant's Exhibit no. 24, in Rockefeller v. Alfred Merritt, *Transcript*, 601–604. The resulting list of Lon's transactions with Wetmore, dated November 1, 1893, is in Merritt Papers. See also pp. 197, 198, below.

[6] Leonidas, undated memorandum no. 5, Merritt Papers; Gates, *Truth*, 3, 14.

[7] George D. Rogers (Rockefeller's secretary) to Rockefeller, July 21, 1893, Rockefeller Archives; Norcross, "Spoliation," 154–156, quoting a 1912 letter to Norcross from John E. Merritt, Merritt Papers.

[8] Gates to Andrus Merritt, October 2, 1893, Rockefeller Archives, quoted in Gates, *Truth*, 13.

[9] Gates, *Truth*, 7; Leonidas, "Answer to Gates," 11, 12, 14. On buying more mines, see Leonidas, undated memorandum, no. 5, and "Outline History," 6, 9 — both in Merritt Papers; Andrus, "Story," 168.

[10] Newton H. Winchell, "Structures of the Mesabi Iron Ore," in Lake Superior Mining Institute, *Proceedings*, 13:189–204 (Ishpeming, Mich., 1908); *Lake Superior Iron Ores*, 322. The quotations below appear in Mussey, *Combination in the Mining Industry*, 114, 136.

[11] On the stock market, see Hoffmann, *Depression of the Nineties*, 62, 114, 248. Leonidas later accused Rockefeller of manipulating the market and preventing the sale of Consolidated stock; Stanley Committee, *Hearings*, 3:1907. But Lewis to Brothers and Nephew, October 1, 1897, [p. 11], Merritt Papers, laid the blame elsewhere charging that "AR [*Andrus*] went out on the street and proclaimed to the world that *he* would make this stock perfectly worthless and the man that had any would wish he had sold for a '*nickle*.' This was a good way to use those he and the rest of us had Indused to joyn us in filling our Contract made with the Consolidated Co," he added ironically.

[12] Gates to Rockefeller, December 11, 1893, Rockefeller Archives. Some paragraphs have been provided for clarity.

[13] Hoyt to Leonidas and Alfred, October 5, December 20, 1893, in American Steel Barge Company v. Merritt, Complaint, 48–50, New York Supreme Court, in Case no. 171, Federal Records Center, microfilm in MHS.

[14] Here and two paragraphs below, see W. C. Farrington to Hill, December 30, 1893; Hill to Jay C. Morse, January 4, 1894, and to H. H. Porter, January 5, 1894; Morse to Hill, January 5, 1894; Porter to Hill, January 8, 1894, all in President's Subject Files, no. 1,655, Great Northern Railway Records. Hill, who was in Washington, D.C., received a telegram stating that the "Merritts have limited our time until tomorrow [*February 1*] noon, unless we can say something definitely in way of an advance, they will positively close to others . . . if they sell to others we are permanently out of it"; S. A. Phillips to Hill, January 31, 1894. The incident does not appear in biographies of Hill, and the James J. Hill Papers in St. Paul are not yet open to researchers.

For the Merritt versions of the negotiations, see Alfred testimony in Stanley Committee, *Hearings*, 3:1875; Andrus, "Story," 213–215; Norcross, "Spoliation," 207–209, and a telegram signed "Pagoda" (Alfred Merritt) to Leonidas, December 23, 1893 — both in Merritt Papers. Using words often beginning with the letter "P" instead of the names of persons involved, the Merritt brothers communicated at times in a simple code. The telegram read: "Can show it to twenty six [*Rockefeller and Gates*] if advisable. Have had no conference yet. Would you think best to follow that line of policy? Wouldnt Pantry [*Hill?*] pay more with that understanding? Wouldnt it cause twenty six to show his hand either for himself or help sale to Pantry?"

On the Wilson-Gorman tariff, see F. W. Taussig, *Tariff History of the United States*, 289, 300 (New York, 1900); Edward Stanwood, *American Tariff Controversies in the Nineteenth Century*, 2:319–328 (Boston and New York, 1903).

[15] Here and below, see Gates, *Truth*, 14; Norcross, "Spoliation," 197, Merritt Papers; Lon, Alfred, John E., C. C., E. T., Thomas A., A. R., Hulett Merritt to Gates, De-

cember 23, 1893, Rockefeller Archives. On the July 15, 1893, loan, see Chapter 7, note 6, above.

[16] Memorandum, Gates to Rockefeller, December 29, 1893, Rockefeller Archives.

[17] Gates, *Truth*, 15; copies of agreements, February 1, 21, 1894, signed by Andrus, John E., Thomas A., Wilbur J., Alfred, Cassius, Leonidas, Joseph B. Cotton as trustee for Wetmore, and Rockefeller, in Rockefeller Archives. The $150,000 loan made to the family in July, 1893, was absorbed in the money they received; Norcross, "Spoliation," 216, Merritt Papers.

[18] Gates, *Truth*, 15. For a possible explanation, see p. 196, below.

[19] Gates, *Truth*, 15; Gates to Hulett, July 6, 1895, August 6, 1896, to Lewis J. Merritt, August 6, 1896, Rockefeller Archives. Andrus, "Story," 206, blamed the split in the family's ranks on a Rockefeller move to "divide the forces he was bent upon destroying." This explanation is not supported by Lewis to Brothers and Nephew, October 1, 1897, Merritt Papers, in which Lewis presented his side of the story and accused the others of treating him unfairly, abusing Hulett, publicly ridiculing him and his family, "filling the news papers with sensational Trash and vile and unmitigated lies. . . . trying to Break Down my Credit," and having "It Preached from the Pulpit calling me a Traiter and a black sheep." The letter is both bitter and specific in its charges. Hulett accused the family of "persecuting" his "poor sick father," who had suffered a stroke in August, 1893; Hulett to "My Dear Aunt" (Mrs. Leonidas), January 21, 1894 (misdated 1893). See also Hulett to Leonidas, August 28, 1893, Merritt Papers.

[20] Alfred believed that Gates tried to prevent him from resigning when he "said that Mr. Rockefeller had great confidence in me. . . . I could reach the fellows on the range, the explorers and mine owners, and could get traffic contracts for the railroad that probably no other fellow could get." Alfred recalled that he told Gates, "I could not retain my manhood and work for Mr. Rockefeller." Stanley Committee, *Hearings*, 3:1860. On Alfred's resignation, see Directors' Minutes, February 6, 1893, in DM&N, "Synopsis," 53. No one immediately replaced him as president.

[21] Norcross, "Spoliation," 217, Merritt Papers. Cotton was a native of Indiana, a graduate of Michigan Agricultural and Mechanical College, who settled in Duluth in 1888. He served one term in the Minnesota legislature in 1893. See Upham and Dunlap, eds., *Minnesota Biographies*, 144.

[22] Andrus, "Story," 222; Gates, *Truth*, 16. Anak Harris was born in Kentucky in 1838, served in the Confederate Army, began to practice law in Tennessee in 1865, and moved to Fort Scott, Kansas, in 1871. He remained there for 22 years, building up "a large and profitable" practice. He was also active in Democratic politics and achieved a reputation for oratory. A "strenuous opponent" of Populism, Harris left Kansas for Duluth in 1893 to find a place where "there were no farmers." He died in Duluth in 1901; his son Henry died of ptomaine poisoning a year later at the age of 34. Obituaries in *Duluth News Tribune*, May 14, 1901, p. 1, May 12, 1902, p. 5; *Topeka Journal*, May 15, 1901.

[23] Complaint, Plaintiff's Bill of Particulars, and Verdict, in Leonidas Merritt *v.* DM&N, *Paper Book*, 5, 6, 20–23, 75. The *Duluth News Tribune* followed the legal action carefully in its issues of May 27, September 14–16, 18–23, 1894. The case was appealed to the Minnesota Supreme Court but was "dismissed and withdrawn" in June, 1895; see *Duluth Commonwealth*, July 2, 1895. It is interesting to note that, while Leonidas sued for payment for his services, the DM&N board paid Alfred $22,000 for his work as president without a fuss; see Directors' Minutes, April 11, 1894, DM&N, "Synopsis," 47.

[24] Leonidas Merritt *v.* Biwabik Mountain Iron Company, Case no. 9497, in the office of the clerk of St. Louis County District Court, Duluth, microfilm copy in MHS; *Duluth News Tribune*, November 13, 15, 16, 17, 1894.

[25] Leonidas Merritt *v.* Missabe Mountain Iron Company, Case no. 9496, in the office of the clerk of St. Louis County District Court, Duluth, microfilm copy in MHS; *Duluth News Tribune*, November 16, 1894.

[26] A copy of the petition, April 18, 1894, and the removal authorization are in Merritt

v. American Steel Barge Company. See also *Duluth News Tribune,* August 4, 1894, March 29, 1895.

[27] Summons and Complaint, Merritt *v.* American Steel Barge Company. The collateral shares included: DM&N 3,050, Mountain Iron Company 1,535, and Missabe Mountain Iron Company 5. Because of the prominent people involved, the case drew comment in the *New York Times,* March 16, 17, 1894. A succinct account appears in *Duluth News Tribune,* March 14, 1894.

[28] The barge company's "Answer of the Defendant," Merritt *v.* American Steel Barge Company, included the letters. See also Chapter 7, above.

[29] Plaintiff's Complaint and Referee's Report, American Steel Barge Company *v.* Merritt, New York Supreme Court. On the stock auction, see *New York Daily Tribune,* November 24, 1894.

[30] Complaint and Judge Nelson's decision, Merritt *v.* American Steel Barge Company. In November, 1895, the Merritts took the case to the Eighth Circuit Court of Appeals, which reaffirmed the decision; see 79 *Federal Reporter* 228 (March 1, 1897). All charges in the action were finally paid by June 3, 1901.

[31] Plaintiff's Complaint, filed October 29, 1894, and Order Dismissing Cause as to Defendant Gates, May 22, 1895, both in Rockefeller *v.* Merritt, *Transcript,* 119, 205; Stanley Committee, *Hearings,* 3:1864; *Duluth Commonwealth,* June 14, 1895. See also Norcross, "Spoliation," 221, Merritt Papers, estimating Alfred's share.

[32] Plaintiff's Complaint, Rockefeller *v.* Merritt, *Transcript,* 115–118.

[33] Petition for Removal to Federal Court, Order Assigning Judge Riner, and Jury Empaneled, in Rockefeller *v.* Merritt, *Transcript,* 166, 205–207. A useful scrapbook compiled by Joseph Cotton in the possession of the Duluth, Missabe and Iron Range Railroad, Duluth, contains newspaper clippings on the trial from the *Duluth Evening Herald, Duluth Commonwealth,* and *Duluth News Tribune,* June 6–16, 1895.

The jury was composed of A. P. Noyes, foreman (Forest Lake), Thomas E. Bowe (Panola), F. W. Carlton (Warren), William G. Chilton (Frazee), Simon Clark (Duluth), A. S. Collins (Windom), P. V. Flynn (Duluth), J. M. Geist (Duluth), E. A. Mattix (Duluth), P. O. Regan (St. Paul), Samuel M. Register (Stillwater), Luke Stannard (Taylors Falls); *Duluth News Tribune,* June 6, 1895.

For biographical information on Cotton, Davis, Shaw, and Washburn, see Upham and Dunlap, eds., *Minnesota Biographies,* 144, 165, 695, 827. On Baldwin, who practiced law in Duluth from 1891 to 1902 when he moved to Chicago, see John C. Munro, *Semi-Centennial Celebration Edition and History of North Star Lodge No. 23,* 106 (St. Cloud, 1907); *Polk's Duluth City Directory,* 1891–92, p. 137, and 1902, p. 106.

[34] Testimony of George D. Rogers, Rockefeller's secretary, on the proposal to cancel the bonds; Charles F. Rand, president of the Aurora and Spanish-American companies and receiver of the Penokee and Gogebic; Samuel P. Ely, who spent five years opening the Cuban mine; Charles L. Colby, president of the Penokee and Gogebic in 1893; Clarence M. Boss, inspector of mines for Gogebic County, and others, all in Rockefeller *v.* Merritt, *Transcript,* 439–467. The final quotation appears in "Comparative Analysis of the Pleadings," in the case stored in the Federal Records Center. See also Gates to Rockefeller, September 23, 1893, Gates to Cotton, January 10, 1894, Rockefeller Archives; Gates, *Truth,* 18, 20.

[35] Rockefeller *v.* Merritt, *Transcript,* 225–229, 245–272, 380–382; Wetmore's deposition appears on pp. 431–438. On the initial stages of the Consolidated and the Merritt-Rockefeller interview, see also Chapter 7, above. A telegram of July 12, 1893, from Leonidas to M. M. Clark saying he held a conference with Rockefeller that day subsequently figured in another lawsuit brought by Alfred Merritt against Anak A. Harris & Son. See p. 192, below.

[36] Gates testimony, Rockefeller *v.* Merritt, *Transcript,* 378, 382. See also Gates, *Truth,* 23.

[37] Murray to Gates, June 15, 1895, Rockefeller Archives. Two major New York City

newspapers reported the verdict and briefly reviewed the proceedings without editorial comment. See *New York Times* and *New York Daily Tribune*, June 14, 15, 1895.

[38] Gates, *Truth*, 24. See also Gates to Rockefeller, August 7, 1896, Rockefeller Archives.

[39] Leonidas, "Answer to Gates," 2, 3; Leonidas testimony, in Rockefeller *v*. Merritt, *Transcript*, 275–277; *Duluth News Tribune*, June 14, 1895; Andrus, "Story," 230, 233.

[40] Hulett to Gates, penciled note, Rockefeller Archives; Andrus, "Story," 230; Gates, *Truth*, 16, 17. Gates noted the circumstances on Hulett's message.

[41] George D. Rogers, memorandum of three conversations with Anak A. Harris, July 10–12, 1894, Rockefeller Archives.

[42] Gates, *Truth*, 17.

[43] Rogers, memorandum of conversations with Harris, July 10, 11, 1894, Rockefeller Archives.

[44] Andrus, "Story," 232.

[45] Here and below, see 76 *Federal Reporter* 909–919 (November 9, 1896). For local comment, see *Duluth News Tribune*, November 11, 1896.

[46] Andrus, "Story," 247–250. Here and two paragraphs below, see copies of the retraction and closing documents in Rockefeller *v*. Merritt, Federal Records Center. The family's statement was also printed in *Duluth News Tribune*, February 28, 1897; Stanley Committee, *Hearings*, 3:1923; and Gates, *Truth*, 31. Cassius Merritt died in 1894, and Hansen Evesmith signed the retraction as "Hansen E. Smith," administrator of his estate, making a total of 21 signatures. On Cassius' death, see note 62, below.

[47] Stanley Committee, *Hearings*, 3:1916. Lon told the committee that he did not remember signing the retraction.

[48] For the Gates version of the settlement, see *Truth*, 24–26; for the Merritt version, see Leonidas, "Answer to Gates," 64–66; Andrus, "Story," 247–258; Stanley Committee, *Hearings*, 3:1859, 1866, 1916–1918.

[49] Norcross, "Spoliation," 228, and Jed L. Washburn to George H. Bridgeman, January 2, 1897 — both in Merritt Papers; Andrus, "Story," 249; Stanley Committee, *Hearings*, 3:1858, 1866, 1922.

[50] Alfred Merritt et al. *v*. A. A. Harris et al., 11th Judicial District, Case no. 13291, in the office of the clerk of St. Louis County District Court, Duluth, microfilm copy in MHS. See also *Duluth News Tribune*, February 24, June 16, 18, October 11, 1897. The Security Land and Exploration Company was incorporated on December 30, 1891, by a group of Duluth men, including [N?] B. Merritt; Articles of incorporation, Book C-2, p. 574.

[51] See note 11, above; Andrus, "Story," 266–268; Stanley Committee, *Hearings*, 3:1857–1865, 1869–1878 (Alfred testimony), 1902–1907, 1916–1926 (Leonidas testimony). For editorial comment, see, for example, *New York Times*, November 25, 1911, p. 12. After 11 months of hearings the committee concluded that U.S. Steel had "secured an unquestioned and an absolute dominance over the production, manufacture, and transportation of its product." The majority recommended an antitrust action, but eight years passed before the federal government brought suit. In 1920 the government attempted to dissolve the steel corporation on the grounds that its economic power eliminated competition. The U.S. Supreme Court ruled, however, that the corporation had attempted to control the steel industry but failed, and thereafter it had not attempted to destroy the operations of independent manufacturers. See Stanley Committee, *Hearings*, 8:1, 210–212; United States *v*. U.S. Steel Corporation, 251 *United States* 417 (March 1, 1920). A reminiscent article by Alta Hepziabeth Merritt, Alfred's daughter, describing her childhood reactions to "What Rockefeller Did to Us" appeared in the *New York World*, February 18, 25, 1912, magazine sec., p. 1, 2, respectively. My thanks to Grant Merritt for bringing it to my attention.

[52] Gates, *Truth*, 15; Gates to Stanley, Rockefeller to Stanley, both dated December 9, 1911, Rockefeller Archives. Both men chose not to appear on the grounds that the issue

had been argued before the courts and that they had nothing new to add to their sworn testimony. "The statements are false," wrote Rockefeller in part. The Merritts' "own signed retraction is before you. I therefore deem it unnecessary to avail myself of the opportunity now offered to appear before your Committee." Gates to Rockefeller, March 12, 1912, Rockefeller Archives, indicated that 113,000 copies of the pamphlet had been printed and distributed to the editors of every daily and weekly newspaper in the United States, every U.S. senator and representative, corporation directors, everybody listed in *Who's Who in America*, "every white minister of all the leading denominations," and all college presidents and professors.

    [53] Leonidas, "Answer to Gates," 1. Andrus, "Story," 267, said that in a "broken condition Lon prepared, but did not push through to publication, a pamphlet answering Gates and the calumny of the Stanley Investigation Committee."

    [54] Stanley Committee, *Hearings*, 3:1871, 1883.

    [55] James C. Merritt to Alfred Merritt, January 8, 1894 (misdated 1893) and memorandum, Gates to Rockefeller, May 20, 23, 1895 (quoted below) — both in Rockefeller Archives. On the Scheide interview, see also Norcross, "Spoliation," 157–161, quoting a 1912 letter from John, in Merritt Papers.

    [56] Cotton to Gates, March 8, 1894, Rockefeller Archives.

    [57] "Loans to C. W. Wetmore," [November 1, 1893], as well as the accounts cited below are in Merritt Papers. Three shares of Mountain Iron stock are also listed to E. H. Hall in the collateral on these loans.

    [58] Lewis Merritt to Brothers and Nephew, October 1, 1897, [p. 12], Merritt Papers.

    [59] Norcross, "Spoliation," 170, Merritt Papers. On Wetmore's later career, see Chapter 7, note 1, above.

    [60] Stanley Committee, *Hearings*, 3:1863. Harvard business historian Fritz Redlich commented that "the whole transaction is the natural outcome of the very low business ethics of the time, and is especially typical of the brutal business methods of John D. Rockefeller." Redlich, *American Business Leaders*, 1:148. Jules Abels, *The Rockefeller Billions: The Story of the World's Most Stupendous Fortune*, 184 (New York, 1965), wrote: "Rockefeller was clubbed with this stick about the Merritts for years."

    [61] Nevins, *Rockefeller*, 2:369; Stanley Committee, *Hearings*, 3:1913.

    [62] Evesmith to St. Louis County Historical Society, November 19, 1929, copy in SLC. The death of Cassius on April 27, 1894, became part of the conspiracy theory. According to Andrus, "Story," 221, the minister who spoke at Cassius' funeral said that "If Rockefeller [*sic*] had walked up behind Cassius Merritt and shot him in the back, it would have been a merciful act compared to the way he caused his death." Lewis Merritt disagreed. "What rite had you to say that [he?] killed Brother C. C. who owed before he had ever heard of JDR over $400,000 [in?] wild speculation. . . . There was one thing you all don[e] which hurts me very much and which was entirely unessary [*sic*] on Elizas [*Cassius' widow*] part as she had $50,000 of Life Insurance left her and it was this She had 700 shears of Consolidated stock. You went to her and advised her to Sell at any price telling her it would be worthless and so she sold for 8¢ [*$8.00?*] a clean los[s] to her and her orfen children of over $54,000 Dolars. I can[n]ot conc[e]ive how any rational person especial[l]y a Brother could hav[e] don[e] this." Lewis to Brothers and Nephew, October 1, 1897, [p. 16, 17], Merritt Papers.

    [63] See Chapter 4, note 9; *Virginia Enterprise*, March 29, 1901, p. 1; *Minneapolis Tribune*, May 10, 1926, p. 1; King, in *Missabe Iron Ranger*, May, 1967, p. 2; Norcross, "Spoliation," 228, Merritt Papers. On the formation of the American Exploration Company, see *Duluth News Tribune*, October 12, 1894. Andrus was also interested in coal lands in Kentucky in which he lost considerable sums of money in the 1890s.

CHAPTER 9 — OF KINGS, PRINCES, AND INDUSTRIAL EMPIRES — *pages 202 to 230*

¹ Rockefeller, *Random Reminiscences*, 120.

² Directors' Minutes, DM&N, "Synopsis," 53; *Duluth News Tribune*, February 7, 21, 1894; Hearding, "Some Recollections," 280, unpublished manuscript, MHS; Leonidas Merritt, memorandum no. 5, Merritt Papers.

³ Directors' Minutes, 1894–95, DM&N, "Synopsis," 41–72; Minnesota Railroad and Warehouse Commission, *Annual Report*, 1895, p. 308; *Duluth News Tribune*, February 7, 1894, February 6, 1895. Rockefeller bought out the remaining Grant-Chase minority stockholders for about $500,000 at the same per share price he paid the Merritts; *Duluth News Tribune*, April 27, 1894; *New York Daily Tribune*, April 28, 1894; Gates memorandum, April 10, 1894, and other documents on this transaction, Rockefeller Archives. Wetmore's financial status with the railroad may be seen in statements of receipts and disbursements in the files of the Duluth, Missabe and Iron Range Railroad's headquarters, Duluth. On Allibone and Philbin, see King, *Missabe Road*, 57, 71; on Olcott, see *Skillings' Mining Review*, vol. 24, May 4, 1935, p. 1.

⁴ The Merritts' option to repurchase 55,000 shares of Consolidated stock did not formally lapse until January 1, 1895; after that Rockefeller assumed full legal control of all Consolidated property. The revised organizational structure of the constituent companies is in Rockefeller Archives, and in *Duluth News Tribune*, February 5, 1895. See also DM&N, "Synopsis," 57; Poor, *Manual of Railroads*, 1895, p. 28.

⁵ Minnesota Railroad and Warehouse Commission, *Annual Report*, 1893, pp. 292, 293, 298, 300; 1894, pp. 262, 271, 273. Poor, *Manual of Railroads*, 1894, p. 252, gives the mileage as 115.3.

⁶ Minnesota Railroad and Warehouse Commission, *Annual Report*, 1895, pp. 315, 316, 325, 327; 1896, pp. 312, 313, 325.

⁷ Hacker, *World of Andrew Carnegie*, 342–345, 350, 362, 371; *Dictionary of American Biography*, 3:499–506. Carnegie is among the more accessible industrialists of this period. He not only wrote his *Autobiography* (Boston, 1920), but he has also been the subject of several voluminous studies which are cited in the footnotes below.

⁸ On Oliver's 1892 leases, see Chapter 5, above. See also Evans, *Iron Pioneer*, 106, 122, 202–208. On Oliver Mining and Standard Ore, see Secretary of State, Articles of incorporation, Book G-2, p. 291, Book F-2, p. 610.

⁹ *Lake Superior Iron Ores*, 253; Evesmith, "Missabe Mountain Mine," 1–3, in "Sketches"; *Duluth News Tribune*, August 18, 1893.

¹⁰ On Rockefeller's purchase of the Lone Jack, see Gates to Rockefeller, March 2, 1895, and a copy of the Oliver lease — both in Rockefeller Archives. See also *Lake Superior Iron Ores*, 253; *Superior Sunday Forum*, January 20, 1895.

¹¹ Evans, *Iron Pioneer*, 4; Wall, *Carnegie*, 478–536. A full-length biography is George Harvey, *Henry Clay Frick: The Man* (New York and London, 1928).

¹² Harvey, *Frick*, 188; Wall, *Carnegie*, 587. The literature on the Homestead strike is extensive. For a succinct account, see Wall, *Carnegie*, 537–582.

¹³ Carnegie to Frick, August 29, 1892, quoted in James H. Bridge, *The Carnegie Millions and the Men Who Made Them*, 259 (New York, 1903). See also Evans, *Iron Pioneer*, 209. Carnegie's lack of enthusiasm for iron mining may have been based upon an unsuccessful venture on the Marquette Range in 1872, according to Wall, *Carnegie*, 589.

¹⁴ Harvey, *Frick*, 189, 191; Wall, *Carnegie*, 598. The "good fellow" quotation is in Evans, *Iron Pioneer*, 210.

¹⁵ Carnegie to Frick, March 16, 1894, quoted in Harvey, *Frick*, 190, and Wall, *Carnegie*, 600.

¹⁶ Stanley Committee, *Hearings*, 3:2360; Wall, *Carnegie*, 596.

[17] There seems to be general confusion as to the exact date Oliver and Frick concluded their arrangements. Only Evans, *Iron Pioneer*, 211, specified May 1, 1894. Harvey, *Frick*, 189, and Wall, *Carnegie*, 596, fail to commit themselves to a specific date. For example, Wall merely wrote, "Frick, in his capacity as chairman of the company, completed the negotiations with Oliver soon after the Homestead Strike was settled" on November 20, 1892. But Carnegie's reluctance, expressed repeatedly in correspondence, delayed final agreement for nearly 18 months.

[18] *New York Herald*, June 7, 1895, and Wall, *Carnegie*, 599, quote a number of sources. Here and below, see Burton Hendrick, *The Life of Andrew Carnegie*, 2:16–20 (Garden City, N.Y., 1932); Nevins, *Rockefeller*, 2:398–401. By October 23, 1899, the *Duluth News Tribune* reported that Rockefeller's fleet numbered 57 vessels. On Pickands Mather, see Havighurst, *Vein of Iron*, 94. According to Anna Youngman, "Tendencies of Modern Combination," in *Journal of Political Economy*, 15:289 (May, 1907), Rockefeller's 1899 purchase of 30 vessels from the American Steel Barge Company gave him "a dominant position in the lake-ore shipping."

[19] Carnegie to Rockefeller, October 30, 1896, in Rockefeller Archives, quoted in Wall, *Carnegie*, 600.

[20] Agreement, December 9, 1896, Rockefeller Archives; Evans, *Iron Pioneer*, 249. See also "The Carnegie-Oliver-Rockefeller Agreement," in *Iron Age*, vol. 59, February 18, 1897, p. 22.

[21] Here and below, see *Lake Superior Iron Ores*, 252–254, 323, 324. It will be remembered that the Biwabik Mine had been leased by the Merritts in 1892 to Peter Kimberly, who managed to hang on to it through the panic of 1893 with the help of the Tod-Stambaugh Company, iron and steel manufacturers of Cleveland, Ohio. Thus the Lake Superior Consolidated collected royalties on the Biwabik but did not control the operation of the mine. See Setnicker, "Development of the Biwabik Mine," 71, 141.

[22] Carnegie Steel Board, Minutes, January 21, 1897, copy in Rockefeller Archives. Carnegie also increased his holdings in the Oliver Mining Company by purchasing an additional 4,000 shares. Oliver retained a 1/6 interest and remained president; see Evans, *Iron Pioneer*, 221. Oliver Mining changed its name to Oliver Iron Mining in 1898 and became a division of U.S. Steel in 1951. See Articles of incorporation, Book G-2, p. 291, Book S-2, p. 611, and Book M-11, p. 1.

[23] *Lake Superior Iron Ores*, 323, 324; Nevins, *Rockefeller*, 2:403. The *Duluth News Tribune*, April 11, 1898, also named the Commodore, Franklin, Penobscot, and Sellers mines as among those being acquired.

[24] Slightly varying figures may be found in Leith, *Mesabi Iron-Bearing District*, 293, and *Lake Superior Iron Ores*, 322.

[25] Wall, *Carnegie*, 602. The full letter, Oliver to Frick, July 27, 1897, is quoted in Bridge, *Carnegie Millions*, 261–266. See also Evans, *Iron Pioneer*, 253.

[26] Wall, *Carnegie*, 604; Harvey, *Frick*, 195; Evans, *Iron Pioneer*, 240–244.

[27] Copy of the Tilden lease, November 1, 1897, Rockefeller Archives. See also Wall, *Carnegie*, 604; *Lake Superior Iron Ores*, 190, 191.

[28] *Lake Superior Iron Ores*, 190, 191. On the complicated Norrie transaction here and below, see Evans, *Iron Pioneer*, 230–244; Wall, *Carnegie*, 605–607. For information on Schwab, see note 55, below.

[29] Carnegie to Carnegie Steel, September 27, 1897, quoted in Wall, *Carnegie*, 607; Carnegie to Oliver, December 24, 1897, quoted in Evans, *Iron Pioneer*, 243. *Iron Age*, vol. 60, October 7, 1897, p. 17, stated that the purchase would help Carnegie gain additional international trade.

[30] On the D&IR and the Pioneer Mine, see Chapter 3, above. Pioneer lease, July, 1897, Rockefeller Archives; *Lake Superior Iron Ores*, 199; *Ely Miner*, January 19, August 10, 1898; Evans, *Iron Pioneer*, 246, 247, 253. For additional local comment, see *Duluth News Tribune*, August 6, 8, 1898.

[31] Wall, *Carnegie*, 609–611; Evans, *Iron Pioneer*, 249–252. *Lake Superior Iron Ores*,

164, 168, 195, 196, 199, 203, gives these shipping dates: Duluth and Franklin, 1893; Pillsbury, 1898; Stephens, 1903; Shaw, 1917; Rocheleau, 1920.

³² Sale agreement, dated only July, 1898, Rockefeller Archives. See also *Lake Superior Iron Ores*, 199; Wall, *Carnegie*, 609; "The Carnegie-Oliver Ore Interests," in *Iron Age*, vol. 62, September 15, 1898, p. 19.

³³ "Carnegie-Oliver," in *Iron Age*, vol. 62, September 15, 1898, p. 19.

³⁴ Articles of incorporation, Book W-2, p. 151; "The Grand Marais Railroad Project," and Dwight E. Woodbridge, "Lake Iron Ore Matters," in *Iron Age*, vol. 62, August 18, 1898, p. 5, vol. 64, September 7, 1899, p. 8. On Congdon's association with Oliver, see Evans, *Iron Pioneer*, 216.

³⁵ Wall, *Carnegie*, 609; *Poor's Manual of Railroads*, 237 (1898). On Minnesota Iron's mines and the D&IR, see Chapters 3, 4, 5, above.

³⁶ For details on the docks at Two Harbors, see *Duluth News Tribune*, December 20, 26, 1892, January 1, 1894. See also U.S. Commissioner of Corporations, *Report on the Steel Industry*, 1:89.

³⁷ Clark, *History of Manufactures*, 2:235, 238, 580; Commissioner of Corporations, *Steel Industry*, 1:87; *Duluth News Tribune*, February 14, 1898; *Commercial and Financial Chronicle*, 60:299 (February 16, 1895). American Iron and Steel Association, *Bulletin*, 23:132 (May 15, 1889), referred to the creation of this firm as "the most important event that had ever taken place in the history of the iron trade."

³⁸ Ida M. Tarbell, *The Life of Elbert H. Gary: The Story of Steel*, 86–92 (New York, 1925); *Dictionary of American Biography*, 7:175.

³⁹ *Commercial and Financial Chronicle*, 67:633, 1008 (September 24, November 12, 1898); *Mineral Industry*, 7:395 (1898); Commissioner of Corporations, *Steel Industry*, 1:87. The *Duluth News Tribune*, September 13, 1898, announced the formation of Federal.

⁴⁰ *Commercial and Financial Chronicle*, 70:684 (April 7, 1900).

⁴¹ Albro Martin, *James J. Hill and the Opening of the Northwest*, 14, 28, 35, 38, 49, 69, 80, 91–107, 132–154, 279, 396, 468 (New York, 1976). For other views of Hill, see Joseph G. Pyle, *Life of James J. Hill* (Garden City, N.Y., 1917); Robert Sobel, *The Entrepreneurs: Explorations Within the American Business Tradition*, 110–147 (New York, 1974). For the Merritt version of Hill's refusal, see Chapter 8, above.

⁴² These somewhat complicated and often confusing corporate arrangements can best be followed in Great Northern Records: "Duluth and Winnipeg Railroad Company, Corporate History and Organization," 1–11; copy of the "Articles of Incorporation of the Duluth-Superior and Western Railway Co."; and "Duluth, Superior and Western Railway Company," 1–30. For Hill's early interest, see *Minneapolis Tribune*, December 16, 1892; on the Duluth and Winnipeg and the Merritts, see Chapter 5, note 7, above.

⁴³ For a detailed outline of the financial maneuvers and extracts from the board's authorization, see "History of Investments Made by the Great Northern Railway Company in Securities of the Duluth and Winnipeg Railroad Company, Duluth Superior and Western Railway Company and Affiliated Companies," 1–15, in Great Northern Records. The subsidiary firms had all been incorporated on August 18, 1891, at $50,000 each by, among others, Leonidas and Daniel H. Merritt. Copies of the articles of incorporation for the Minawa, Minosin, Nibiwa, Wabigon, and Wenona iron companies are in North Star Iron Company, Great Northern Records. They had been organized to hold title to lands belonging to North Star Iron, a firm originally incorporated in Maryland on March 25, 1890; it continued to operate under a West Virginia statute dated May 18, 1892. Articles of incorporation, Book M, p. 210, Secretary of State's office, Charleston, W. Va. Each iron company issued 500 shares with North Star holding 495 and the directors one share each. According to "Brief History of Organization of North Star Iron Company of West Virginia," Great Northern Records, Alfred, Daniel, Leonidas, and Lewis J. Merritt held a total of 757 shares in the parent North Star company.

⁴⁴ Wright resigned as a member of the Winnipeg board of directors in 1893. See Ammi

W. Wright to the President, February 18, 1893, Great Northern Records; Articles of incorporation, Book F-2, p. 89; *Duluth News Tribune*, May 5, 1892; *Mississippi Valley Lumberman*, vol. 23, February 3, 1893, p. 7; Franklin A. King, "Logging Railroads of Northern Minnesota," in Railway and Locomotive Historical Society, *Bulletin*, 93:101 (October, 1955). Wright and Davis offered to sell Weyerhaeuser the land itself, but his associates did not trust mineral prospecting, and the firm took timber rights only for $1,200,000. They then tried unsuccessfully to sell the land for $3.00 an acre, but "couldn't get anyone to 'nibble'"; Van Brunt, *Duluth*, 2:549. See also Larson, *White Pine Industry*, 233; Ralph W. Hidy, Frank E. Hill, Allan Nevins, *Timber and Men: The Weyerhaeuser Story*, 107 (New York, 1963).

[45] Copy of the Mahoning Mine lease in Louis W. Hill Papers, Northwest Area Foundation, St. Paul; *Lake Superior Iron Ores*, 253; W. C. Agnew, *Recollections of the Early History of the Mahoning Ore and Steel Company Prior to 1909*, 2, 7 (n.p., [1909?]), MHS; W. C. Agnew, "Early History of Mahoning Iron Mine is Told by its Discoverer," in *Skillings' Mining Review*, vol. 11, May 12, 1923, pp. 1, 2, 6, 17; "The Mahoning Iron Mine," in *Iron Age*, vol. 64, November 9, 1899, pp. 1–3.

[46] On the acquisition by Eastern, including a copy of the formal agreement, see Duluth, Mississippi River and Northern Railway, Minute Book, May 1, 1899, p. 33, minutes of "A Meeting of the Stockholders," 64, and "A Meeting of the Directors," 64–66, in Great Northern Records. Hill mistakenly testified it was about 25,000 acres; Stanley Committee, *Hearings*, 4:3155, 3172. See also "Report of the Committee on Investigation of the Great Northern Ore Lands," in Minnesota, *House Journal*, 2:1659 (April 20, 1907).

[47] For the subsequent story of this organization, see "History of the Great Northern Ore Properties," in *Skillings' Mining Review*, vol. 53, November 28, 1964, pp. 1, 4, 5. See also Martin, *Hill*, 576; Louis W. Hill, Memorandum to Great Northern Board of Directors, January 9, 1946, Great Northern Records.

[48] Copy of the lease and articles of incorporation for the Allouez Bay Dock Company, Great Northern Records; Great Northern Railway Company, *Fourteenth Annual Report*, 10 (1902) and *Twentieth Annual Report*, 17 (1908), both in the Louis W. Hill Papers. Louis also supervised the construction of the Great Northern's Allouez Bay dock at Superior in 1900. The largest ore dock built up to that time, it was 62 feet wide, extended over 1,500 feet from shore, and stood 73 feet above the water. Its superstructure consisted of 13,216 pilings of Pacific fir driven "to refusal" 40 to 60 feet into the lake bed. See Dwight E. Woodbridge, "The Eastern Minnesota Ore Dock," in *Iron Age*, vol. 65, June 21, 1900, pp. 11–13, especially a chart on p. 12 comparing the size of all Lake Superior ore docks in 1900.

[49] Although incorporated in Marquette, Michigan, the Lake Superior Company filed articles of incorporation in Minnesota on February 15, 1900; Dwight E. Woodbridge, "The Hill Iron Ore Properties," in *Iron Age*, vol. 77, January 4, 1906, pp. 26–28; the other 10 shares went to Robert Farrington, one of Hill's subordinates. A copy of the contract, October 20, 1899, between the Great Northern and Lake Superior companies is in Louis W. Hill Papers. During January and February, 1913, the executive committee, board of managers, and stockholders took the necessary steps to dissolve the firm; formal papers filed with the state of Michigan, April 19, 1913 (copies in Great Northern Records). See also *Duluth News Tribune*, January 28, 1899; Stanley Committee, *Hearings*, 4:3161, 3162, 3169. Minnesota, *General Laws*, 1887, p. 324, allowed no corporation except those building or running a railroad to "acquire, hold or own over 5,000 acres of land."

[50] Woodbridge, in *Iron Age*, vol. 77, January 4, 1906, p. 28; Wirth, *Minnesota Iron Lands*, 162–169; "The Hill Ore Holdings on the Mesabi Range," in *Iron Trade Review*, vol. 36, February 19, 1903, p. 39. In September, 1906, the Great Northern, acting through the Lake Superior Company, Limited, announced its lease (formally dated January 2, 1907) of 48,317 acres of ore land to the United States Steel Corporation. Hill

struck a hard bargain. He received a royalty of $.85 per ton of ore mined during the first year, plus an annual increase of 4 per cent in succeeding years. In addition, the agreement guaranteed Hill's transportation network the right to carry all iron ore mined on these lands at the usual rate of $.80 per ton, divided between railroad and dock charges. When he was asked by the Stanley committee how he had negotiated such a profitable contract, Hill replied, "That was the simplest thing in the world. I did not have the slightest difficulty. I told them that was the price they could have it at, and it was a loss of their time and a waste of mine to ask for any other figure." U.S. Steel accepted this rate in order to increase its control of the Mesabi Range from 52 to about 75 per cent but voided the lease in 1912 (effective January 1, 1915) for fear of a federal suit against its monopolistic position. Hill said, "I think they were frightened to death. . . . They had buck fever." Stanley Committee, *Hearings*, 4:3204, 3233–3242; copy of resolutions from the board of directors, stockholders, and board of managers and "Synopsis" of the lease in Louis W. Hill Papers. See also "The Hill Iron Ore Lands," in *Engineering and Mining Journal*, 82:693 (October 13, 1906).

⁵¹ Great Northern Railway Company, *Twenty-third Annual Report*, 24 (1911).

⁵² On the growth of stock sales after 1897, see Thomas R. Navin and Marian V. Sears, "The Rise of a Market for Industrial Securities, 1887–1902," in *Business History Review*, 29:129–138 (June, 1955). Ralph L. Nelson, *Merger Movements in American Industry, 1895–1956*, 6, 37, 116–126 (Princeton, N.J., 1959), suggested that the sale of securities was the most important factor promoting industrial consolidation and pointed out that the number of companies eliminated increased from 43 in 1895 to 303 in 1898 to 1,208 in 1899, and then averaged 380 a year from 1900 to 1903. For the composition of mergers among manufacturers of finished and semifinished steel products, see Eliot Jones, *The Trust Problem in the United States*, 189–196 (New York, 1921). On pools, see Temin, *Iron and Steel*, 174–189. For other points of view on the merger movement, see W. Elliot Brownlee, *Dynamics of Ascent*, 201 (New York, 1974); Glenn Porter, *The Rise of Big Business, 1860–1910*, 71–84 (New York, 1973); Alfred D. Chandler, Jr., *Strategy and Structure: Chapters in the History of the Industrial Enterprise*, 29–51 (Cambridge, Mass., 1962). On the capitalization of U.S. Steel, see Commissioner of Corporations, *Steel Industry*, 1:167–179.

⁵³ A readable biography is Frederick Lewis Allen, *The Great Pierpont Morgan*, 10–13, 38, 81–125 (New York, 1949).

⁵⁴ More than three companies produced unfinished steel, but none seriously challenged these major firms. Among the others were Jones & Laughlin Steel Company, Bethlehem Steel Company, and Republic Iron & Steel Company. For lists see *Commercial and Financial Chronicle*, 68:899 (May 13, 1899); *Iron Age*, vol. 67, February 14, 1901, p. 29. On the formation of National Steel, see *Commercial and Financial Chronicle*, 68:672, 69:230, 285 (April 8, July 29, August 5, 1899); Commissioner of Corporations, *Steel Industry*, 1:89. On William Moore, see *Who Was Who in America, 1897–1942*, 862. For the Bessemer Steamship Company and Federal Steel's formation, see pp. 209, 217, above. On the formation of Pittsburgh Steamship, see Waldon Fawcett, "The Struggle for Lake Shipping," in *Iron Age*, vol. 64, October 19, 1899, p. 30; Bridge, *Carnegie Millions*, 272–274; Wall, *Carnegie*, 623; Evans, *Iron Pioneer*, 252, 253.

⁵⁵ Abraham Berglund, *The United States Steel Corporation*, 64–66 (New York, 1907); Harold C. Livesay, *Andrew Carnegie and the Rise of Big Business*, 182–186 (Boston, 1975); Robert Hessen, "Charles M. Schwab, President of United States Steel, 1901–1904," in *Pennsylvania Magazine of History and Biography*, 96:204–212 (April, 1972).

⁵⁶ Kolko, *Triumph of Conservatism*, 32. The story of these events and Schwab's prominent role in them has been told many times. See, for example, Frederick Lewis Allen, *The Lords of Creation*, 1–27 (New York, 1935); Robert Hessen, *Steel Titan: The Life of Charles M. Schwab*, 114–116 (New York, 1975); Wall, *Carnegie*, 784–789; Hendrick, *Life of Carnegie*, 2:128–143. For Schwab's recollections, see Stanley Committee,

Hearings, 2:1276. Carnegie personally received $225,639,000 in U.S. Steel first mortgage, 5 per cent gold bonds in return for his transfer of $86,145,000 in bonds and $92,639,000 in stock of Carnegie Steel.

[57] A copy of the Rockefeller-Morgan transaction is in Rockefeller Archives. See also Tarbell, *Life of Gary*, 118–120; Nevins, *Rockefeller*, 2:417–422; Rockefeller, *Random Reminiscences*, 131. Note that the quotation expresses the same solicitous attitude reflected in Rockefeller's view of the Merritt acquisitions discussed in Chapter 8, above.

[58] Here and below, see "United States Steel Corporation Charter," in *Iron Trade Review*, 34:18–20 (February 28, 1901); Berglund, *United States Steel Corporation*, 70, 72; "The Great Consolidation," in *Iron Age*, vol. 67, February 14, 1901, pp. 28–31. U.S. Steel's Great Lakes ore fleet, with a total carrying capacity of 12,000,000 tons, was composed of 59 ships from the Bessemer Steamship Company, 22 from Federal Steel, 13 from American Steel & Wire, 12 from Pittsburgh Steamship, and 6 from National Steel. On Gary's position, see Tarbell, *Gary*, 123, 127, 135. For a comparison of securities issued by U.S. Steel in 1901 with those of constituent companies and for a breakdown of its capacity, see Commissioner of Corporations, *Steel Industry*, 1:113, 365. U.S. Steel realized its error in overlooking Hill and moved to correct it in 1906; see note 50, above.

[59] Under the heading, "United States Steel Corporation," *Iron Age*, vol. 67, March 7, p. 39, April 4, p. 41, 1901, was cautiously neutral. For sample press reaction in major eastern newspapers as well as abroad, see Sullivan, *Our Times*, 2:351–355.

[60] Gleed, "The Steel Trust and Its Makers," and Ely, "An Analysis of the Steel Trust," in *Cosmopolitan*, 31:32, 429 (May, August, 1901). A satirical view by John B. Wallace, "The World's Greatest Revolution," appeared in the same magazine, 30:677–680 (April, 1901).

[61] Similar fears of monopoly were voiced by the *Duluth Evening Herald*, February 20, p. 3, March 4, p. 4, 1901.

[62] The United States both exported and imported ore in the 1890s, but its exports were small. Imports, however, increased steadily throughout the decade, and more than half of the imported ore came from Cuba and Sweden. See American Iron and Steel Association, *Bulletin*, 29:139, 213 (June 20, September 20, 1895); 30:66 (March 20, 1896); 33:139 (November 15, 1899); 35:53 (April 10, 1901).

[63] *St. Paul Pioneer Press*, March 22, 1901, p. 4; *Minneapolis Journal*, March 21, 1901, p. 2; Minnesota, *Senate Journal*, 1901, p. 579; *House Journal*, 1901, p. 866; Minnesota, *General Laws*, 1901, pp. 269–271. The *Duluth News Tribune* of March 2, 1901, p. 3, also called for action to encourage manufacturers to locate in the city, and on March 6, p. 1, bemoaned the failure of John E. Searles of New York who had proposed to build an iron and steel plant there. In 1914 the Minnesota Steel Company, a subsidiary of U.S. Steel, built a plant in the Morgan Park area of the city. See Francis N. Stacy, "Pittsburgh Moving West," in *World's Work*, 27:328 (January, 1914).

[64] Calculated from statistics in *Lake Superior Iron Ores*, 122–133.

[65] *Lake Superior Iron Ores*, 133, 173, 191, 199, 252–254; John Birkinbine, "Mines and Mining — Iron Ore," in 52 Congress, 1 session, *House Miscellaneous Documents*, no. 340, part 1, p. 14 (serial 3008); U.S. Bureau of the Census, *Mines and Quarries, 1902*, 403 (Washington, 1905), provided the following rank of selected states as producers of iron ore from 1850 to 1902.

1850	*1860*
1. Pennsylvania	1. Pennsylvania
2. Ohio	2. Ohio
3. Maryland	3. New York
4. Tennessee	4. New Jersey
5. Kentucky	5. Michigan
16. Wisconsin	13. Wisconsin
18. Michigan	

1870	*1880*
1. Pennsylvania	1. Pennsylvania
2. Michigan	2. Michigan
3. Ohio	3. New York
4. New York	4. New Jersey
5. Maryland	5. Ohio
11. Wisconsin	15. Wisconsin

1889	*1902*
1. Michigan	1. Minnesota
2. Alabama	2. Michigan
3. Pennsylvania	3. Alabama
4. New York	4. West Virginia
5. Minnesota	Virginia
6. Wisconsin	5. Tennessee
	7. Wisconsin

CHAPTER 10 — THE WESTERN MESABI AND THE CUYUNA —
pages 231 to 259

[1] Leith, *Mesabi Iron-Bearing District*, 4:25, 46–49; Van Barneveld, *Iron Mining in Minnesota*, 175. An account of Ramsey's iron ore venture appears in *Grand Rapids Herald-Review*, April 5, 1922, p. 1.

[2] Donald L. Boese, *John C. Greenway and the Opening of the Western Mesabi*, 2–5 (Grand Rapids, Minn., 1975); [Ellis], *Iron Ranges of Minnesota*, [36]; Mesabi Range township maps of Alfred and Leonidas Merritt, MHS; "The Development of Itasca's Mining Industry," in *Grand Rapids Herald-Review*, December 21, 1921, p. 3; Nute, ed., *Mesabi Pioneer*, 23, 35, 77.

[3] On Hartley, see J. A. A. Burnquist, *Minnesota and Its People*, 4:490–494 (Chicago, 1924); Toensing, *State Officials*, 50; "G. G. Hartley of Duluth is Summoned," in *Skillings' Mining Review*, vol. 10, January 21, 1922, p. 5. On Cole, see Evans, *Iron Pioneer*, 293; *Who Was Who*, 4:188; Woodbridge and Pardee, eds., *St. Louis County*, 2:742; Boese, *Greenway*, 5–9, 15–17, 132. For the Canisteo Mining Company, see Articles of incorporation, Book G-3, p. 482.

[4] Nute, ed., *Mesabi Pioneer*, 29; Boese, *Greenway*, 16, 70, 87–89. On the ores and the Trout Lake concentrator here and below, see Van Barneveld, *Iron Mining in Minnesota*, 177, 179; Woodbridge and Pardee, eds., *St. Louis County*, 2:745–747; James E. Rottsolk, *Pines, Mines, and Lakes: The Story of Itasca County, Minnesota*, 39 (n.p., 1960); *Lake Superior Iron Ores*, 255.

[5] For biographical data on Greenway and a good account of early Coleraine, see Boese, *Greenway*, 43, 45–57, 85–111, 205. Because of his experience in Coleraine, Greenway also became involved in the building of the U.S. Steel town of Gary, Indiana, in 1906.

[6] The quotations appear in Boese, *Greenway*, 106, 107.

[7] W. S. Montgomery, "Duluth and Winnipeg Railroad Company," 12, in Great Northern Railway Company Records; Rottsolk, *Pines, Mines, and Lakes*, 11, 12, 43–45. Pokegama Falls was dammed in 1901 to create power for a paper mill. See also Minnesota Railroad and Warehouse Commission, *Annual Report*, 1892, p. 416; *Minnesota State Census*, 1895, p. 24, and 1905, p. 19. On village incorporations here and below, see Harvey Walker, *Village Laws and Government in Minnesota* (University of Minnesota, Bureau of Research in Government, *Publication*, no. 6 — Minneapolis, 1927).

[8] Rottsolk, *Pines, Mines, and Lakes*, 36, 54; Railroad and Warehouse Commissioners, *Annual Report*, 1903, sec. 15, p. 53; Woodbridge and Pardee, eds., *St. Louis County*, 2:725–727. Small logging railroads had earlier penetrated the pinelands of Itasca County; the Duluth, Mississippi River and Northern, the old Wright-Davis road, reached Nashwauk before 1899. See Larson, *White Pine Industry*, 361; Rottsolk, *Pines, Mines, and Lakes*, 23.

[9] Woodbridge and Pardee, eds., *St. Louis County*, 2:749; Boese, *Greenway*, 22, 25–41; *Grand Rapids Herald-Review*, May 7, 1904, p. 8; *Minnesota State Census*, 1905, p. 19; *Lake Superior Iron Ores*, 249; Railroad and Warehouse Commissioners, *Annual Report*, 1907, p. 393. On Charles Bovey, see Upham and Dunlap, eds., *Minnesota Biographies*, 67.

[10] [*Ellis*], *Iron Ranges of Minnesota*, [87, 90, 91]; *Lake Superior Iron Ores*, 256; Woodbridge and Pardee, eds., *St. Louis County*, 2:749, 751, 752; Prosser, *Rails to the North Star*, 43.

[11] Bridge, *Carnegie Millions*, 259.

[12] "Field Notes of the Survey of the Subdivision and Meander Lines," T45N, R29, 30W, Crow Wing County, in Secretary of State's Office.

[13] Roland D. Irving and Charles R. Van Hise, "The Penokee Iron-Bearing Series of Michigan and Wisconsin," in U.S. Geological Survey, *Monographs*, no. 19, plate 1 (Washington, D.C., 1892); Winchell and Winchell, *Iron Ores of Minnesota*, 154.

[14] Robert G. Schmidt, *Geology and Ore Deposits of the Cuyuna North Range Minnesota*, 3 (Washington, D.C., 1963); P. W. Donovan, "Some Aspects of Exploration and Drilling on the Cuyuna Range," and Carl Zapffe, "A Survey of the Developments and Operations in the Cuyuna Iron Ore District," in Lake Superior Mining Institute, *Proceedings*, 20:125, 137 (Ishpeming, Mich., 1915); [Anna Himrod], *The Cuyuna Range: A History of a Minnesota Iron Mining District*, 3 (St. Paul, 1940). The present chapter relies heavily on the latter study, which was prepared by Himrod under the Minnesota Historical Records Survey Project, a division of the Work Projects Administration (WPA), and on background material assembled by Himrod in the MHS manuscripts collection, cited in note 20, below.

[15] E. C. Harder and A. W. Johnston, *Preliminary Report on the Geology of East Central Minnesota Including the Cuyuna Iron-Ore District*, 133, 134 (Minneapolis, 1918); Carl Zapffe, "Matters of Interest to Operators Regarding the Cuyuna District," and J. Wilbur Van Evera, "A Brief History of the Cuyuna Range," in Lake Superior Mining Institute, *Proceedings*, 20:192, 23:89 (Ishpeming, Mich., 1920, 1923).

[16] On Fort Ripley and Old Crow Wing, see Holmquist and Brookins, *Minnesota's Major Historic Sites*, 68, 69.

[17] Eugene V. Smalley, *History of the Northern Pacific Railroad*, 323 (New York, 1883); Carl Zapffe, *Brainerd, Minnesota, 1871–1946*, 2, 6, 11, 39, 76 (Minneapolis, 1946); WPA, "Historical Sketch, Crow Wing County," 10, 11, 37 (n.d.), MHS; A. J. Crone, *Pioneers of Deerwood*, 4–6 (Deerwood, 1923); *Minnesota State Census*, 1885, p. 15; *U.S. Census*, 1900, *Population*, 217.

[18] Here and below, see [Himrod], *Cuyuna Range*, 4, 6, 10; *Brainerd Journal*, February 23, 1893. Pajari's recollections appear in *Deerwood Enterprise*, May 9, 1930, p. 1.

[19] Here and below, see Crone, *Deerwood*, 4, 45; *Skillings' Mining Review*, vol. 21, December 3, 1932, p. 3; Neil M. Clark, "How the Needle of a Compass Pointed the Way to Fortune," in *American Magazine*, February, 1922, pp. 16, 88–93. On Adams' farm, see Hiram M. Drache, *The Day of the Bonanza*, 4, 74, 118 (Fargo, N.D., 1964). Letters from Adams to trustees in the East are in Jay Cooke Papers, Harvard University, copies in MHS. See, for example, Adams to F. O. French, special trustee, February 13, 1881.

[20] [Himrod], *Cuyuna Range*, 14, 150; Durand A. Hall to Anna Himrod, March 21, 1938, WPA Historical Records Survey Papers, 1937–39, MHS; Articles of incorporation, Book E-3, p. 152.

[21] Hemstead to Himrod, January 1, 1938, quoted in [Himrod], *Cuyuna Range*, 15.

²² Here and below, see Kirby Thomas, "A Promising Lake Superior Iron District," in *Mining World*, 21:446–448 (November 5, 1904); Harder and Johnston, *Preliminary Report*, 98; Van Evera, in Lake Superior Mining Institute, *Proceedings*, 23:85; [Himrod], *Cuyuna Range*, 17, 18, 105, 133. Jamison to Himrod, January 28, 1938, WPA Historical Records Survey Papers, indicated that the Orelands Mining Company sold stock "on the strength of what the U.S. Steel were doing," and suggested that that corporation began drilling near Dam Lake south of Kimberly in Aitkin County because the formation was "similar to the Missaba."

²³ On naming the Cuyuna, see Robert M. Adams to Himrod, December 7, 1937, in WPA Historical Records Survey Papers; C. K. Leith, "A Summary of Lake Superior Geology," in American Institute of Mining Engineers, *Transactions*, 36:101 (1906). Leith's diagram is in *Mesabi Iron-Bearing District*, 203.

²⁴ Van Evera, in Lake Superior Mining Institute, *Proceedings*, 23:86; [Himrod], *Cuyuna Range*, 26. Hobart Iron Company was incorporated in 1900; Articles of incorporation, Book Y-2, p. 1.

²⁵ [Himrod], *Cuyuna Range*, 32–36; *Mining World*, 34:1187 (June 10, 1911).

²⁶ [Himrod], *Cuyuna Range*, 36; Articles of incorporation, Book Q-3, p. 517; "Cuyuna Range, Minnesota," in *Engineering and Mining Journal*, vol. 90, December 17, 1910, pp. 12–14; Adams to J. A. Ferguson, October 15, 1909, in Adams Family Papers, MHS; *Duluth News Tribune*, October 19, 1908, p. 1; Railroad and Warehouse Commissioners, *Annual Report*, 1911, p. 490. In 1958 the state of Minnesota sued Cuyler Adams' heirs for ownership of the iron ore near Rabbit Lake. See Minnesota *v.* Adams *et al.*, 251 Minnesota 521 (1957).

²⁷ Here and below, see Upham and Dunlap, eds., *Minnesota Geographic Names*, 156–158; *Lake Superior Iron Ores*, 29, 31, 33, 172, 180; Northern Minnesota Editorial Association, *Program and Souvenir*, [2] (Crosby, 1920); Burnquist, *Minnesota and Its People*, 4:618; *Brainerd Dispatch*, August 11, 1911, p. 6; *Crosby-Ironton Courier*, September 27, 1961, p. 1; Van Evera, in Lake Superior Mining Institute, *Proceedings*, 23:86.

²⁸ Adams to Howard Elliott, August 8, 1910, Adams Family Papers; [Himrod], *Cuyuna Range*, 49, 50, 54, 107; "Certificate of Incorporation" (copy) in Northern Pacific Railway Company Records; Articles of incorporation, Book U-3, p. 524. Cuyuna Northern, Board of Directors, Minutes, June 18, 1914, Northern Pacific Records, say the railroad was sold to Northern Pacific for $258,597.86.

²⁹ *Brainerd Dispatch*, April 14, p. 7, May 4, p. 5, May 12, p. 7, May 26, p. 7, 1911. On the ore docks, below, see "Northern Pacific Ore Dock: Handling Ore from the Cuyuna Range," in *Iron Trade Review*, 50:729–733 (October 23, 1913); *Lake Superior Iron Ores*, 276, 316. The mines shipping by 1913 were the Armour No. 1 and No. 2, Barrows, Thompson, Ironton, Kennedy, Louise, and Pennington.

³⁰ *Lake Superior Iron Ores*, 276; Carl Zapffe, "Manganiferous Ores of the Cuyuna District, Minnesota," in American Institute of Mining and Metallurgical Engineers, *Transactions*, 71:372–385 (1925); [Himrod], *Cuyuna Range*, 53; Gar A. Roush, *Strategic Mineral Supplies*, 31–69 (New York, 1939); George F. Brightman, "Cuyuna Iron Range," in *Economic Geography*, 18:282 (July, 1942). Early in the production of Cuyuna iron it was believed there was "no market for manganiferous ores"; *Engineering and Mining Journal*, 93:1025 (May 25, 1912).

³¹ Donovan and Van Evera, both in Lake Superior Mining Institute, *Proceedings*, 20:138, 23:86; Percy W. Donovan, "Churn-Drill Angle Holes on the Cuyuna," in *Engineering and Mining Journal*, 96:1117 (December 13, 1913); *Lake Superior Iron Ores*, 274; Harder and Johnston, *Preliminary Report*, 103. On the development of individual mines, see Harder and Johnston, 101. On hydraulic stripping, see Edward P. McCarty, "Hydraulic Stripping at Rowe and Hillcrest Mines on the Cuyuna Range," in Lake Superior Mining Institute, *Proceedings*, 20:162–173.

³² Adams to James T. Hale, July 15, 1908, Adams Family Papers; *U.S. Census*, 1900, pp. 215, 217, and 1920, *Population*, pp. 509, 511.

[33] Unsuccessful attempts to unionize the Minnesota ranges by the Western Federation of Miners in 1907 and the International Workers of the World in 1916 resulted in two widespread strikes. Not until 1943 did Oliver Mining recognize the United Steelworkers of the CIO, which had successfully unionized the steel industry six years earlier. See Neil Betten, "Strike on the Mesabi — 1907" and "Riot, Revolution, Repression in the Iron Range Strike of 1916," in *Minnesota History*, 40:340–347, 41:82–93 (Fall, 1967, Summer, 1968); Hyman Berman, "Education for Work and Labor Solidarity: The Immigrant Miners and Radicalism on the Mesabi Range," typed paper, 1963, copy in MHS; Donald G. Sofchalk, "Organized Labor and the Iron Ore Miners of Northern Minnesota, 1907–1936," in *Labor History*, 13:214–242 (Spring, 1971).

[34] The statistics here and two paragraphs below are in William D. Trethewey, *Mining Directory Issue, Minnesota, 1973*, 213, 218 (University of Minnesota, *Bulletin* — Minneapolis, 1973); *Iron Ore, 1977*, 10 (Cleveland, 1977); *Lake Superior Iron Ores*, 41, 219, 227 (1952).

[35] Edward W. Davis, who died in 1973, described the work referred to here and below in *Pioneering with Taconite* published in 1964. See also Frederick T. Witzeg, "A Geographical Study of the Taconite Industry of Northeastern Minnesota," Ph.D. thesis, University of Illinois, 1957. On present taconite plants, see William D. Trethewey, *Mining Directory of Minnesota, 1978*, 252–257 (Minneapolis, 1978).

[36] Rossman, "An Inspiring Chapter in the Story of Iron," in Iron Mining Industry of Minnesota, *Facts About Minnesota Iron Mining*, 40 (n.p., 1958), copy in MHS.

INDEX

Printed in the USA
CPSIA information can be obtained
at www.ICGtesting.com
JSHW021435221024
72172JS00002B/13

9 780873 514910